Bioinorganic Chemistry—II

Kenneth N. Raymond, EDITOR

University of California, Berkeley

A symposium co-sponsored
by the Division of Inorganic
Chemistry and by the Division
of Biological Chemistry at
the 171st Meeting of the
American Chemical Society,
New York, N.Y.,
April 7–9, 1976

ADVANCES IN CHEMISTRY SERIES 162

AMERICAN CHEMICAL SOCIETY
WASHINGTON, D. C.   1977

Library of Congress CIP Data

Bioinorganic chemistry—II.
  (Advances in chemistry series; 162 ISSN 0065-2393)

  Includes bibliographical references and index.

  1. Metals—Physiological effect—Congresses. 2. Bio-
logical transport—Congresses. 3. Electron transport—
Congresses. 4. Biological chemistry—Congresses.
  I. Raymond, Kenneth N., 1942-   . II. American
Chemical Society. Division of Inorganic Chemistry. III.
American Chemical Society. Division of Biological Chem-
istry. IV. Series: Advances in chemistry series; 162.
[DNLM: 1. Biochemistry—Congresses. 2. Chemistry—
Congresses. QD151.2 B613 1976]

QD1.A355 no. 162 [QP532]   540'.8s   [574.1'921]
ISBN 0-8412-0359-8   77-22225   ADCSAJ 162 1-448

Advances in Chemistry Series

Robert F. Gould, *Editor*

FOREWORD

Advances in Chemistry Series was founded in 1949 by the American Chemical Society as an outlet for symposia and collections of data in special areas of topical interest that could not be accommodated in the Society's journals. It provides a medium for symposia that would otherwise be fragmented, their papers distributed among several journals or not published at all. Papers are refereed critically according to ACS editorial standards and receive the careful attention and processing characteristic of ACS publications. Papers published in Advances in Chemistry Series are original contributions not published elsewhere in whole or major part and include reports of research as well as reviews since symposia may embrace both types of presentation.

CONTENTS

PREFACE

Although the term "bioinorganic chemistry" is relatively new, this area of research can be traced back at least as far as the late 1800's to the discovery of iron-containing compounds in blood. However, the new name has come out of renewed interest and activity in the field that in turn has developed from an accumulated body of information about the many and varied biochemical roles played by metal-containing biomolecules. Often this role played by the metal is the leading one, with the remaining parts of what may be a large protein molecule acting as supporting cast. Examples are cobalt in vitamin B_{12}, iron in myoglobin and hemoglobin, the iron–sulfur cluster in the bacterial ferridoxins, and so on. The confluence of the new coordination chemistry of the last two decades, the biochemical characterization of many metal-containing systems, and the application of new and powerful physical tools has made the research area described as "bioinorganic" one of the fastest moving in chemistry. Since the papers given at the Centennial Meeting of the American Chemical Society were to be representative of the vanguard of chemical research in 1976, the organizers chose to include a symposium on bioinorganic chemistry.

For this symposium, I took as my charter the presentation of a timely, representative account of the state of research in this field. It was impossible, within the time constraints of the symposium, to describe all of the problems which are now being pursued most actively, and other recent symposia have presented progress reports for several of these problems. For both of these reasons, I chose to present only three areas which I think are important, which had not been covered extensively by earlier symposia, and which seem to represent the research in the general area: biological transport of iron, biological electron transport and copper-containing biomolecules, and molybdenum-containing biomolecules. This volume is also organized into these three divisions.

It was also hoped that this symposium would stimulate exchanges of ideas, information, and viewpoints by bringing together investigators with very diverse backgrounds, but who are working on closely related problems. Thus chemists, biochemists, plant physiologists, and medical researchers were among the speakers in the program. The presentations included review and background material as well as reports of original research. It is hoped that this balance and the diverse backgrounds of

vii

the authors will make this collection instructive as well as informative for students and research workers in the field.

Financial support in the form of a grant from the Petroleum Research Fund of the American Chemical Society provided travel funds which made it possible for several speakers from Europe to participate in the conference. I also owe thanks to a great many people who helped in various ways either with the symposium itself or the publication of the manuscripts. First of all, I must thank the authors, who were so prompt in meeting deadlines and so tolerant of my correspondence when the deadlines neared. I would also like to thank Jerry Zuckerman, Dick Holm, and the others who were instrumental in arranging the symposium. The editorial assistance of many of the authors, my research colleagues, and June Smith was invaluable in the latter stages of preparing the manuscripts for publication.

Chemistry Department   KENNETH N. RAYMOND
University of California
Berkeley, Calif. 94720
August 1976

Biological Transport of Iron

Siderophores: Biochemical Ecology and Mechanism of Iron Transport in Enterobacteria

J. B. NEILANDS

Department of Biochemistry, University of California, Berkeley, CA 94720

Siderophores protect Escherichia coli *and* Salmonella typhimurium *from certain phages, bacteriocins, and antibiotics by two mechanisms. The first is adsorption competition for outer membrane receptors. Thus ferrichrome competes with T1, T5, Φ80, colicin M, and albomycin for a common site (tonA) in* Escherichia coli *and with phage ES18 and albomycin in* Salmonella. *Ferric enterobactin similarly antagonizes colicin B. In the second mechanism siderophores nonspecifically protect against the B group colicins in an event requiring use of siderophore iron. Ferric enterobactin and cognate membrane receptors are overproduced at low levels of iron. Experiments with ^{55}Fe and tritiated ligand and with the isostructural chromic analog show that ferrichrome rapidly delivers its iron while the ligand more slowly, although again as the iron complex, penetrates the cell.*

Until recently the problem of iron assimilation in microbial species could be likened to a quiet, pastoral scene with a few "low powered" microbiologists and biochemists patiently tending their crop. Just in the past two years, however, this scene has been changed radically by the sudden shift of emphasis from structural chemistry and fungi to membrane physiology and enteric bacteria. Things happen fast in enteric bacteria, and we conclude that progress is directly proportional to the generation time of the species. It was the finding of a common locus for attachment of microbial iron transport compounds (siderophores), phages, and colicins at the outer membrane of *Escherichia coli,* first reported in 1974 (*1*), that changed the tempo from subsistence farming to agribusiness.

Accordingly, this chapter deals mainly with some of the very recent advances in knowledge of iron transport systems in enteric bacteria. I review critically and place in historic perspective current understanding of the competition phenomenon between phages, colicins, and sidero-phores for outer membrane receptors, the resistance to certain colicins imparted by iron, and, finally, the mechanism of siderophore-mediated iron uptake.

The general properties of siderophores have been described extensively (*2*), and up-to-date lists of the individual compounds, their sources from aerobic and facultative aerobic species, and their properties have been published (*3, 4, 5*). (Porphyrin Products, P. O. Box 31, Logan, UT 84321, sell a limited number of siderophores.) The earlier literature on iron assimilation by microbes, including enteric species, may be found elsewhere (*6, 7*). For information on the chemical constitution and physiological role of the outer membrane of enteric bacteria, the reader is referred to Nakae and Nikaido (*8*).

Roles of Iron in Microbial Physiology and Processes for Its Assimilation

It is well to recall that the theoretical basis for this research resides in the omnipresent role of iron in microbial physiology (*9*). Iron compounds, from the ferredoxins to cytochrome oxidase, transfer electrons over a redox potential spanning the better part of one volt. The ferredoxins thus supply the low potential electrons required for reduction of nitrogen and carbon dioxide in nitrogen fixation and photosynthesis, respectively, while the cytochromes enable respiration and concomitant conservation of chemical energy. The significance of iron storage and oxygen transport proteins in microbes is largely unknown, although one mold species contains ferritin, and hemoglobin-like compounds occur in yeast. The iron-containing oxygenases and hydroperoxidases play essential roles in oxygen and hydrogen peroxide metabolism. One form of superoxide dismutase contains iron (*10*). The biological reduction of nitrogen requires, in addition to ferredoxin, an iron protein and an iron–molybdenum protein. Finally, in *E. coli* and in all other microbes not utilizing the functionally equivalent vitamin B_{12} system, iron is a component of ribotide reductase. Thus iron has a key role in the synthesis of the DNA of microorganisms, including the enteric species.

In view of the many crucial biofunctions of iron, it might be expected that microbes would be equipped with diverse systems for acquiring the metal. These can be designated low affinity (nonspecific) or high affinity (specific). The former can be blocked by use of a chelator that is not assimilated, e.g., citrate in the case of *Salmonella typhimurium.* The high affinity system involves specific carrier molecules (siderophores) and a cognate membrane receptor transport system. The two components of the high affinity system can be altered by mutation, and they are hence accessible to the experimental methods of biochemical genetics. This chapter will be restricted to the high affinity system, the main contours of which are depicted in Figure 1.

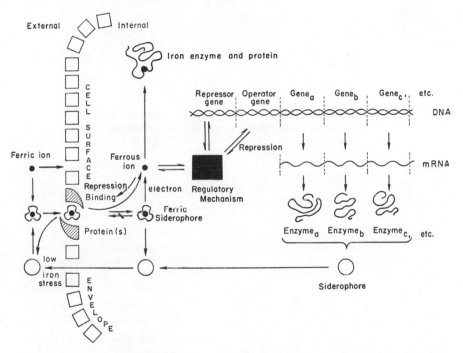

Figure 1.   Schematic of the two iron transport systems of microorganisms.
The high affinity system is comprised of specific carriers of ferric ion (siderophores) and their cognate membrane bound receptors. Both components of the system are regulated by iron repression through a mechanism which is still poorly understood. The high affinity system is invoked only when the available iron supply is limiting; otherwise iron enters the cell via a nonspecific, low affinity uptake system. Ferrichrome apparently delivers its iron by simple reduction. In contrast, the tricatechol siderophore enterobactin may require both reduction and ligand hydrolysis for release of the iron.

Siderophores

Practically all aerobic and facultative aerobic microbial species critically examined for their presence excrete siderophores. The latter have also been called iron transport compounds and siderochromes, but siderophore, first proposed by Lankford (6), is preferred, since it suggests the

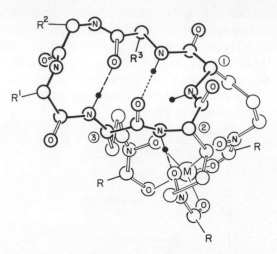

Figure 2. Crystal and solution structure of the ferrichrome siderophores as determined by x-ray diffraction (11) and high resolution NMR (12). The ferrichrome peptides differ in the nature of the acyl substituent at the metal hydroxamate (R) and in the side chains of the three small, neutral, spacer amino acids (R^1, R^2, and R^3). Ferrichrome: M = Fe; R = CH_3; R' = R^2 = R^3 = H (see also Figure 6 and Refs. 3 and 4).

function and is consonant with the terminology used for the alkali metal cation complexing agents, the ionophores. In contrast to the latter, however, the natural role of the siderophores in producing strains has been established by the rigorous methodology of biochemical genetics. The carrier-mediated flux of monovalent ions across membranes, while explaining the uncoupling and antibiotic properties of ionophores, does not solve the problem of the natural mechanism for alkali metal ion transport.

In siderophores the liganding atoms are oxygen, except in mycobactin, where a single heterocyclic nitrogen participates in bonding to the iron. Most of the compounds characterized to date are hydroxamic acids, with ferrichrome (Figure 2) as the prototype. The hydroxamate-type siderophores are abundant in molds and fungi and are less common, apparently, in prokaryotes. Schizokinen has been isolated from *Anabaena* sp., a blue-green alga (*13*). *Aerobacter aerogenes* forms aerobactin, a hydroxamate-type siderophore. Otherwise, only catechol-type members of the series have been described thus far from enteric organisms. The siderophore common to all of these species is enterobactin

(Figure 3). With ferric ion these ligands form thermodynamically stable, high spin, exchangeable complexes which are virtually specific for Fe^{III}.

In systematic surveys of aerobic microbial species for siderophores, it has usually been reported that some do not form these compounds. The reason for this is unclear, but several rationalizations can be advanced. It is possible that the assay has been too insensitive or otherwise defective. If the substance lacked charge transfer bands, it would not be detected by the colorimetric tests. Among the exceedingly sensitive bacterial tests, *S. typhimurium enb-7* is to be preferred over *Arthrobacter* JG-9, since citrate can be used to suppress low affinity iron transport in the former and, in addition, the latter does not respond to phenolate type siderophores (15). Organisms being tested for siderophore production are usually grown in glucose-salts media, and the possibly tenuous assumption is made that adventitious iron concentration is low enough to derepress siderophore synthesis but high enough to allow that level of growth which will maximize product yield. Finally, the iron binding activity may remain associated with the cell envelope even under conditions of low iron stress.

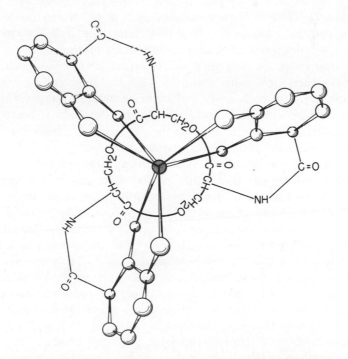

Journal of the American Chemical Society

Figure 3.   Ferric enterobactin, the ferric complex of cyclotri-2,3-dihydroxy-N-benzoyl-L-serine (14)

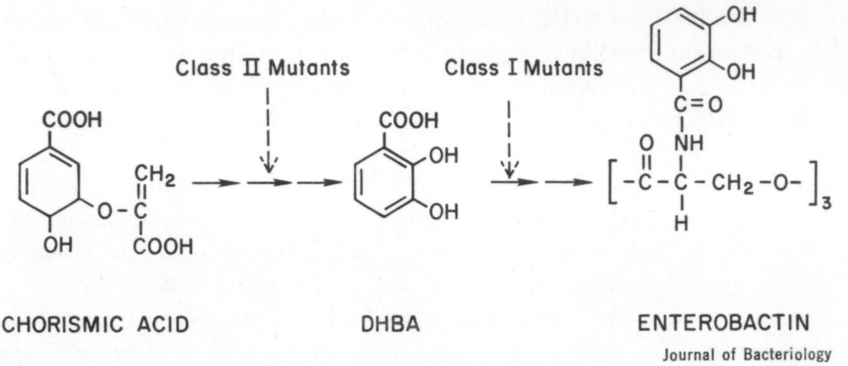

CHORISMIC ACID   DHBA   ENTEROBACTIN

Journal of Bacteriology

Figure 4. Biosynthetic pathway from chorismate to enterobactin showing nature of the class I and class II mutations in Salmonella typhimurium LT-2
(25)

Advantages of Enteric Bacteria for Studies of Microbial Iron Assimilation

Enteric bacteria cannot achieve the cell mass/unit volume of culture fluid seen with either more strictly aerobic prokaryotes or with yeasts. Nonetheless the former have a number of features which render them attractive for the experiments described here, the most important of which is the substantial level of understanding of their genetics. This facilitates the collection, mapping, and interpretation of the biochemistry of mutants. They grow rapidly and are attacked by a variety of phages and by antibiotics of both lower and higher molecular weight, the latter proteinaceous in character and collectively designated bacteriocins. These tools make the enteric bacteria the organisms of choice for investigating the mechanism of bacterial iron assimilation.

Thus, a mutation in the aromatic pathway between chorismate and enterobactin will impair the ability of the organism to scavenge iron when this element is in limiting concentration in the environment. Such *enb* mutants of *S. typhimurium* fail to grow in citrate minimal media without added siderophore. *E. coli,* in contrast, possesses an inducible transport system for ferric citrate. However, *ent* mutants of this organism grow poorly on low iron glucose salts media and fail to grow when succinate is used as energy source. All of these mutants thrive on nutrient broth or agar or on minimal media containing reasonable levels of available inorganic iron. Mutants defective in the transport of ferric enterobactin will be internally iron deficient on these special media and will give culture fluids with a deep wine color, provided some iron is present. These observations neatly illustrate the side-by-side existence of high and low affinity iron transport systems and afford simple means of acquiring mutants in which to probe the mechanism of the former system.

Siderophore—Phage—Colicin Interaction

The recent history of high affinity iron transport in enteric bacteria can be traced to the detection, in 1961, of 2,3-dihydroxybenzoic acid as a product of tryptophan metabolism by cell suspensions of *A. aerogenes* (*16*). The glycine conjugate of this phenolic acid had earlier been isolated from low iron cultures of *Bacillus subtilis* (*17*). Brot et al. (*18*) characterized 2,3-dihydroxy-N-benzoylserine as an extracellular metabolite of a methionine—vitamin B_{12} auxotroph of *E. coli*, and in the same year Corwin et al. (*19*), working with *S. typhimurium*, showed that sensitivity to chromic ion is linked to the *trp* operon and should map close to the *tonB* ("T-one") mutation in *E. coli* K-12. Wang and Newton (*20, 21, 22*) then demonstrated that deletion through the *tonB—trp* region of *E. coli* imparted Cr^{III} sensitivity and a high iron requirement. (Cr^{III} is believed to copolymerize with iron and to make it unavailable for growth of the microbes.) Wang and Newton defined two types of mutants, namely, a class lacking the *tonB* function and unable to grow even in the presence of ferric citrate—and hence transport defective—and a second class, blocked in 2,3-dihydroxybenzoyl serine synthesis, which could utilize ferric citrate. Dihydroxybenzoyl serine was isolated from *E. coli* and *A. aerogenes* by O'Brien et al. (*23*), who also obtained the substance by synthesis, and from *S. typhimurium* by Wilkins and Lankford (*24*).

An important contribution was made by B. N. Ames in the course of his work on the mapping of the chromosome of *S. typhimurium*. Using

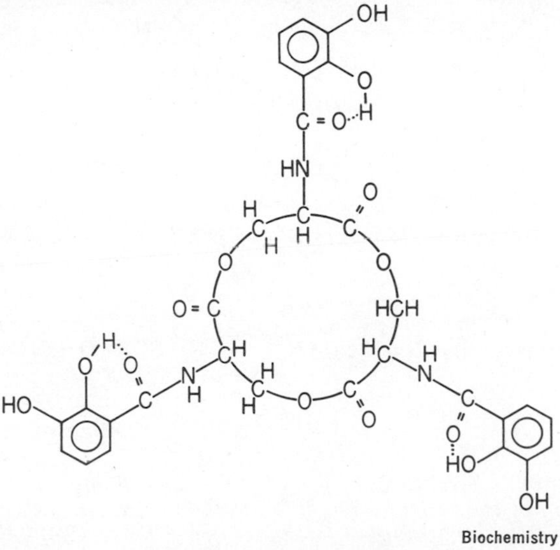

Biochemistry

Figure 5.   Structure of enterobactin, as deduced by PMR and ^{13}C *NMR* (*27*)

a growth medium which happened to contain high levels of citrate (Vogel–Bonner medium), he isolated a series of mutants which would only grow in the presence of salts of inorganic iron. Pollack et al. (25) showed these iron mutants to be blocked before (class II) and after (class I) dihydroxybenzoyl serine (Figure 4). Class I, but not class II, mutants excreted catechols, and both failed to grow on medium E without addition of a chelator sufficiently powerful to remove iron from ferric citrate. The wild type produced, at the terminus of the biosynthetic pathway, an ether-soluble catechol characterized as the cyclic trimer of 2,3-dihydroxy-N-benzoyl-L-serine, named enterobactin (26) (Figure 5).

Working independently, O'Brien and Gibson (28) isolated the identical compound from *E. coli* and named it enterochelin. The Australian

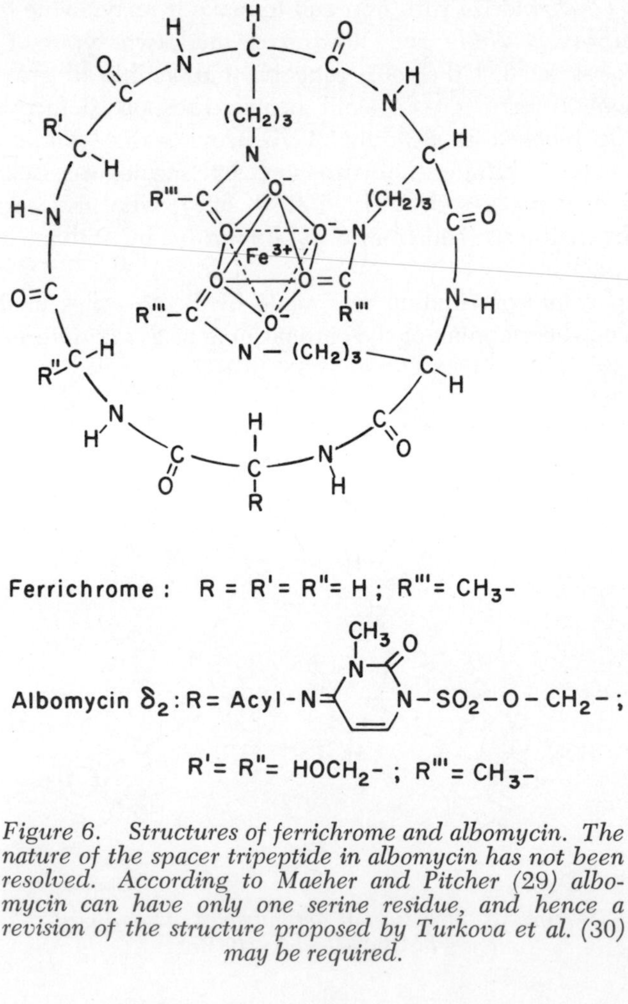

Ferrichrome :   R = R'= R"= H ; R'''= CH$_3$-

Albomycin δ_2: R= Acyl-N= [N(CH$_3$)-C=O ring] N-SO$_2$-O-CH$_2$- ;

R'= R"= HOCH$_2$- ; R'''= CH$_3$-

Figure 6.   Structures of ferrichrome and albomycin. The nature of the spacer tripeptide in albomycin has not been resolved. According to Maeher and Pitcher (29) albomycin can have only one serine residue, and hence a revision of the structure proposed by Turkova et al. (30) may be required.

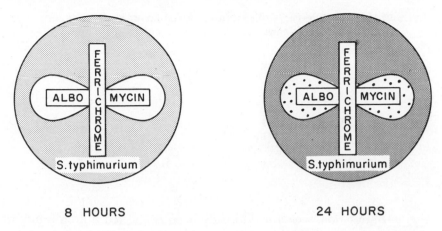

Figure 7. Antagonism between ferrichrome and albomycin for membrane receptor in Salmonella typhimurium *and the appearance of mutants resistant to the antibiotic (2). Several of these mutants owe their resistance to a defect in ferrichrome transport. The nature of this effect as competition for a surface site was elucidated by Zimmerman and Knusel (32).*

workers objected to the term enterobactin on the grounds that the suffix "bactin" should be reserved for dihydroxamates, such as mycobactin. However, since the names of a number of other dihydroxamates were already in the literature, e.g., rhodotorulic acid, schizokinen, and mycelianamide, and since it has become important to specify the microbial origin of this substance, we adhere to the original designation, namely, enterobactin.

There is no published evidence that *S. typhimurium* LT-2 or *E. coli* K-12 form any high affinity iron carrier other than enterobactin. In studying the nutritional requirements of the *enb* mutants of *Salmonella,* Pollack et al. (*25*) made the remarkable observation that the mutants responded even better to ferrichrome than to their native siderophore, enterobactin (minimum optimal concentration 0.1 and $0.5 \mu M$ for 50-min generation time, respectively). The existence of a unique antibiotic, albomycin (Figure 6) which is a close structural analog of ferrichrome, suggested the intriguing possibility of using the former antimetabolite as a probe for the transport of the latter siderophore, and it was quickly shown that albomycin-resistant mutants (*sid*) were unable to transport radioactive ferrichrome (*31*) (Figure 7). Luckey et al. (*33*) extended this work and showed that most albomycin-resistant mutants mapped near *panC.* However, all attempts to demonstrate defective inner membranes in the *sid* mutants failed.

At this point the recently published work (*34*) on the vitamin B_{12}– colicin E receptor focused our attention on the outer membrane of *E. coli.*

Table I.   Phage, Colicin, and Antibiotic Resistance in
ton Mutants of *Escherichia coli* K-12

Strain	Phage	Resistance Colicin	Antibiotic
tonA	T1 T5 Φ80 Φ80h	M	albomycin
tonB	T1 Φ80	M B V I	albomycin

The *tonB* mutation, which had already been connected to iron transport (*see* above), also affected the outer membrane. Inspection of the region of the *E. coli* linkage map analogous to the *sid* locus in *S. typhimurium* revealed that *tonA*, the structural gene for the T5 outer membrane receptor, maps near *pan*. These considerations raised the possibility that the *tonA* product could be related to ferrichrome and a *sid* function in *E. coli*. The characteristics of the *ton* mutations known at that time, with the exception of albomycin resistance, are recorded in Table I.

Competition at the Outer Membrane. FERRICHROME RECEPTOR. *Escherichia coli.* It was easily demonstrated that *tonA* and *tonB* mutants were also resistant to albomycin (*1, 35*). The first experiments to test competition were performed by simply adding ferrichrome to wild type cells plus phage in a plate assay. Later an adsorption assay was used to

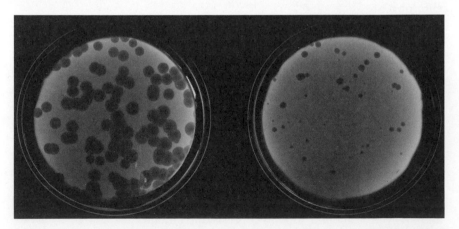

Journal of Bacteriology

Figure 8.   Protection of Escherichia coli K-12 from Φ80 vir by ferrichrome. (Left) no ferrichrome; (right) 50-nmol ferrichrome added during preadsorption of phage to cells prior to plating on a lawn of E. coli (35)

quantitate the results further. Phage Φ80 was chosen since the *tonA* and *tonB* mutations inactivate its receptor, and, in addition, it is more easily handled than agents with an identical host range, such as phage T1 and colicin M. Figure 8 shows the effect of adding 50 nmol of ferrichrome to *E. coli* K-12 CGSC856 plus Φ80 vir at the preadsorption stage prior to plating. This concentration of ferrichrome substantially diminished both the number and the size of the plaques.

Control experiments showed that exposure of Φ80 to ferrichrome for up to 15 hr had no effect on the viability of the phage, indicating the siderophore does not irreversibly inactivate the latter. We next examined a number of phage strains. As seen in Table II, λ and a λ–Φ80 hybrid with the host range of the former were unaffected by ferrichrome. However, a λ–Φ80h hybrid with Φ80 host range was powerfully antagonized

Table II.   Effect of Ferrichrome[a] on Plaque Formation by Phage Strains (35)

Strain	Inhibition[b] (%)
Φ80h	100
λcI	< 5
h$+$λ$_{att}$80$_{imm}$80	< 5
h$+$80$_{att}$80$_{imm}$λ	100

[a] Ferrichrome added during preadsorption to give 0.1 μmol/30 ml plate.
[b] Average of three determinations.

Journal of Bacteriology

by ferrichrome. Phage Φ80h, a host range mutant of Φ80 forming plaques on *tonB*, but not on *tonA*, strains of *E. coli* was inhibited in plaque production by ~ 60% at 1 nmol added ferrichrome per plate! The adsorption assay was used to test the chromic and aluminum analog of ferrichrome, chromichrome, and alumichrome, respectively, both of which were shown capable of some competition with Φ80. This proved that iron per se was not necessary for the effect since the chromic analog of ferrichrome, owing to its high crystal field stabilization energy, is kinetically inert and does not exchange with environmental iron. Ferrichrome A, rhodotorulic acid (which is active in the plate assay), a number of siderophores structurally distinct from ferrichrome, and synthetic ferric complexing agents such as EDTA did not influence adsorption of Φ80. The activity of rhodotorulic acid in the plate assay may possibly be ascribed to a receptor-repression phenomenon triggered by iron supply to the cells. The sum of these experiments indicated, as expected, that within the siderophore series, the receptor can only accommodate molecules closely related to ferrichrome, e.g., albomycin and the metal ion analogs. Ferrichrome protection against colicin M was verified. As seen in Figure 9, competition of Φ80 adsorption by ferrichrome yielded an S-shaped curve, indicating that more than one molecule of ferrichrome may enter the re-

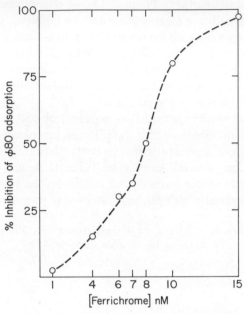

*Figure 9. Quantitative effect of ferrichrome
in preventing adsorption of Φ80 to Escherichia coli K-12 (35)*

ceptor complex. These data lead to the scheme depicted in Figure 10, where phage, ferrichrome or albomycin, and colicin M all compete for a common site on the outer membrane of *E. coli*. This interpretation of the biochemical function of the *tonA* protein drew support from the experiments of Hantke and Braun (*36*), who detected lack of ferrichrome-stimulated [55]Fe transport in *tonA* mutants.

Since the isolated T5 receptor binds the virus, it was of interest to test for ferrichrome competition in vitro (*37*). Log phase cells of *E. coli* JC6724 grown in L broth and M9 minimal media showed ~ 30% difference in their plaque-forming abilities. The recent work of McIntosh and Earhart (*38*) suggests that the latter medium can be further deprived of iron in order to boost synthesis of certain siderophore receptors. These workers have since demonstrated loss of outer membrane proteins in strains defective in siderophore transport (*39*) (Workers in Japan were apparently the first to note the appearance of bands of large polypeptides in outer membrane SDS gel profiles from *Escherichica coli* grown in low iron media (*40*)). Ferrichrome was unable to inactivate the partially purified receptor by exposure for periods of up to 22 days at 4°C. With the cell-free receptor preparation, ferrichrome was again highly potent in competing for it with T5, and neither deferrioxamine B nor rhodo-torulic acid were active. Chromichrome, alumichrome, ferricrocin, and the tri- and dimethyl esters of ferrichrome A were similarly effective, but the monomethyl ester was only weakly active. Attempts to measure binding of radioactive ferrichrome, even quite high specific activity material, to this receptor have not been successful, indicating the complex is quite highly disassociated.

Thus the following picture of ferrichrome protection mechanism against phages T1, T5, Φ80, and colicin M emerges. The phenomenon occurs at the outer membrane or at the Bayer adhesion zones (*41*). Iron is not required for the event since among the isostructural metal ion analogs of ferrichrome, both kinetically labile (Al^{3+}) and stable (Cr^{3+}) ions will substitute for Fe^{3+}. Ferrichrome A is inactive, but progressive removal of the negative charge by esterification imparts activity. Within the cyclohexapeptide moiety of ferrichrome, serine-for-glycine substitu-tions (e.g., ferricrocin) are tolerated, but siderophores of substantially different structure, even rhodotorulic acid, do not accommodate to the contours of the receptor.

Salmonella typhimurium. This organism is sensitive to albomycin and, like *E. coli*, should possess a ferrichrome receptor. The observation by Stocker (*42*) that a correlation exists between sensitivity to albomycin and phage ES18 provided the hint that the outer membrane binding site for the latter may be the ferrichrome receptor in *S. typhimurium* LT-2. Phage ES18 attacks smooth and rough strains not lysogenic for *Fels2*; ES18.h1, a host range mutant, will propagate on strains carrying *Fels2* (*43*). Using *S. typhimurium enb7*, ferrichrome at μmolar concentrations prevented the adsorption of phages ES18 and ES18.h1 (*44*). Three *sid* mutants, resistant to ES18.h1, adsorbed the phage poorly. All three resistant *sid* mutants co-transduced with high frequency at *panC*.

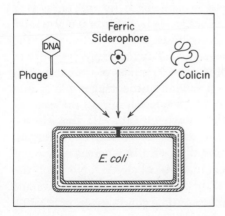

Figure 10. Diagrammatic sketch of the competition between phage coli-cin and siderophore for the outer membrane receptor in the enteric bacteria. (A) Escherichia coli K-12. (1) Ferrichrome receptor: phage = T1, T5, Φ80; colicin = M; sidero-phore = ferrichrome. (2) Ferric en-terobactin receptor: phage = ?; coli-cin = B; siderophore = ferric entero-bactin. (B) Salmonella typhimurium LT-2. (1) Ferrichrome receptor: phage = ES18; bacteriocin = ?; sid-erophore = ferrichrome. (2) Ferric enterobactin receptor: phage = ?; bacteriocin = ?; siderophore = ferric enterobactin.

Thus, *S. typhimurium* appears to have a common surface receptor for ferrichrome, albomycin, and ES18 phages which is the genetic equivalent of the ferrichrome-T1, T5, Φ80-colicin M receptor in *E. coli*.

FERRIC ENTEROBACTIN RECEPTOR. *Escherichia coli*. Guterman (*45, 46, 47*) demonstrated that enterobactin could protect *E. coli* against colicin B but did not define the mechanism of the interaction. However, since enterobactin prevented adsorption of colicin B, it seemed most probable that there is a common membrane receptor for ferric enterobactin and colicin B. Using a ferric enterobactin utilization mutant AN272 (*fes⁻*), it was possible to show that this siderophore specifically protects against colicin B (*48*) in the absence of iron supply from the complex. These data support the previous proposal (*35*) that colicin B and ferric enterobactin compete for a common receptor involved in the uptake of the latter siderophore. The situation is as depicted schematically in Figure 10 except that a phage has not yet been detected in this system. The relation of this receptor to the *fep* (ferric enterobactin permease) mutation needs to be established, initially at least by accurate genetic mapping. The *fep⁻* strain AN270 is fully sensive to colicin B (*49*). Colicin B-resistant strain RWB18 lacks the outer membrane component designated 95p (*50*), i.e., the ferric enterobactin receptor.

Salmonella typhimurium. No work has been reported on the ferric enterobactin receptor in *S. typhimurium*. The system presently lacks the convenient phage and colicin probes, although an agent such as colicin B might bind to *S. typhimurium*. Thus Guterman et al. (*51*) showed that rough strains of *S. typhimurium* were sensitive to coliphage BF23 and adsorbed the E colicins; this is the putative vitamin B_{12} receptor (*see* "Discussion," below).

Nonspecific Siderophore Protection against the "B" Group Colicins. From the preceding it is clear that the function of the *tonA* gene is for biosynthesis of the outer membrane receptor for ferrichrome recognition. The facts that *tonB* mutants are tolerant to colicins B, V, and Ia, as well as resistant to Φ80 and defective in utilization of iron compounds, prompted us to test the effect of siderophores on the activity of the B-group colicins (*48*). For this purpose the following assay system was devised. A filter paper disc containing 20–100 nmol of the test substance was applied to the surface of a colicin-nutrient agar plate, and the radius of exhibition of growth of the seeded tester strain was recorded. It is apparent from the results shown in Table III that siderophores in general—even ferric citrate—protect the *tonA⁺* enterobactin⁻ strain of *E. coli* RW193 from the colicins B, V, and Ia whereas only ferrichrome or its chromic analog protect against colicin M. None of the siderophores tested inhibited colicins K or E1. Synthetic chelating agents were active, but only as the iron complexes. The *tonA⁻* strain of AN193,

Table III.   Siderophore Nutrition and Protection
of *E. coli* K-12 Strain RW193 *(48)*

Protective Potency[a]

Siderophore	Colicin M	Colicin B	Colicin V	Colicin Ia	Nutrition[a,b]
Ferric enterobactin	0	++	++	++	++
Ferrichrome	++	++	++	++	++
Ferric schizokinen	0	++	++	++	++
Ferric rhodotorulate	0	+	+	+	+
Iron(III) citrate	0	+	+	+	+
Chromium(III)-deferriferrichrome	+	0	0	0	

[a] The symbols indicate growth relative to growth of cells in the absence of siderophore. ++ indicates an ~ 10–20-mm radius zone of growth about the disc; + indicates an ~ 1–9-mm zone; 0 indicates no growth.

[b] "Nutrition" refers to siderophore-dependent growth in the absence of colicin, observed on NB-apoferrichrome A plates.

Journal of Bacteriology

as expected, was not protected against colicins B, V, and Ia by ferrichrome. A *SidA* mutant which was unable to use hydroxamates on M9 minimal plates was not protected. In a *fes⁻* mutant, lacking ferric enterobactin esterase, enterobactin protected against colicin B but not against colicins Ia or V.

The B group colicins are believed to act by blocking the energy metabolism of the cell. Accordingly, the siderophores were tested for ability to block the action of the poisons azide, 2,4-dinitrophenol, or CCCP; no protection was found.

A triple layer plate technique (52) capable of distinguishing receptor and tolerant mutants, the latter insensitive to colicin but retaining surface receptor, was applied in order to ascertain which colicin receptors were involved in the transport or utilization of specific siderophores (53). This study confirmed the earlier (35, 48) postulates that colicin B and ferric enterobactin share a common surface receptor in *E. coli* RW193. Other receptors exist for colicins V and Ia. The mechanism of protection by enterobactin against colicin B is thus analogous to that observed for the colicin M–ferrichrome pair (Figure 10). These and other siderophores, as well as a variety of iron compounds, protect against other B-group colicins, but by a noncompetitive, nonspecific mechanism which may be ascribed to either repression of the membrane receptors or interdiction of the action of the colicins at a stage following their binding to the membrane. In any event, the siderophores do not directly inactivate the colicins since a series of mutants were obtained which were not protected by the siderophores.

Guterman and Dann (47) obtained *E. coli* mutants, mapping at about 57 min on the chromosome, which were colicin B resistant by virtue of excretion of enterobactin (*exbB* function). We confirmed this effect by use of enterobactin⁻ strain, RW193, and its *exbB*⁻ derivative. The latter strain was fully sensitive to colicin B in the absence of the enterobactin precursor, 2,3-dihydroxybenzoic acid (*54*). Hantke and Braun (*55*) defined a second locus at 65 min on the *E. coli* genetic map, designated *feu* (ferric enterobactin uptake), which is believed to be involved in ferric enterobactin utilization and the action of colicins B, V, and I. However, it is not known if the *feu* mutation affects the colicin receptors or the colicin targets, nor is it known how many membrane proteins are altered by this mutation. The function of this gene is unclear since other mutants lacking the colicin I and V receptors grow normally on enterobactin, and only colicin B appears to interact specifically with the siderophore in competition for a surface-binding site (*53*). Frost and Rosenberg (*56*) have proposed that *tonB*⁻ strains lack a membrane component required for Φ80 adsorption and ferric enterobactin transport since the strains are defective in both of these functions. However, enterobactin does not block adsorption of Φ80. As indicated above, a component of the Φ80 attachment site is the ferrichrome receptor. The *tonB* mutation invokes a number of changes in membrane proteins and blocks utilization of a host of structurally unrelated siderophores. Additionally, certain mutants in the *tonB* region are unable to utilize enterobactin but remain sensitive to Φ80 (*55*). Accordingly, the biochemical function of the *tonB* gene product remains undefined.

Mechanisms of Siderophore-Mediated Iron Uptake

The kinetic lability of ferric siderophores requires that transport experiments be performed with molecules bearing separate radioactive labels in the metal and ligand moieties. As coordination compounds the siderophores are thermodynamically stable and kinetically labile. The formation constants are typically 10^{30}. In the case of ferrichrome the exchange half time at pH 6.3 and 37° is about 10 min (*57*). Published work (*58, 59*) with doubly labeled ferric schizokinen in *Bacillus megaterium* and ferric aerobactin in *A. aerogenes* as well as a study of ferric enterobactin in *E. coli* (*60*) in each instance suggests a synchronous uptake mechanism for iron and ligand.

Ferrichrome iron transport in the enteric bacteria affords an experimentally feasible model for mechanistic studies. The ligand is rugged and can be labeled to high specific activity by microwave discharge activation of tritium gas. The labile ferric ion can be replaced with chromium to yield the kinetically stable isostructural chromic complex, chromichrome (*61*). Assuming this analog has the transport properties of the iron complex, its ³H counts would be an index of the behavior of the intact coordination compound. Both *E. coli* K-12 and *S. typhimurium* LT-2 have potent systems for utilization of ferrichrome iron, there is no evidence that they synthesize any high affinity carrier other than entero-bactin, mutants defective in the synthesis of the latter siderophore are readily obtained, and, finally nonspecific iron transport can be suppressed with NTA or citrate, respectively.

  E. coli. FERRICHROME. The uptake rates of ^{55}Fe-ferrichrome, ^{3}H-ferrichrome, and Λ-*cis*-^{3}H-chromichrome, all in separate cultures of RW193, are shown in Figure 11 (*62*). The essentially identical rates for uptake of the ^{55}Fe and ^{3}H of the chromichrome suggests rapid transport of the intact coordination compound with concomitant rejection of the ligand. The latter does slowly penetrate the cell via a second mechanism, apparently again as the iron complex since addition of excess iron to the

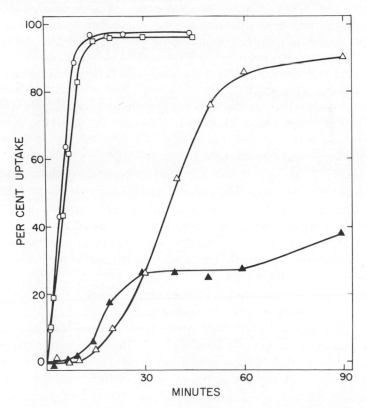

Figure 11. Uptake of the radioactive label of ^{55}*Fe-ferri-chrome* (○), *a cis-chromic* ^{3}*H-deferriferrichrome* (□), ^{3}*H-fer-richrome* (△), *and* ^{3}*H-ferrichrome with excess ferric NTA* (▲) *in* Escherichia coli *RW193. Cell optical density at 650 nm was 1.2, and siderophore concentration was 0.5* μ*M (62).*

system does not enhance significantly the rate of ^{3}H uptake. The exceedingly fast rate of iron removal indicates that the mechanism is via reduction, and it is obvious that the Λ-cis coordination isomer is active and that isomerization of dissociation of the ligand cannot occur during transport.

ENTEROBACTIN. Langman et al. (63) studied iron uptake by non-enterobactin synthesizing strains of E. coli K-12 which contained additional mutations. In the presence of added enterobactin, mutants lacking the ferric enterobactin permease (fep) transported very little iron, and modest amounts were taken up by a mutant lacking the ferric enterobactin esterase (fes) while the parent strain vigorously transported iron. There is excellent genetic evidence that the fes gene product is required for utilization of enterobactin iron although this esterase has not been studied in any depth as a biochemical entity. The Australian workers claim that the preferred substrate is ferric enterobactin, but Bruce and Brot (64) found only the free ligand to be hydrolyzed. Frost and Rosenberg (60) studied the behavior of ^{55}Fe- and ^{14}C-labeled enterobactin in E. coli K-12 mutants not synthesizing endogenous enterobactin and containing fes and fep mutations. In the former they showed that iron and ligand accumulated within the cells at approximately equal rates whereas in the parent strain, radioactivity was leaked to the medium. The fep mutations are relatively (40%) closely linked to the six or seven ent genes on the E. coli chromosome.

Recently Frost and Rosenberg (56) demonstrated that strains of K-12 blocked in aromatic biosynthesis and also tonB$^-$ could fabricate enterobactin from precursors such as 2,3-dihydroxybenzoic acid. The siderophore synthesized endogenously from 2,3-dihydroxybenzoic acid was pictured as carrying iron from the periplasmic space across the cytoplasmic membrane. Evidently ferric enterobactin could not traverse the outer membrane in these strains.

Salmonella typhimurium. FERRICHROME. Ferrichrome iron transport in enb7 proceeds by two concurrent mechanisms (Figure 12) (62). In one, the iron is snatched out rapidly with effective accumulation of the free ligand in solution. With added iron, the latter enters the cell at a rate identical to that of chromichrome. Again the Λ-cis isomer is active without dissociation, and reductive release is virtually certain.

FERRIOXAMINE. Experiments in enb7 with ^{3}H- and ^{55}Fe-labeled ferrioxamine B revealed that the iron of this siderophore, as in the case of ferrichrome, is rapidly removed by the cells (62). The ligand penetrates the organism at a substantially slower rate, once more as the iron chelate. Interestingly, there was zero transport of either the cis or trans isomers of chromioxamine B, the Cr^{III} analog of ferrioxamine B.

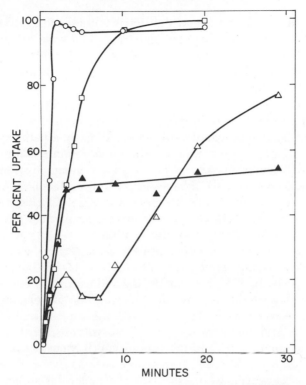

Figure 12. Uptake of the radioactive label of
^{55}*Fe-ferrichrome* (○), *Λ-cis-chromic-*^{3}*H-deferrifer-*
richrome (□), ^{3}*H-ferrichrome* (△), *and* ^{3}*H-ferri-*
chrome with excess ferric citrate (▲) *in* Salmonella
typhimurium *LT-2. Cell optical density at 650 nm*
was 1.2, and siderophore concentration was 0.5 μM
(62).

Discussion

The specific siderophore common to a large number of enteric bacteria, enterobactin, feeds iron to *E. coli* and also protects the latter against
the B group colicins. Ferrichrome, a siderophore elaborated by higher
microorganisms such as species of *Ustilago, Neurospora, Aspergilli,* and
Penicillia (4), is efficiently utilized by enteric bacteria. The latter organisms use this siderophore as a source of iron and to compete with and/or
protect against a series of phages, colicins, and the antibiotic, albomycin.

Clearly, protection of enteric bacteria against phages and colicins
occurs by two distinct processes—one involving specific direct adsorption
competition for common outer membrane receptors and a second, nonspecific, noncompetitive mechanism involving a cell-mediated event invoked by iron.

The existence of ferrichrome and ferric enterobactin receptors in the outer membrane of enteric bacteria confirms the discovery, first reported for vitamin B_{12} (34), for a genuine transport role for this segment of the cell envelope. The properties of the four analogous systems known at the present time are shown in Table IV.

E. coli does not synthesize vitamin B_{12} and grows well without this nutrient. However, in certain circumstances there is a marginal benefit to the cell to be able to manufacture methionine by a pathway involving the vitamin. For this purpose the required apoprotein is synthesized, and the receptor protein is inserted in the membrane in anticipation of finding B_{12} in the environment (70). Phages BF23 and the E colicins have somehow managed to adapt to the B_{12} receptor (71). Besides the surface receptor, a second protein component in the periplasmic space may be necessary for further transport of the vitamin.

A similarly rational argument can be made for the evolutionary retention of the ferrichrome receptor. The earth is 4.5–4.6 billion years old. During the prebiotic era, iron was in the reduced form and reasonably soluble. Similar conditions prevailed during the development of the anaerobic, prokaryotic cells. However, with the appearance of blue-green bacteria some 3 billion years ago, true photosynthesis began, oxygen was placed in the atmosphere, and the iron promptly went out of solution as $Fe(OH)_3$ ($K_{sol} = < 10^{-38}M$). In response to this crisis, cells synthesized siderophores. However, it appears that the molecular weights of these substances may have exceeded the free diffusion limit of the outer membrane, which may be in the range of 700 daltons (8). This outer membrane is freely permeable to amino acids, monosaccharides, and mono-

Table IV. Outer Membrane

Nutrilite	Mol Wt	Gene	Locus (min)	Mol Wt (SDS)
		Receptor		
		Escherichia coli K-12		
Vitamin B_{12}	1357	_btuB_ (_bfe_)	88[d]	60,000[e]
Ferrichrome[a]	740	_tonA_	3	85,000[f]
Ferric enterobactin[a]	788[b]	_feuB_[c]	72.5	ca. 95,000[g]
Maltose and maltodextrins	342+	_lamB_	90	—
		Salmonella typhimurium LT-2		
Ferrichrome	740	_sid_	9	—

[a] These siderophores also protect E. coli against the B-group colicins, but by a mechanism involving cell-mediated utilization of their ferric ion rather than by an adsorption competition for a cell surface receptor.
[b] Trisodium salt.
[c] (65, 66).

mers in general, but not to siderophores. This in turn required the elaboration of a cell surface receptor to facilitate uptake of such larger molecules. Finally, microbes discovered how to make antibiotic analogs, e.g., albomycin, of the nutrient ligands, and in the final stage of molecular parasitism, the receptors were taken over by phages and colicins. Presumably the pervasively important role of iron in microbial physiology has prompted the retention of the siderophore receptors in spite of the fact that they have become the port of entry to phages and antibiotics both large (colicins) and small (albomycin, ferrimycin).

The maltose molecule is too small to be fitted conveniently into the molecular barrier hypothesis, but the receptor for this substance functions well with maltodextrins which do have higher molecular weights. The receptor is induced with maltose, a substrate which is doubtless commonly encountered in the diet of enteric bacteria. Biologically the receptor serves for transport of maltose and for chemotaxis to this substrate (72). The data presented in Table IV, especially that pertaining to the ferrichrome receptor, demonstrate convincingly that phage receptors were designed for nutritious substances.

The stoichiometry and mechanism of ligand binding to any of the four receptors listed in Table IV, following their isolation from the outer membrane, has yet to be demonstrated. Competition with phage proceeds well with the T5 receptor in vitro even though the complexes have rather high dissociation constants. It is not necessary that the ligands, phages, and colicins have structural similarities since each may bind to a different locus of the receptor.

Receptors in Enteric Bacteria

	Noxious Agents	
Phage(s)	*Antibiotic*	*Colicin*
	Escherichia coli K-12	
BF23	—	E1,E3
T1,T5,Φ80	albomycin	M
—	—	B
λ	—	—
	Salmonella typhimurium LT-2	
ES18	albomycin	—

[a] Loci are from the recalibrated linkage map of *Escherichia coli* K-12 (67).
[e] (34).
[f] (68).
[g] (69).

The extreme potency of siderophores in counteracting the B-group colcins suggests an important biological role for such molecules in a mechanism which resembles, but which is fundamentally different in nature from, the outer membrane competition phenomenon.

Figure 13 summarizes the present state of knowledge regarding interactions of siderophores, phages, and colicins with membrane systems in the enteric bacteria. Over the eons of evolutionary time, noxious agents have acquired the capacity to exploit these receptors as a means of penetrating the cell envlope. It would be important to ascertain if this analogy extends to plant and animal virus receptors.

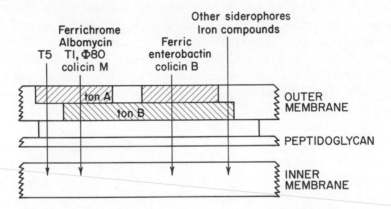

Figure 13.   Schematic of the organization and properties of siderophore receptors in Escherichia coli *K-12.*

As seen in Table I, colicins V and Ia require the tonB gene, but the receptors for these agents have not been correlated with any specific siderophore or other nutrient substance. The binding of ferric enterobactin has been defined as the biochemical function of the colicin B receptor (53). Iron supply to the cell interferes with adsorption of colicin Ia, thus suggesting this receptor is designed for a siderophore, the specific nature of which is still unkown (73).

The molecular events associated with the noncompetitive protection mechanism remain to be elucidated. That the two mechanisms by which siderophores protect against phages and colicins are related can be seen in the response of E. coli to siderophores such as rhodotorulic acid. This substance is active in protecting against Φ80 in the plate assay but is unable to prevent adsorption of the virus, and it is totally inactive in competing for T5 receptor in vitro (35, 37). These data would be consistent with repression of the *tonA* receptor or some component specified by *tonB*, promoted by iron nutrition, in much the same way that siderophores nonspecifically protect against the B-group colicins. Binding experiments by Mizushima et al. (74) and Hollifield et al. (75) as well as work with a colicin B resistant strain RWB18 by McIntosh and

Earhart (39) provide the evidence that in SDS gels the ferric entero-bactin receptor behaves as a component with mol wt ∼ 95,000. Both parts of the high affinity iron transport system endogenous to *E. coli* have now been shown to be under iron repression. Gilchrist and Konisky (76) observed lower levels of colicin Ia receptor in a heme requiring mutant of *E. coli*, a result which might be ascribed to an intracellular accumulation of iron. The latter represses formation of the Ia receptor (66).

The *exbB* mutants (47) are derepressed in enterobactin synthesis and produce this siderophore in iron-containing media. The means whereby iron represses enterobactin synthesis is still obscure. Several years ago it was noted that growth of *E. coli* on low iron media led to changes in various *t*RNAs (77). In *E. coli* K-12 aromatic amino-acid synthesizing enzymes are also derepressed in low iron media, possibly because of the diversion of the chorismate pool to enterobactin (78).

Experience to date would seem to mandate that all known siderophores should be screened against the phage and colicin receptors for which biochemical functions have not yet been defined. Thus the colicin V and Ia receptors may well be assigned to siderophores which have not yet been examined.

The transport experiments with ^{55}Fe- and ^{3}H-labeled siderophores are, to say the least, confusing. A minimum of three mechanisms are possible:

(1) Transfer of just the naked metal ion to the cell surface

(2) Transport of the intact chelate compound followed by internal delivery of the iron

(3) Separation of the metal and ligand with simultaneous uptake of each and rejection of the ligand (62).

Ferrichrome transport in *E. coli* adheres to mechanism 2, since the nondissociable chromic complex enters the cell at the same rate as the iron. It is somewhat mysterious that the ligand is so efficiently rejected in the more rapid phase of siderophore iron assimilation. The ligand—again as the chelate compound—enters the cell at a substantially slower rate.

In *S. typhimurium* a combination of two concurrent processes, either mechanisms 1 and 3 or mechanism 2, would explain the observed results. The former would account for the rapid phase of iron uptake while the latter would explain the slower penetration rate of the intact chelate. In this organism the iron of ferrioxamine B is again rapidly assimilated, and the chelate enters at a slower pace, but no information on the mechanism can be gleaned from the chromic complexes as they are not transported. It is possible that the active isomer has not been tested in this system. An inspection of molecular models implies that four trans isomers might exist (79).

We conclude that plural mechanisms exist for siderophore iron utilization in the enteric bacteria. The iron may be rapidly removed with (*E. coli*) or without (*S. typhimurium*) effective transport of the ligand. The rate of this process is such that labilization of the iron by reduction appears most likely. The intact ferric chelates also pass the cell envelope, but by a generally slower mechanism. An estimate of the number of atoms of iron acquired per bacterium indicates true uptake rather than adsorption to the cell surface has taken place. In both organisms the Λ-cis chromic coordination isomer of ferrichrome is active, indicating that dissociation and/or isomerization is not obligatory for its transport.

Provided reduction of the iron is necessary, mechanisms 1 and 3 would enable the cell to exclude certain contaminating ions, e.g., aluminum, which might find their way into siderophores. It would also obviate the accumulation of a possibly toxic ligand and the need to back-transport this substance. Unlike the situation which obtains with vitamin B_{12} and maltose, only the iron atom of the siderophores needs to be built into the cell protoplasm. On the other hand, a certain level of the chelate within the cell could trigger regulatory signals or relay information on the ecological composition of the extracellular environment. Interestingly, in *fes⁻* mutants enterobactin stimulates the use of hydroxamate siderophores (*80*).

To the differences between *E. coli* and *S. typhimurium* in siderophore iron utilization we have recorded here must be added the observation that the former has an inducible system for uptake of ferric citrate while the latter cannot utilize this siderophore. The opposite transport specificity exists for free citrate in these organisms (*81*). The receptor for ferrichrome in *S. coli* is functional for T1, T5, Φ80, and colicin M while the corresponding receptor in *S. typhimurium* accommodates phage ES18. The later organism wears a lipopolysaccharide coat, substances which have also been shown to bind ferrous and ferric iron (*82*). The lipophilic surface of the cell may elevate the redox potential of the iron of the hydrophilic siderophores thus rendering the metal ion easier to reduce. The T5 receptor of *E. coli* is believed to be a lipoprotein (*83*). Clearly, ferric enterobactin, with its triple negative charge, will accept an electron only with considerable difficulty. Ferrichrome, which has zero charge, and all of the other known siderophores, with the exception of the mycobactins, are quite hydrophilic.

The active transport aspect of siderophore iron utilization remains largely unexplored, although such a process probably exists—in addition to facilitated diffusion—in view of the high affinity of the cells for siderophores and the likely participation of an energized state of the membrane in their transport. Experiments with whole cells must perforce be performed with mutants lacking the ability to make these carrier molecules.

This, however, introduces a number of complications, and for information on the molecular mechanics of transport it would be desirable to work with a simpler system, such as membrane vesicles, and such experiments are in progress (*84*). Many unresolved questions in siderophore transport await resolution.

Siderophore formation is believed to be under a repressive type of control since addition of iron does not immediately block the biosynthetic pathway through poisoning of preexisting enzymes. In regard to regulation of the total system (siderophore + receptor), there is a possibility of coordinate control by iron. The operon is repressed by media concentrations of the element in the $0.1–1.0\,\mu M$ range—available rather than absolute level is the important consideration—and the precise molecular form of the repressive iron is unknown. For solubility reasons it is unlikely to be just ferric ion, but it might conceivably be free ferrous ion. The latter is believed to activate cyclic AMP phosphodiesterase. The *exb* mutants of Guterman and Dann (*47*) which resist colicin B by virtue of overproduction of enterobactin in iron-rich media are of special interest. Wayne (*80*) showed that the *exbB* mutation is very closely linked and is possibly identical to the *metK* locus (S-adenosylmethionine synthetase) and that the methionine requirement in these mutants could be ascribed to a block in the conversion of homoserine to cystathionine. He showed that spermidine and certain other polyamines repressed enterobactin synthesis. The enzymes involved in conversion of S-adenosylmethionine to spermidine appeared normal in an *exbB* strain as did also the level of iron-dependent thiomethylation of its tyrosine *t*RNA. The interpretation of these observations is not immediately apparent, but it is certain that the regulation of the high affinity iron operon will shortly be scrutinized in detail in several laboratories. While the work described above clearly implicates the *tonA* gene in the synthesis of the outer membrane ferrichrome receptor, the nature of the *tonB* mutation still rquires elucidation.

A third problem still outstanding is the exact mechanism of action, in terms of molecular mechanics, of the outer membrane siderophore receptors. Many of these proteins are large enough to span the envelope, but whether they actually do so has not been verified experimentally. Based on the B_{12} model, it is generally assumed that they facilitate transport of larger hydrophilic molecules across the outer membrane to the inner membrane or to the cytoplasm. There are several advantages of studying the ferrichrome system. The wide range of analogs available in the ferrichrome series enabled delineation of the specificity parameters of the outer membrane receptor (*37*). In addition, the stable non-exchangeable Cr^{3+} complex could be used to demonstrate transport of the intact complex, presumably into the cytoplasm (*62*). These properties of ferrichrome facilitated interpretation of the mode of competition of this particular siderophore with T1, T5, Φ80, albomycin, and colicin M for the tonA receptor and in this way elucidated the biochemical function of the latter. The Cr^{3+} coordination analog of ferric enterobactin can be prepared, but upon exposure to air the chromic ion rapidly oxidizes the ligand (85). The enterobactin system, however, has advantages of its own, such as the repression of both siderophore and receptor by iron. Since the siderophore is endogenous, mutational analysis can be applied to both components of the high affinity transport system.

Finally, the mechanism of assimilation of iron by the low affinity pathway should be explored. Here the mutational analysis so productive in the specific siderophore-mediated pathway is of limited value, except in the somewhat negative sense that the latter system can be eliminated by simple genetic manipulation.

A number of additional recent developments with a bearing on iron metabolism in enteric bacteria should be recorded. Miles and Khimji (86) surveyed over 80 strains, drawn in part from Escherichia, Klebsiella, Salmonella, Proteus, and Shigella species, by testing for stimulation of low density seedings of indicator organisms on agar containing ethylene diaminediorthohydroxyphenyl acetic acid. They concluded that iron chelators from the bacteria in these genera were functionally exchangeable. No correlation between virulence and chelator production could be found. On the other hand, Yancey et al. (87) noted a pronounced drop in virulence in S. typhimurium mutants blocked in enterobactin production. Wake et al. (88) showed that P^+ plague strains of Yersinia pestis contained more siderophore-producing organisms than P^- strains. The topic of iron and infection was reviewed by Weinberg (89).

Guinea pigs fed a synthetic diet fare poorly without addition of certain factors, one of which may be enterobactin. Both Kincaid and Odell (90) and Briggs (91) have obtained spectacular weight gains with a supplement of 5 ppm enterobactin, but the work has been confounded by an erratic response.

Grady et al. (92) tested a number of iron-chelating drugs for their ability to remove iron from the iron-loaded rat. They found 2,3-dihydroxybenzoic acid, the metal-binding center of enterobactin, and rhodotorulic acid to be the most promising, especially the latter. The objective of these experiments is to find a nontoxic chelating drug for controlling the transfusion-induced siderosis associated with the long term therapy for thalassemia, aplastic anemia, and related blood disorders.

The nature of the tonB mutation required for assimilation of the iron of all known siderophores remains elusive in spite of a continuing discussion of the subject (73, 93, 94). It has recently been reported that tonB mutants are defective in the energy-dependent step in B_{12} transport, although the surface receptors for the vitamin and for maltose are unaffected by the mutation (95).

The synthesis of the colicin Ia receptor is clearly derepressed at low iron (73, 96), but a specific siderophore has not been assigned to this large polypeptide constituent, which is programmed by the *cir* gene at 43 min on the chromosome map. *FeuB* is the specific locus for the colicin B–ferric enterobactin receptor (66).

A high molecular weight, citrate-inducible, outer membrane polypeptide has been detected by slab gel electrophoresis (65). This is believed to be the receptor for the ferric citrate transport system earlier characterized by Frost and Rosenberg (60).

According to one report (97) the colicin M receptor is overproduced at low iron. On the other hand, Braun et al. (93) reviewed the functional organization of the outer membrane and stated that the *tonA* component is not increased in iron deficiency.

An outer membrane pore protein of molecular weight 25,000 is required for nucleoside uptake as well as for T6 and colicin K resistance. It is believed to be directed by the *tsx-nup* gene locus at 9–10 min on the *E. coli* chromosomal map (98, 99).

A transport-defective mutant of *E. coli* is advocated for the convenient preparation of ferric enterobactin (100). In *S. typhimurium,* as well as in *E. coli,* several proteins of the outer membrane are regulated by the iron content of the medium (101).

While most workers report the outer membrane siderophore receptors to have molecular weights in the 75–95K range, some variation in the magnitude of these numbers may be attributed to the preparative and analytical methods as well as to the particular standards used. Since enterobactin will rapidly remove iron from ferrichrome, the transport of the latter must perforce be studied in mutants lacking the former. However, such mutants often display multiple lesions. Additionally, isogenic strains have seldom been used and variations in media and cultural conditions will further confound attempts to compare results reported from different laboratories.

Acknowledgment

Research reported herein from the author's laboratory was performed by J. R. Pollack, M. Luckey, R. Wayne, K. Frick, and J. Leong and was supported by USPHS grants AI 04156 and AM 17146.

Literature Cited

1. Wayne, R., Neilands, J. B., "Abstracts of Papers," 168th National Meeting, American Chemical Society, Atlantic City, NJ, September 1974, MICR 3.

2. Neilands, J. B., in "Inorganic Biochemistry," G. Eichhorn, Ed., p. 167, Elsevier, Amsterdam, 1973.
3. Rodgers, G., Neilands, J. B., in "Handbook of Microbiology," A. I. Laskin and H. A. Lechevalier, Eds., Vol. II, p. 823, C. R. C. Press, Cleveland, 1973.
4. Diekmann, H., in "Handbook of Microbiology," A. J. Laskin and H. A. Lechevalier, Eds., Vol. III, p. 449, C. R. C. Press, Cleveland, 1973.
5. Emery, T., *Adv. Enzymol.* (1971) **35**, 135.
6. Lankford, C. E., *Crit. Rev. Microbiol.* (1973) **2**, 273.
7. Rosenberg, H., Young, I. G., in "Microbial Iron Metabolism," J. B. Neilands, Ed., p. 67, Academic, New York, 1974.
8. Nakae, T., Nikaido, H., *J. Biol. Chem.* (1975) **250**, 7359.
9. Neilands, J. B., "Microbial Iron Metabolism," Academic, New York, 1974.
10. Yamakura, F., *Biochim. Biophys. Acta* (1976) **422**, 280.
11. Zalkin, A., Forrester, J. D., Templeton, D. H., *J. Am. Chem. Soc.* (1966) **88**, 1810.
12. Llinas, M., *Struct. Bonding* (1973) **17**; 135.
13. Simpson, F., Neilands, J. B., *J. Phycol.* (1976) **12**, 44.
14. Isied, S. S., Kuo, G., Raymond, K. W., *J. Am. Chem. Soc.* (1976) **98**, 1763.
15. Lankford, C. E., personal communication.
16. Pittard, A. J., Gibson, F., Doy, C. H., *Biochim. Biophys. Acta* (1961) **49**, 485.
17. Ito, T., Neilands, J. B., *J. Am. Chem. Soc.* (1958) **80**, 4645.
18. Brot, N., Goodwin, J., Fales, H., *Biochem. Biophys. Res. Commun.* (1966) **25**, 454.
19. Corwin, L. M., Fanning, G. R., Feldman, F., Margolin, P., *J. Bacteriol.* (1966) **91**, 1509.
20. Wang, C. C., Newton, A., *J. Bacteriol.* (1969) **98**, 1135.
21. *Ibid.* (1969) **98**, 1142.
22. Wang, C. C., Newton, A., *J. Biol. Chem.* (1971) **246**, 2147.
23. O'Brien, I. G., Cox, G. B., Gibson, F., *Biochim. Biophys. Acta* (1969) **246**, 2147.
24. Wilkins, T. D., Lankford, C. E., *J. Infect. Dis.* (1970) **212**, 129.
25. Pollack, J. R., Ames, B. N., Neilands, J. B., *J. Bacteriol.* (1970) **104**, 635.
26. Pollack, J. R., Neilands, J. B., *Biochem. Biophys. Res. Commun.* (1970) **38**, 989.
27. Llinas, M., Wilson, D. M., Neilands, J. B., *Biochemistry* (1973) **12**, 3836.
28. O'Brien, I. G., Gibson, F., *Biochim. Biophys. Acta* (1970) **215**, 393.
29. Maeher, H., Pitcher, R. G., *J. Antibiot.* (1971) **24**, 830.
30. Turkova, J., Mikes, O., Sorm, F., *Collect. Czech. Chem. Commun.* (1964) **29**, 280.
31. Pollack, J. R., Ames, B. N., Neilands, J. B., *Fed. Proc.* (1970) **29**, 801.
32. Zimmerman, W., Knusel, F., *Arch. Mikrobiol.* (1969) **68**, 107.
33. Luckey, M., Pollack, J. R., Wayne, R., Ames, B. N., Neilands, J. B., *J. Bacteriol.* (1972) **111**, 731.
34. DiMasi, D. R., White, J. C., Schnaitman, C. A., Bradbeer, C., *J. Bacteriol.* (1973) **115**, 506.
35. Wayne, R., Neilands, J. B., *J. Bacteriol.* (1975) **121**, 497.
36. Hantke, K., Braun, V., *FEBS Lett.* (1975) **49**, 301.
37. Luckey, M., Wayne, R., Neilands, J. B., *Biochem. Biophys. Res. Commun.* (1975) **6**, 687.
38. McIntosh, M. A., Earhart, C. F., *Biochem. Biophys. Res. Commun.* (1976) **70**, 315.
39. McIntosh, M. A., Earhart, C. F., personal communication.
40. Uemura, J., Mizushima, S., *Biochem. Biophys. Acta* (1975) **413**, 163.

41. Bayer, M. E., *J. Virol.* (1968) **2**, 346.
42. Stocker, B. A. D., personal communication.
43. Kuo, T., Stocker, B. A. D., *Virology* (1970) **42**, 621.
44. Luckey, M., Neilands, J. B., *J. Bacteriol.* (1976) **127**, 1036.
45. Guterman, C. K., *Biochem. Biophys. Res. Commun.* (1971) **44**, 1149.
46. Guterman, S. K., *J. Bacteriol.* (1973) **114**, 1217.
47. Guterman, S. K., Dann, L., *J. Bacteriol.* (1973) **114**, 1225.
48. Wayne, R., Frick, K., Neilands, J. B., *J. Bacteriol.* (1976) **126**, 7.
49. Hollifield, W., personal communication.
50. McIntosh, M. A., personal communication.
51. Guterman, S. K., Wright, A., Boyd, D. H., *J. Bacteriol.* (1975) **124**, 1351.
52. Davies, J. K., Reeves, P., *J. Bacteriol.* (1975) **123**, 96.
53. Wayne, R. R., Neilands, J. B., *Fed Proc.* (1976) **35**, 1453.
54. Wayne, R., personal communication.
55. Hantke, K., Braun, V., *FEBS Lett.* (1975) **59**, 277.
56. Frost, G. E., Rosenberg, H., *J. Bacteriol.* (1975) **124**, 704.
57. Lovenberg, W., Buchanan, B. B., Rabinowitz, J. C., *J. Biol. Chem.* (1963) **238**, 3899.
58. Davis, W. B., Byers, B. R., *J. Bacteriol.* (1971) **107**, 491.
59. Arceneaux, J. E. L., Davis, W. B., Downer, D. M., Haydon, A. H., Byers, B. R., *J. Bacteriol.* (1973) **115**, 919.
60. Frost, G. E., Rosenberg, H., *Biochim. Biophys. Acta* (1973) **330**, 90.
61. Leong, J., Raymond, K. N., *J. Am. Chem. Soc.* (1974) **96**, 6628.
62. Leong, J., Neilands, J. B., *J. Bacteriol.* (1976) **126**, 823.
63. Langman, L., Young, I. G., Frost, G. E., Rosenberg, H., Gibson, F., *J. Bacteriol.* (1972) **112**, 1142.
64. Bryce, G. F., Brot, N., *Biochemistry* (1972) **11**, 1708.
65. Hancock, R. E. W., Hantke, K., Braun, V., *J. Bacteriol.* (1976) **127**, 1370.
66. Konisky, J., personal communication.
67. Bachmann, B. J., Low, K. B., Taylor, A. L., *Bacteriol. Rev.* (1976) **40**, 116.
68. Braun, V., Schaller, K., Wolff, H., *Biochim. Biophys. Acta* (1973) **323**, 87.
69. McIntosh, M. A., Earhart, C. F., personal communication.
70. White, J. C., DiGirolamo, P. M., Fu, M. L., Preston, Y. A., Bradbeer, C., *J. Biol. Chem.* (1973) **248**, 3978.
71. Bradbeer, C., Woodrow, M. L., Khalifah, L. I., *J. Bacteriol.* (1976) **125**, 1032.
72. Schwartz, M., *J. Mol. Biol.* (1975) **99**, 185.
73. Konisky, J., Soucek, S., Frick, K., Davies, J. K., Hammond, C., *J. Bacteriol.* (1976) **127**, 249.
74. Mizushima et al., personal communication.
75. Hollifield, W., et al., personal communication.
76. Gilchrist, M. J. R., Konisky, J., *J. Bacteriol.* (1976) **125**, 1223.
77. Griffiths, E., *FEBS Lett.* (1972) **25**, 159.
78. McCray, J. W., Jr., Herrmann, K. M., *J. Bacteriol.* (1976) **125**, 608.
79. Leong, J., Raymond, K. N., *J. Am. Chem. Soc.* (1975) **97**, 293.
80. Wayne, R., Doctoral dissertation, University of California, Berkeley, CA, 1976.
81. Luckey M., Doctoral dissertation, University of California, Berkeley, CA, 1976.
82. Wang, W. S., Korczynski, M. S., Lundgren, D. G., *J. Bacteriol.* (1970) **104**, 556.
83. Braun, V., personal communication.
84. Negrin, R., personal communication
85. Raymond, K. N., personal communication.
86. Miles, A. A., Khimji, P. L., *J. Med. Microbiol.* (1975) **8**, 477.

87. Yancey, R. J., Jr., Breeding, S. A. L., Lankford, C. E., "Abstracts of Papers," American Society of Microbiology Meeting, Atlantic City, NJ, May 2–7, 1976.
88. Wake, A., Misawa, M., Matsui, A., *Infect. Immun.* (1975) **12**, 1211.
89. Weinberg, E. D., *Science* (1974) **184**, 952.
90. Kincaid, R. L., O'Dell, B. L., *Fed. Proc.* (1974) **33**, 666.
91. Briggs, G., personal communication.
92. Grady, R. W., Graziano, J. H., Akers, H. A., Cerami, A., *J. Pharmacol. Exp. Ther.* (1976) **196**, 478.
93. Braun, V., Hancock, R. E. W., Hantke, K., Hartmann, A., *J. Supramol. Str.* (1976) **5**, 37.
94. Pugsley, A. P., Reeves, P., *J. Bacteriol.* (1976) **126**, 1052.
95. Bassford, P. J., Bradbeer, C., Kadner, R. J., Schnaitman, C. A., *J. Bacteriol.* (1976) **128**, 242.
96. Hancock, R. E. W., Braun, V., *FEBS Lett.* (1976) **65**, 208.
97. Pugsley, A. P., Reeves, P., *Biochem. Biophys. Res. Commun.* (1976) **70**, 846.
98. Hantke, K., *FEBS Lett.* (1976) **70**, 109.
99. McKeown, M., Kahn, M., Hanawalt, P., *J. Bacteriol.* (1976) **126**, 814.
100. Young, I. G., *Prep. Biochem.* (1976) **6**, 123.
101. Bennett, R. L., Rothfield, L. I., *J. Bacteriol.* (1976) **127**, 498.

RECEIVED July 26, 1976.

Kinetically Inert Complexes of the Siderophores in Studies of Microbial Iron Transport

KENNETH N. RAYMOND

Department of Chemistry, University of California, Berkeley, CA 94720

The compounds called siderophores (earlier called sidero-chromes) are low-molecular weight chelating agents which are manufactured by microbes and are involved in their cellular iron transport. Kinetically inert complexes of the siderophores have been prepared by replacing the native ferric ion, which is kinetically labile in biological systems, with the kinetically inert chromic ion. The metal–substituted complexes and related model compounds have then been used as chemical probes, using vis-uv and circular dichroism spectroscopy, to elucidate the coordination geometries of siderophores, and as biological probes, using the kinetic inertness of the chromic siderophore complexes, to study the mechanisms of cellular iron transport in several microbial species. The siderophores studied include the hydroxamate-containing ferrichromes and ferrioxamines and the catechol-containing compound enterobactin.

The preceding and following chapters amply illustrate the reasons why microbial iron transport compounds are worthy of our attention—both from the biochemical and medical points of view. However, one might wonder what this has to do with coordination chemistry. The obvious answer is that these are, after all, coordination compounds. But more than that, when viewed from the perspective of a coordination chemistry, new experiments or new approaches suggest themselves. This is always the exciting potential of interdisciplinary research. This chapter is the result of a research project which has involved extensive collaboration between J. B. Neilands' laboratories and my own. Many of the details of the transport studies of kinetically inert, metal-substituted siderophores in microbial systems were presented in the previous chapter. I will focus here on the coordination chemistry of these compounds and how metal-substituted siderophore complexes can be used both as chemical probes (using spectroscopic techniques) for the structures of these materials, and as biological probes in membrane transport studies.

The compounds called siderophores (earlier called siderochromes) are low-molecular weight materials which are manufactured by microbes and are involved in their cellular iron transport. The biochemistry of the siderophores has been discussed in the previous paper and has been reviewed extensively and recently (1). The siderophores are all chelating ligands which form extremely stable octahedral complexes with high-spin ferric iron. Two important classes of these compounds—the ferrichromes and ferrioxamines—are trihydroxamic acids which (except for those containing charged substituents) form neutral complexes using three bidentate hydroxamate monoanions. These complexes of Fe(III) are all kinetically labile. Even the large hexadentate ligands such as ferrichrome, which completely enclose the ferric ion with an octahedral cavity, have exchange rates on the order of several minutes at physiological conditions of pH and temperature. In contrast, complexes in which chromic ion is substituted for ferric ion, although structurally the same, are kinetically inert. This has been demonstrated for model hydroxamate complexes (2), desferriferrichromes (3), and ferrioxamines (4). Subsequent transport studies have been carried out using several of these kinetically inert complexes.

Another common ligand functional group found in the siderophores is catechol (o-dihydroxybenzene). Catechol is similar to hydroxamates in being a bidentate ligand which coordinates through two oxygen atoms, but is a dianion. Except for the oxygen sensitivity of the catechol complexes (because of the ease of oxidation of the ligand), they are very similar in kinetic and spectroscopic properties to the hydroxamate complexes.

Structure and Properties of Ferric Complexes in Siderophores

General Chemistry of Iron Chelates. The aqueous chemistry of Fe(III) is dominated by its Lewis acidity. Several pH units below that of physiological solutions, hydrolysis and polymerization reactions of ferric ion take place. At physiological pH ferric ion is quantitatively insoluble as the hydroxide. The K_{sp} for $Fe(OH)_3$ is 2×10^{-39} (5) while the K_{sp} for ferrous hydroxide, $Fe(OH)_2$, is 8×10^{-16} (6). The biological consequences of these numbers are profound because, since this planet produced an oxidizing atmosphere, the ultimate source of iron for all biological systems has been inorganic Fe(III). Even the complexation of ferric ion is not always enough to make it useful to biological systems, since the hydrolysis of such complexes often produces such high-molec-ular weight hydroxy-bridged polymers that transport across cell mem-branes is impossible (7).

During the last 10–15 years a number of low-molecular weight compounds of natural origin have been found to bind Fe(III) specifically and transport it in biological systems (1). Most if not all of the com-pounds of this type include hydroxamate or phenolate groups as ligands. Upon loss of the proton, the anion is a very strong chelating agent with an amazing specificity for Fe^{3+}. The general chemistry of the hydroxamic acids forms a part of classical organic chemistry. The reaction with ferric ion is a standard test for the hydroxamate functional group. The acid dissociation of the hydroxamic acids typically gives pK_as on the order

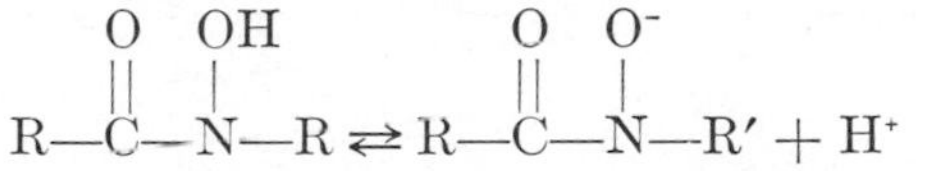

$$\underset{R}{\underset{|}{R}}\;\overset{O}{\overset{||}{C}}-\overset{OH}{\overset{|}{N}}-R \rightleftarrows \overset{O}{\overset{||}{R-C}}-\overset{O^-}{\overset{|}{N}}-R' + H^+$$

of 9. The subsequent reaction with ferric ion gives a very stable five-membered ring (Figure 1). Above very acid pH, three hydroxamic acids

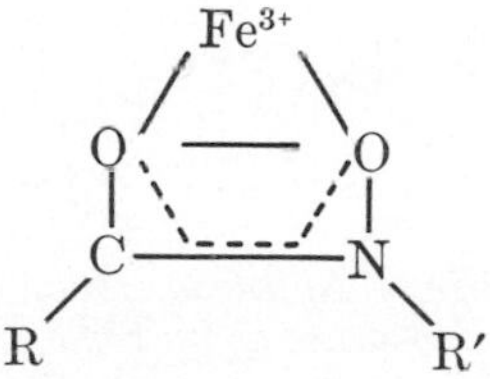

Figure 1.   Ferric hy-droxamate complex

will bind to form a neutral, octahedral complex of Fe^{3+}. The formation constants for even the simple monohydroxamic acids are very large and quite specific for Fe^{3+}. For acetohydroxamic acid ($R = CH_3$, $R' = H$) the pK_a is 9.35, and the logarithms of the stepwise formation constants K_1, K_2, and K_3 are 11.42, 9.68, and 7.2, for an overall formation constant, β_3, of the tris complex of 2×10^{28} (8, 9). In contrast, the overall forma-tion constant, β_2, for the bis complex of ferrous ion is only 3×10^8. That this sensitivity is caused more by the size of the ion than its charge can be seen in the β_3 values for the tris complexes of Al^{3+} (3×10^{21}) and La^{3+} (8×10^{11}) (9). The great disparity between the complexing strength of the hydroxamic acids for Fe^{3+} and Fe^{2+} is probably their most important property for iron transport, since the reduction of the ferric ion complex within the cell provides a ready means of releasing the complexed iron and freeing the ligand for another shuttle trip back to pick up Fe^{3+}.

The stabilities of the naturally occurring trishydroxamic acid com-plexes are among the greatest known. For example, the widely used and

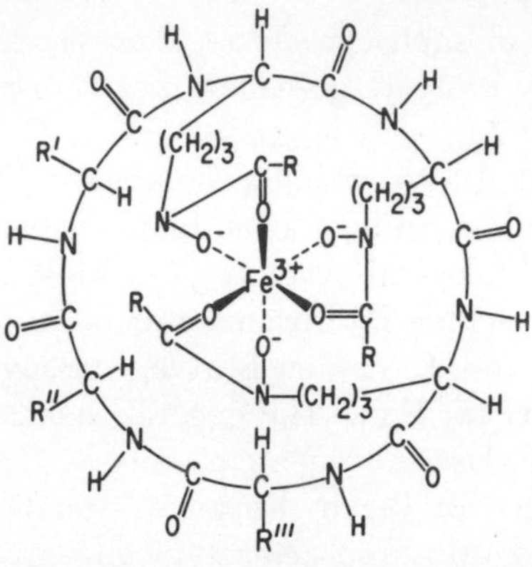

Siderochrome	R'	R''	R'''	R
Ferrichrome	H	H	H	CH_3
Ferrichrysin	CH_2OH	CH_2OH	"	"
Ferricrocin	H	"	"	"
Ferrichrome C	"	CH_3	"	"
Ferrichrome A	CH_2OH	CH_2OH	"	$-CH=C(CH_3)-CH_2CO_2H$ (*trans*)
Ferrirhodin	"	"	"	$-CH=C(CH_3)-CH_2CH_2OH$ (*cis*)
Ferrirubin	"	"	"	$-CH=C(CH_3)-CH_2CH_2OH$ (*trans*)
Albomycin δ_1	$-CH_2OSO_2-N$... 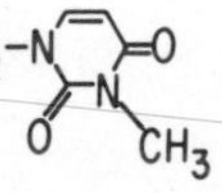	"	CH_2OH	CH_3

Figure 2.   Structure of the ferrichromes. The basic structural feature is a cyclic hexapeptide with the three hydroxamic acid linkages provided by a tripeptide of δN-acyl-δN-hydroxyl-l-ornithine. The Λ-cis coordination isomer is shown in each case.

very powerful hexadentate chelate EDTA has a formation constant log K of 25.1 while that for desferriferrichrome (Figure 2) is 29.1 and for desferriferrioxamine E (Figure 3) is 32.4 (*10*). The tris(hydroxamate) complexes typically are water-soluble, neutral compounds. In all of these complexes the iron is high-spin Fe(III) and, in contrast to the iron in the heme-containing proteins, is readily exchanged. This lability is, of course, expected for high-spin d^5 complexes, although the kinetics of exchange for these hexadentate ligands is much slower than, for example, tris bidentate complexes. The ferric ion can be removed from the complexes of the trihydroxamic acids by treating with dilute base or reduction of Fe(III) to Fe(II).

The structure of the simple hydroxamate complex tris(benzohydroxamato)iron(III) ($R = \phi$, $R' = $ H in Figure 1) has shown the most stable crystalline form of the solid to be the racemic cis isomer (*11*). (The convention for symbols of absolute configurations Δ and Λ are those of the IUPAC Proposal (*12*). The cis isomer is defined as the isomer which has C_3 point symmetry. *See* Ref. 3 for further discussion). Since both the cis and trans isomers of the chromic complex have been isolated (2), the similar geometries of the chromic and ferric compounds (vide infra) would indicate that the cis and trans ferric complexes are probably in solution in approximately the same proportions (60%, 40%, respectively), and it is the predominant cis isomer which crystallizes out. In the ferric complexes, the rapid isomerization of the complexes in solution therefore leads exclusively to crystallization of the cis isomer.

Most of the naturally occurring hydroxamic acids have three hydroxamic acid groups per molecule. The iron complexes of these trihydroxamic acids have a characteristic broad absorption band at 420–400 nm, and therefore originally were given the generic name siderochromes (1). The three hydroxamate groups are linked either as side arms from a cyclic peptide (as in the ferrichromes, Figure 2) or as part of a linear or cyclic chain (as in the ferrioxamines, Figure 3). Those with growth-

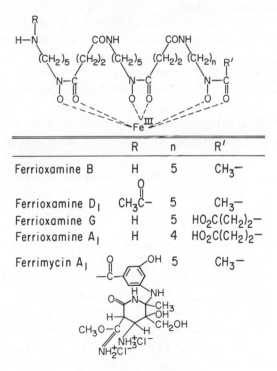

Figure 3. Structure of the linear ferrioxamines. The basic structural feature of the ferrioxamines is repeating units of 1-amino-5-hydroxyaminopentane and succinic acid. Ferrioxamine E is cyclic with n = 5 and an amide linkage such that there are no R or R′ substituents, but just a C–N bond.

promoting activity were named sideramines, and those that are antibiotics were named sideromycins.

The Hydroxamate-Containing Siderophores—Ferrichromes and Ferrioxamines. The ferrichromes (Figure 2) are all trihydroxamic acids produced by fungi such as *Ustilago sphaerogena* (*1*). The ferrioxamines are produced by several species of *Nocardia* and *Streptomyces* (*1*). In contrast to the ferrichromes, linear and cyclic ferrioxamines (Figure 3) have the three hydroxamate groups part of a polyamide chain like beads on a string. One other major difference is that the ligands themselves are not optically active. Only if a substituent group has an optical center, as in the ferrimycins, is there optical activity for the molecule. The ferrichromes have a natural optical activity associated with the ligand. Except in those cases where the ligand is optically inactive (in which case the complexes are racemic mixtures), the previous siderochrome complexes have been found to have a Λ-cis absolute configuration (*see* also Figure 2 in the previous chapter). Thus, while ferrioxamine E is racemic (*13*), x-ray structure analyses of ferrichrome A (*14*) and ferrichrysin (*15*) have shown both to be Λ-cis isomers. A recent structure analysis of the mixed hydroxamate-β-phenol imide siderophore manufactured by mycotic bacteria, mycobactin, has shown that ferric mycobactin also has Λ-cis absolute configuration (*16*). The other physical properties of the ferrichromes have been studied using several techiques. The NMR spectra of Al(III) and Ga(III) derivatives, as compared with the free ligand, have shown that a profound conformation change accompanies complex formation (*17*).

However, despite large differences in ligand molecular structure, all of the hydroxamate siderophores whose structures have been determined to date have been found to be cis complexes with a coordination geometry about the ferric ion which is substantially identical to the simple tris-(benzhydroxamato)–Fe(III) complex. Thus, while ferrioxamine E is racemic but with a cis geometry (*13*), x-ray structure analyses of ferrichrome A (*14*) and ferrichrysin (*15*) have shown both to be Λ-cis isomers.

The Catechol-Containing Siderophore—Enterobactin. The isolation and characterization of the cyclic triester 2,3-dihydroxy-*N*-benzoyl-*l*-serine, a tricatechol siderophore (Figure 4), were independently reported by both Pollack and Neilands (*18*) and O'Brien and Gibson (*19*). The ligand was isolated from cultures of *Salmonella typhimurium* and *Escherichia coli* and given the names enterobactin and enterochelin, respectively. Enterobactin is an efficient cellular transport agent but, unlike ferrichrome, intracellular release of the iron involves enzymatic hydrolysis of the enterobactin to the monomer, 2,3-dihydroxy-*N*-benzoyl-*l*-serine (*1*).

It can be seen from molecular models that two diastereoisomers are possible for the ferric enterobactin complex, Λ-cis and Δ-cis. These are not mirror images because of the optical activity of the ligand. The similarity of the roles played by the ferrichromes and enterobactin lent additional speculative interest to the preferred absolute configuration of the iron complex (*20*). The structural studies of the tris catechol complexes (vide infra) and the spectroscopic properties of the chromic

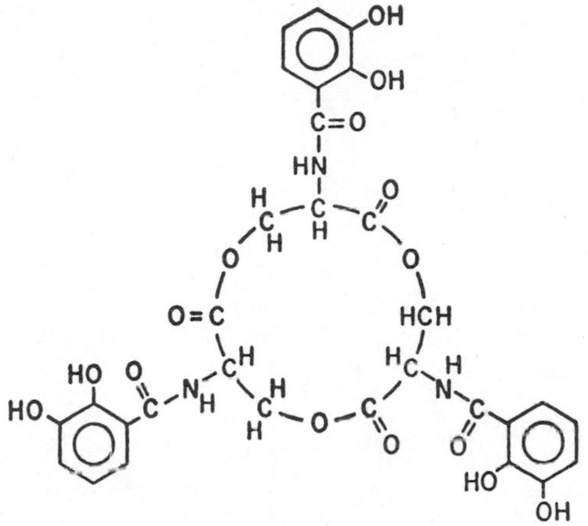

Figure 4. Structural diagram of enterobactin

enterobactin complex have led to an assignment of geometry for the most stable isomer of the ferric enterobactin complex (Figure 5).

Replacement of Ferric Ion by Chromic Ion in Siderophores

Hydroxamate Siderophores. GEOMETRIC ISOMERS. Many of the questions regarding the structure–function relationship of the siderophores could not be answered in detail because of the kinetic lability of these high-spin Fe(III) complexes. This lability always left ambiguous, for example, whether or not metal transport occurs via uptake of the intact molecular complex. Surprisingly, the coordination chemistry of the siderophore ligands with metal ions other than ferric was largely unknown. (A brief report of the CD spectrum of the Cr(III) complex of desferriferrichrysin has appeared (*21*). However, the complex apparently was not isolated, and the CD spectrum was not interpreted.) We therefore began to investigate the coordination geometries of siderophore ligands or their ligand moieties with kinetically inert trivalent metal ions such as Co(III) and Cr(III). Since hydroxamic acids are unsymmetrical bidentate ligands, there are both geometric and optical isomers in tris-

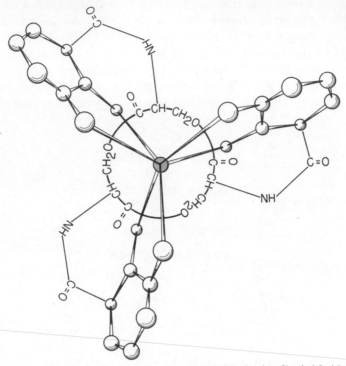

Figure 5. A schematic of the Δ-cis isomer of chromic and ferric enterobactin. The metal lies at the center of a distorted octahedron formed by the oxygen atoms of the three catechol dianions.

(hydroxamate) complexes. As noted earlier for an octahedral complex formed with three equivalent optically active hydroxamate anions, there are two geometric isomers possible—trans and cis. Each geometric isomer consists of Δ and Λ optical isomers (*12*). Often these are diastereoisomers because of the ligand optical activity, in which case there are four possible isomers—Λ-cis, Λ-trans, Δ-cis, and Δ-trans.

Preliminary exploratory research was directed toward preparing and characterizing Cr(III) or Co(III) complexes. These are d^3 and low-spin d^6 metal ions, respectively, which have the greatest possible ligand field stabilization energy and hence are kinetically inert toward ligand substitution and isomerization reactions. This is in contrast to the high-spin d^5 ferric ion which has zero ligand field stabilization energy (*22*). Thus, in contrast to the ferric siderophore complexes, chromic or cobaltic-substituted complexes should be kinetically inert.

MODEL HYDROXAMATE COMPLEXES. Attempts to prepare tris(hydroxamate) complexes of Co(III) with benzohydroxamic acid or its

N-methyl derivative resulted in oxidation of the ligand with concomitant reduction of Co(III) to Co(II). The preparation of tris(benzohydroxamato)chromium(III), Cr(benz)$_3$, was successful and resulted in the separation and characterization of its two geometric isomers (2). The half-lives for isomerization of these complexes near physiological conditions is on the order of hours. To facilitate the separation of all four optical isomers of a simple model tris(hydroxamate)chromium(III) complex, we prepared (using l-menthol as a substituent) the optically active hydroxamic acid, N-methyl-l-menthoxyacethydroxamic acid (men). This resulted in the separation of the two cis diastereoisomers of tris(N-methyl-l-menthoxyacethydroxamato)chromium(III) from the trans diastereoisomers and their characterization by electronic absorption and circular dichroism spectra.

Thin layer chromatography of the tris(benzohydroxamato)chromium(III) complex resulted in two green bands, corresponding to the cis and trans isomers, whose elution R_{st} values bracketed that of the one broad reddish-brown band of the Fe(III) complex. As just described, the geometric isomers of the Fe(III) complex are in rapid equilibrium in solution, and as a result, the mixture of these isomers elutes as one band with an R_{st} value that is a weighted average of the two individual isomers.

The tris(N-methyl-l-menthoxyacethydroxamato)chromium(III) and -iron(II) complexes, Cr(men)$_3$ and Fe(men)$_3$, were also purified by thin layer chromatography. The iron complex gives one broad reddish-brown band whose elution R_{st} value is bracketed by the bluish-green bands of the cis and trans isomers of the Cr(III) complex (2). As with the tris(benzohydroxamate) complexes, this behavior is caused by the rapid equilibration of the kinetically labile ferric complex.

The isomers of Cr(men)$_3$ isomerize with half-lives (several hours) similar to the Cr(benz)$_3$ complex. The rate of isomerization of the tris-(hydroxamate) complexes is therefore not particularly sensitive to the substituent of the hydroxamate nitrogen atom, since the men ligand contains an alkylated nitrogen atom, and the benz ligand contains an unsubstituted nitrogen atom. In the absence of an induced strain, the corresponding siderophore complexes must isomerize much more slowly because of the steric constraints of the ligand.

Although four diastereoisomers (Λ-cis, Λ-trans, Δ-cis, and Δ-trans) are expected for Cr(men)$_3$, thin layer chromatography of the complex yielded only three bluish-green bands. Two of these are the resolved Λ-cis (10%) and Δ-cis (21%) isomers, and the third (69%) is an unresolved mixture of the Λ-trans (31%) and Δ-trans (38%) isomers.

One other key difference between the chromic and ferric ions is their spectroscopic properties. Since ferric ion is a high-spin d^5 ion in the siderophore complexes, there are no spin-allowed d–d electronic transitions. Thus the vis-uv absorption spectra of the siderophores are not all caused by metal chromophore centers but rather are from ligand–metal or ligand–ligand transitions (largely charge transfer) which vary enormously from one compound to another, even though the coordination geometry about the Fe(III) may be the same. In contrast, octahedral (or nearly octahedral) complexes of Cr(III) have two well established d–d absorption bands that are localized on the metal chromophore and thus are insensitive to changes in the metal–ligand complex which is outside the immediate coordination sphere of the metal.

CHROMIC FERRICHROME COMPLEXES. The spectra for the model chromic hydroxamate complexes are reproduced in Figure 6. Since the visible and CD spectra of the isomers are wholly dominated by the metal complex chromophore, these data can be used to characterize and to identify coordination isomers of complexes formed by the siderophores. The preparation and characterization of the chromic complexes of desferriferrichrome and desferriferrichrysin have been reported (3). Although an examination of molecular models for both complexes shows two coordination isomers are possible (Λ-cis and Δ-cis), both chromic complexes consist exclusively of the Λ-cis isomer. These results agree with x-ray crystallographic investigations which have shown that both ferrichrysin and ferrichrome A crystallize as only the Λ-cis isomer (14, 15). Both chromic complexes have identical CD spectra which are the same as the Λ-cis Cr(men)$_3$ spectrum (Figure 6).

CHROMIC FERRIOXAMINE COMPLEXES. The preparation and characterization of chromic complexes of ferrioxamine B (see Figure 3) have been reported (4). From an examination of molecular models, the five geometric isomers (one cis and four trans) shown in Figure 7 are possible. Each of these isomers exists as a racemic mixture, and the separation of the cis geometrical isomer was accomplished. A second fraction was isolated which consists of one or more trans isomers. The geometries of these isomers were assigned on the basis of their vis-uv spectra (Figure 8) which are superimposable upon those of the cis-and trans-Cr(men)$_3$ complexes (Figure 6).

Both the cis and trans geometrical isomers of chromic ferrioxamine B isomerize to equilibrium solutions with half-lives of several days at room temperature. This is considerably slower than that found for the simple tris hydroxamate complexes such as Cr(men)$_3$ and is caused by the steric constraints of the ferrioxamine B ligand and its hexadentate chelation.

Catecholate Siderophores. SIMPLE CATECHOL COMPLEXES. As noted earlier, the common siderophore for enteric bacteria is the tricatechol, enterobactin (Figure 4). In order to perfect synthetic and

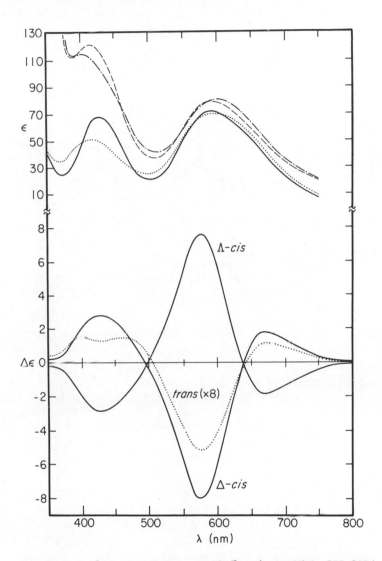

Figure 6. Absorption spectra of Cr(benz)₃ in 17% CH₃OH/ CHCl₃ solution and both absorption and CD spectra of Cr(men)₃ in 3% CH₃OH/CHCl₃ solution.

cis-Cr(benz)₃, (– – –); trans-Cr(benz)₃, (– · –); cis-Cr(men)₃, (———); trans-Cr(men)₃, (· · ·). *The CD spectrum of the mixture of trans isomers (31% Λ, 38% Δ) has multiplied by eight since the net optical activity of the Λ, Δ mixture is small. The CD bands near 415 nm are assigned as the high energy ⁴A₂ → ⁴E transition (point group C₃) which come from the ⁴A₂g → ⁴T₁g absorption band in octahedral symmetry. The large bands near 570 nm are assigned as the low energy ⁴A₂ → ⁴E transition, and the bands near 670 nm are assigned as the ⁴A₂ → ⁴A₂ transition. Both of these transitions come from the ⁴A₂g → ⁴T₂g absorption band in octahedral symmetry.*

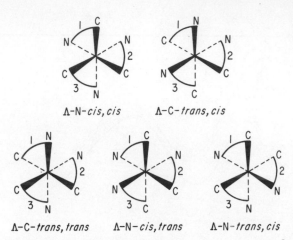

Figure 7. The five enantiomeric geometrical isomers of ferrioxamine B. The oxygen donor atoms of each hydroxamate group have been omitted for clarity. The Λ optical isomer is shown in each case. See Ref. 4 for nomenclature of these geometrical isomers.

separation techniques to be used with the small amounts of enterobactin available, simple catechol complexes were prepared as model compounds. Spectroscopic data of the simple model compounds then could be used in assigning geometries for enterobactin isomers. The previous chemical literature of tris(catechol) complexes of transition metal ions is sparse.

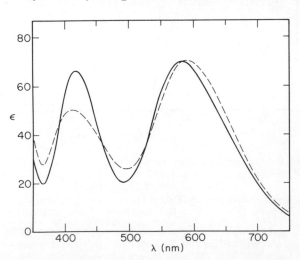

Figure 8.   Absorption spectra of the cis isomer and trans isomers of chromic desferriferrioxamine B in aqueous solution.   Cis, (———); trans, (– – –).

The only reference to a chromic complex reported that it was rapidly hydrolyzed in dilute aqueous solution (23). This, of course, would preclude separation of optical isomers of the tris chelates. Nevertheless, these complexes were reinvestigated before preparing the chromic enterobactin complex. It was found that the complexes are very stable in the absence of oxygen (24). The usual oxygen sensitivity of the catechol dianion was found to be substantially increased in the chromium complex. (The ease of oxidation of coordinated catechol and related ligands has been demonstrated for a series of metal complexes (Ni^{2+}, Cu^{2+}, Zn^{2+}, etc.) by Holm et al. in relation to the 1,2-benzenedithiolato analogs (25)). It is this oxidation of the chromium complex that causes the green-to-red color changes reported previously as hydrolysis. All preparations and handling of the chromium catechol complex were therefore carried out under inert atmosphere conditions.

Although only partial resolution of solutions of $[Cr(cat)_3]^{3-}$ was achieved at neutral pH, complete resolution was attained at pH 13 and 5°C. The rate of loss of optical activity for resolved $[Cr(cat)_3]^{3-}$ was found to depend strongly on hydrogen ion concentrations, varying from half-times of several minutes to several hours between pH 7 and pH 13 (24).

COMPARISON WITH CHROMIC ENTEROBACTIN. The visible and circular dichroism spectra of $[Cr(cat)_3]^{3-}$ and $[Cr(enterobactin)]^{3-}$ complexes are shown in Figures 9 and 10. The absorption spectra are similar except that the ligand-localized transition occurs at lower energy in the enterobactin complex, thus masking the $^4A_{2g} \rightarrow {}^4T_{1g}$ (for D_h symmetry) d–d transition, which appears as a shoulder on the edge of the more intense $\pi \rightarrow \pi^*$ ligand transitions. This is apparently caused by the fact that enterobactin contains ortho-acyl-substituted catechol rings.

Thus the vis-uv spectra of $[Cr(cat)_3]^{3-}$ and $[Cr(enterobactin)]^{3-}$ are too dissimilar to allow detailed comparisons and confident prediction of structure based on such comparisons. However, there is a dramatically different situation found in comparing the CD spectra of $[Cr(cat)_3]^{3-}$ and $[Cr(enterobactin)]^{3-}$, which are found to be substantially identical (Figure 10). This is because the interfering charge transfer band is not associated with the chiral center and hence does not contribute to the optical activity.

The crystal and molecular structure of a salt of $[Cr(cat)_3]^{3-}$ and the known $[Cr(cat)_3]^{3-}$ absolute configurations give the following assignment: the predominant isomer of the chromic enterobactin monomeric complex has a Δ-cis absolute configuration (Figure 5). The similarity of the chromic and ferric complexes allows this assignment to be made for the ferric complex as well. This is the opposite absolute configuration of the other optically active siderophores characterized to date. The opposite

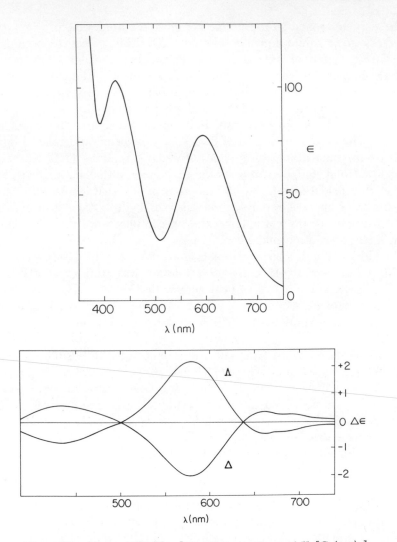

*Figure 9.   (a) (top) Visible absorption spectrum of $K_3[Cr(cat)_3]$
in water.   (bottom) Circular dichroism spectra of Δ- and Λ-
$K_3[Cr(cat)_3]$ solutions.*

absolute configurations of chromic enterobactin and chromic ferrichrome
can be seen clearly in comparing their CD spectra (Figure 10b). The
role of the siderophores as cellular permeases for ferric ion therefore
does not depend on the complex always having a Λ-cis configuration,
although this configuration or others may be specifically transported in
individual microbial-ligand systems.

The molecular structure of enterobactin has not as yet been established by diffraction techniques and, although coordination of ferric ion

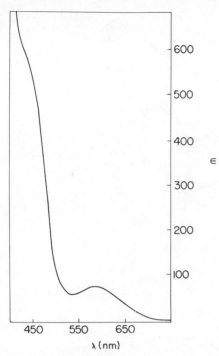

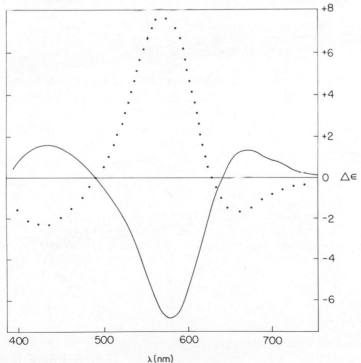

Figure 10.   (a) (left) Visible absorption spectrum of $[NH_4]_3[Cr(enterobactin)]$. (b) (below) Circular dichroism spectra of Δ-$[NH_4]_3[Cr(enterobactin)]$ (———) and chromic ferrichrome ($\cdots\cdots$) (the latter from Ref. 3).

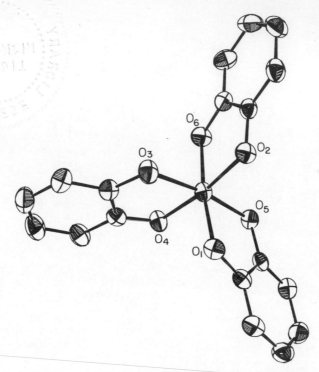

Figure 11. A perspective drawing of the [M(O₂C₆-H₄)₃]³⁻ anions. M = Cr, Fe, as viewed down the molecular threefold axis.

by enterobactin previously had been assumed to be an octahedral complex which involves only the catechol moieties of the ligand, no firm structural evidence for this was available. Furthermore, the use of Cr-(III) in place of Fe(III) to enable transport studies of optically active

Table I.   Structural Parameters

Charge, $n =$	P^a
	1
Average M–O distance (Å)	1.723 (4)
Average ring O–M–O angle (°)	91.4 (2)
Average O–O ring distance (Å)	2.466 (6)
Ligand bite[e]	1.431
Trigonal twist angle (°)[f]	58.9
Plane-to-plane distance (Å)[g]	1.940

[a] Ref. *27*.
[b] Ref. *28*.
[c] Ref. *29*.
[d] Ref. *26*.
[e] Ratio of the O–O ring distance to M–O distance. *See* Ref. *30*.

siderochrome complexes has been justified here and in the previous chapter on the basis that such complexes would be isostructural. High-spin Fe(III) and Cr(III) are within 0.03 Å in ionic radius of one another, but the crystal field stabilization energy (CFSE) for the chromic complex (12 Dq) is considerably greater than that for high-spin ferric ion (0 Dq). Any shift by the chromic complex towards octahedral from trigonal prismatic coordination, as evidenced by the trigonal twist angle, may be attributed to this crystal field effect.

GEOMETRY OF METAL CATECHOL COMPLEXES. The coordination geometries of $[Fe(cat)_3]^{3-}$ and $[Cr(cat)_3]^{3-}$ have been determined by single crystal studies of the salts $K_3[M(cat)_3] \cdot 1.5 \, H_2O$ ($M = Cr, Fe$) in order to explore the crystal field effect of chromic ion on the coordination geometry and, indirectly, to determine the coordination geometry of enterobactin itself (26).

The $[M(cat)_3]^{3-}$ complexes (Figure 11) are distorted from octahedral geometry with approximately D_3 molecular point symmetry. The structural parameters of the tris(catechol) complexes reported to date are compared in Table I. The ligand bite (ratio of the O–O ring distance to the M–O distance), the trigonal twist angle, and the trigonal plane-to-plane distance vary smoothly across the table as ionic radii increase. The final geometry represents a balance between distortions of the O–M–O angle and O–O ring distance and variations of the twist angle from octahedral to trigonal prismatic. In comparing the chromic and ferric catechol structures, the difference in M–O bond length is not large enough to cause the nearly six-degree difference in twist angle. This must be attributed to the difference in crystal field stabilization energy (ΔCFSE) between octahedral and trigonal-prismatic geometries. Although significant in terms of the precision of the structure determinations, the ferric and chromic complexes are close enough in geometry to regard similar

for $[M(cat)_3]^{n-}$ Complexes (26)

Si^b	As^c	Cr^d	Fe^d
2	1	3	3
1.784(18)	1.843(5)	1.986(4)	2.015(6)
88.7(2)	88.2(5)	83.56(14)	81.26(7)
2.490(6)	2.565(7)	2.646(6)	2.625(2)
1.396	1.392	1.333	1.303
55.9(5)	55.2(10)	50.5(6)	44.7(10)
2.093	2.194	2.247	2.303

f This angle is defined by viewing the complex in projection down the molecular three-fold axis. It is then the rotation required to bring the top and bottom planes (of three oxygen atoms each) into coincidence. This angle is 60° for octahedral and 0° for trigonal prismatic coordination.

g Plane-to-plane distance for the two trigonal oxygen atom planes described in *f*.

Journal of the American Chemical Society

chromic-substituted siderophore complexes as structurally identical to the natural ferric complexes for biological purposes.

FERRIC–CATECHOLATE FORMATION CONSTANTS. The very high affinity for ferric ion which all siderophores display is essential to their role in obtaining iron for the microorganism using the ligand. This is always accomplished in an environment which also contains many other strong complexing agents for ferric ion. Hence a very high formation constant for the siderophore complex is essential for survival in the competitive world of the microorganism. As described earlier and reviewed elsewhere (1), the hydroxamate siderophores have formation constants for reactions of the type:

$$Fe^{3+} \;+\; \left[\underset{N-C}{\overset{O^- \; O}{}} \right]_3 \;\rightleftharpoons\; Fe\left[\underset{O-C}{\overset{O-N}{}} \right]_3$$

which range between 10^{30} and 10^{32}. These values are only two to three orders of magnitude greater than the overall formation constant, β_3, for the simple tris(monohydroxamate) complexes.

In contrast to the hydroxamate siderophores, little or nothing is known about the stability constant for the catechol siderophore, enterobactin. Prior to determining the formation constant of enterobactin (for which hydrolysis of the ligand presents special problems), the reaction of catechol itself with ferric ion has been investigated (31).

Catechol is a very weak acid and hence at low pH is a poor ligand. The kinetics and equilibria of its reactions with ferric ion under acidic conditions have been investigated (32). Under such conditions, even with excess catechol, ferric ion forms only a transient 1:1 complex which eventually undergoes a redox reaction to give ferrous ion and orthoquinone as products. This reaction has a redox potential just greater than zero at pH 1. At higher pH's the extremely large formation constant of the tris catechol ferric complex strongly reverses the potential, such that ferrous ion will reduce orthoquinone to form the tris catechol ferric complex. In the absence of air, both the chromic and ferric tris catechol complexes are stable indefinitely in basic aqueous solution (24).

The equilibrium constants involved in the reaction $Fe^{3+} + 3\,cat^{2-} \rightleftharpoons Fe(cat)_3{}^{3-}$ were determined as follows. An aqueous solution of Fe^{3+} ($5.5 \times 10^{-3}M$) and catechol ($1.48 \times 10^{-2}M$), initially made basic with the addition of KOH, was titrated with $1.24M$ HCl under an oxygen-free atmosphere at 22° and ionic strength (KCl) 0.16–0.22M (Figure 12). The acid dissociation constants for catechol were determined independently (under similar experimental conditions) to be $pK_{a1} = 9.38$ and

$pK_{a2} = 13.28$. (All stability and association constants mentioned are corrected to ionic strength of $0.1M$.) Using these constants and literature values for the hydrolysis constants of Fe(III) ($FeOH^{2+}$, $Fe(OH_2^+)$, $Fe_2(OH)_2^{4+}$), a classical Bjerrum $\bar{n}$ vs. pL plot produced approximate values of the metal-ligand stability constants. Least squares refinement of the cumulative stability constants converged at the values log $\beta_1 =$

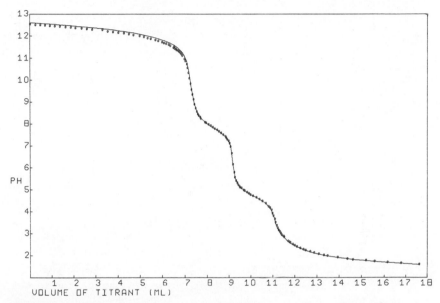

Figure 12.   Titration curve.   An initially basic aqueous solution of Fe^{3+} (5.5×10^{-3}M) and catechol (1.84×10^{-2}M) is titrated with 1.24M HCl at 22° under an oxygen-free atmosphere.   (———), least squares fit to the observed data (discrete points).   Data past pH 10 were given zero weights.

21.5, log $\beta_2 = 36.6$, and log $\beta_3 = 45.9$. The inclusion of iron hydrolysis in the refinement model primarily affected the calculated value of β_1. The distribution of the various species in solution as a function of pH is shown in Figure 13.

The weak acidity of catechol makes its effective formation constant much less than $10^{45.9}$ near physiological pH. However, any chelate effect should tend to make the formation constant for enterobactin in larger than β_3 for catechol. Thus 10^{45} can be regarded as a lower bound for the reaction $Fe^{3+} + ent^{6-} \rightleftarrows Fe(ent)^{3-}$.

Summary

This paper has focused on the coordination chemistry of the siderophores. At this stage in our studies of metal-substituted siderophores we have established the following:

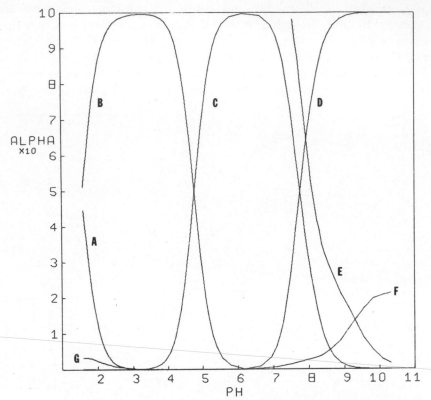

Figure 13.   Distribution curve, as a function of pH, of the various species
formed in the ferric-catechol titration experiment.
(A) free Fe^{3+}; (B) $Fe(cat)^+$; (C) $Fe(cat)_2^-$; (D) $Fe(cat)_3^{3-}$; (E) H_2cat; (F) $(H\ cat)^-$;
(G) ferric hydrolysis products ($FeOH^{2+}$, $Fe_2(OH)_2^{4+}$, $Fe(OH)_2^+$). Cat, catecholate
dianion; ALPHA, concentration of the particular species divided by the total iron
concentration.

(1) The chromic-substituted siderophore complexes can be prepared
and, in contrast to the naturally occurring ferric complexes, are kinetically inert to isomerization or ligand substitution.

(2) The visible and circular dichroism spectra of the chromic siderophore complexes are closely related to the corresponding spectra of simple
model complexes of hydroxamate or catecholate ligands. This provides
a spectroscopic probe for structure in assigning the geometries of the
siderophore complexes.

(3) The structure and bonding of the chromic and ferric complexes
(despite their differences in kinetic properties) are sufficiently alike to
regard them as identical for biological systems.

(4) The chromic-substituted siderophores can be used to study the
mechanisms of microbial iron transport. These studies rely on the kinetic
inertness of the chromic complex and would be impossible to carry out
using other techniques or probes.

Acknowledgment

I am pleased to acknowledge my co-workers, past and present, whose efforts have been summarized here. They are John Leong, Stephan Isied, Alex Avdeef, Frank Fronczek, Leo Brown, Jim McArdle, Hunter Nibert, and Gilbert Kuo. The collaboration of J. B. Neilands continues to be a seminal influence. This research has been supported by USPHS grant AI-11744.

Literature Cited

1. Neilands, J. B., Ed., *Microbial Iron Metabolism*, Academic Press, New York, N.Y. 1974.
2. Leong, J., Raymond, K. N., *J. Am. Chem. Soc.* (1974) **96**, 1757.
3. *Ibid.* (1974) **96**, 6628.
4. *Ibid.* (1975) **97**, 293.
5. Biedermann, G., Schindler, *Acta Chem. Scad. II* (1957) 731.
6. Leussing, D. L., Kolthoff, I. M., *J. Am. Chem. Soc.* (1953) **75**, 2476.
7. Spiro, T. G., Saltman, P., *Struct. Bonding* (1969) **6**, 116.
8. Schwarzenbach, G., Schwarzenbach, K., *Helv. Chim. Acta* (1963) **46**, 1390.
9. Anderegg, G., L'Eplattenier, F., Schwarzenbach, G., *Helv. Chim. Acta* (1963) **46**, 1400.
10. *Ibid.* (1963) **46**, 1409.
11. Lindner, Von H. J., Gottlicher, S., *Acta Cryst.* (1969) **B25**, 832.
12. *Inorg. Chem.* (1970) **9**, 1.
13. Poling, M., Van der Helm, D., Book of Abstracts, American Crystallographic Association, Spring Meeting, Berkeley, 1974, abstract Q7, p. 111.
14. Zalkin, A., Forrester, J. D., Templeton, D. H., *J. Am. Chem. Soc.* (1966) **88**, 1810.
15. Bränden, C. I., private communication, 1974.
16. Hough, E., Rogers, D., *Biochem. Biophys. Res. Commun.* (1974) **57**, 73.
17. Llinás, M., *Struct. Bonding* (1973) **17**, 135.
18. Pollack, J. R., Neilands, J. B., *Biochem. Biophys. Res. Commun.* (1970) **38**, 989.
19. O'Brien, I. G., Gibson, F., *Biochim. Biophys. Acta* (1970) **215**, 393.
20. Neilands, J. B., "Structure of Microbial Iron Transport Compounds" in "Structure and Function of Oxidation Reduction Enzymes," Å. Akeson and A. Ehrenberg, Eds., pp. 541–547, Pergamon, Oxford, 1972.
21. Buerer, T., Gulyas, E., *Proc. 9th Int. Conf. Coord. Chem., St. Moritz-Bad* (1966) 512.
22. Basolo, F., Pearson, R. G., "Mechanisms of Inorganic Reactions," 2nd ed., Wiley, New York, 1967.
23. Weinland, R., Walter, E., *Z. Anorg. Allg. Chem.* (1923) **126**, 141.
24. Isied, S. S., Kuo, G., Raymond, K. N., *J. Am. Chem. Soc.* (1976) **98**, 1763.
25. Röhrscheid, F., Balch, A. L., Holm, R. A., *Inorg. Chem.* (1966) **5**, 1542.
26. Raymond, K. N., Isied, S. S., Brown, L. D., Fronczek, F. R., Nibert, J. H., *J. Am. Chem. Soc.* (1976) **98**, 1767.
27. Allcock, H. R., Bissell, E. C., *J. Am. Chem. Soc.* (1973) **95**, 3154.
28. Flynn, J. J., Boer, F. P., *J. Am. Chem. Soc.* (1969) **91**, 5767.
29. Kobayashi, A., Ito, T., Marumo, F., Sacto, Y., *Acta Crystallogr., Sect. B* (1972) **28**, 3446.
30. Wentworth, R. A. D., *Coor. Chem. Rev.* (1972) **9**, 171.

31. Avdeef, A., Sofen, S. R., Bregante, T. L., Raymond, K. N., unpublished data.
32. Mentasti, E., Pelizzetti, E., Saini, G., *J. Chem. Soc., Dalton Trans.* (1973) 2605, 2609.

RECEIVED July 26, 1976.

Role of Iron in the Regulation of Nutritional Immunity

IVAN KOCHAN

Department of Microbiology, Miami University, Oxford, OH 45056 and
Wright State University, School of Medicine, Dayton, OH 45431

*Iron and iron-binding bacterial products, called sidero-
phores, promote the growth of avirulent and virulent strains
of* Mycobacterium tuberculosis *and* Escherichia coli *in
mammalian sera and in mice. Without iron, serum-exposed
bacteria die quickly. This effect can be neutralized with
exogenous iron or with siderophores which supply bacteria
with iron of transferrin–iron complexes. The iron starvation
imposed by serum is strengthened in infected animals by
developing generalized and localized hypoferremia and by
fever. Unlike most bacteria, highly virulent bacteria can
obtain iron and multiply in normal mammalian sera because of the outer membrane-associated lipopolysaccharide
which participates in the acquisition of iron by bacterial
cell. The injection of infected animals with iron or with
siderophores promotes bacterial growth and the development of infectious disease.*

In addition to factors which cause direct injury to microbial cells, an animal body possesses a more subtle mechanism for defense against microbial parasitism. Animals can starve parasites by limiting the availability of nutrilites which are essential for microbial growth and multiplication. An acute competition between parasites and their hosts for nutrilites essential for the welfare of their cells is most evident in cases in which a given nutrilite cannot be synthesized either by the host or by the parasite. Only after the death of the host does the competition for various nutrilites cease, because dead tissues provide the microorganisms with the essential nutrilites.

Through the centuries of coexistence, animals and their microbial parasites have developed elaborate mechanisms to satisfy their requirements for growth-essential nutrilites. These competitive host–parasite interactions became evident only recently, when techniques were developed to study the behavior of molecules in living systems. The effectiveness with which animals can make nutrilites unavailable for the invading parasites may determine their resistance to initiation of microbial infection. Experimental findings and clinical observations of the nutritional starvation of microbes in animals led to formulation of the concept of nutritional immunity (1, 2, 3). By definition, nutritional immunity expresses the native and acquired abilities of an animal to restrict the availability of growth-essential nutrilites for use by parasitic microorganisms. Only a limited effort has been made, and that in only a few bacterial infections, to determine the effects of the nutritional depletion on the development and the progression of infectious diseases.

Iron seems to be the most clearly defined of the requirements for essential nutrilites needed for the growth of microorganisms in an animal host. Several studies indicate that an acute competition for iron between iron-binding materials of host and parasites characterizes the host–parasite relationship (1, 4, 5, 6, 7). In an animal body, iron is associated with iron-binding proteins and, therefore, it is usually unavailable for use by microbial parasites. To satisfy their need for iron, microorganisms produce strong iron-chelating products, siderophores, which remove the metal from iron-containing proteins and supply it to invading microbes. The effectiveness with which microorganisms produce the iron-providing siderophores determines the speed of their multiplication in animals and may be an indicator of microbial virulence (8, 9, 10).

The purpose of this chapter is to describe the competition for iron between iron-binding proteins of the animal and the siderophores of bacterial parasites. This discussion will be limited to two bacterial species—a slow-growing organism *Mycobacterium tuberculosis* and a fast-growing organism *Escherichia coli*. Both organisms produce specific siderophores which have been defined chemically and physically. Mycobactin, the siderophore of *M. tuberculosis*, because of its hydrophobic nature, is associated mostly with the lipoidal cell wall of the tubercle bacillus (11) whereas enterochelin (enterobactin), the siderophore of *E. coli* and *Salmonella typhimurium*, is soluble in water and is rapidly lost by the bacterial cell into the surrounding medium (12, 13).

Experimental Methods

Since iron carriers of the host become ineffective in dead tissues, the retention of physiological conditions in isolated fluids and tissues is of crucial importance in studies of the microbial quest for iron. Usually, the maintenance of excised tissues or blood at 3°C during the preparation of cell lysates or the separation of serum and the subsequent adjustment of the processed materials to pH 7.5 preserve the iron-binding activity of iron carriers and the stability of their complexes with the metal. It is much more difficult to avoid contamination of the processed material with exogenous iron. Traces of iron are present on glassware, in water, and in various inorganic or organic compounds used to prepare microbial media. Exogenous iron interferes with the measurements of the iron-starvation of bacteria exposed to iron-binding substances of animal tissues and fluids. Also, exogenous iron obliterates the competition for iron between iron-chelating substances of host and microbial origin. There is no easy method to remove contamination iron from media. It has been done with some success by repeated extractions of the medium with 8-hydroxyquinoline and chloroform (*14*) and by passage of the medium through columns of Chelex resin (*15*).

Our investigation of the competition for iron between iron-binding substances of animal and microbial origin has been facilitated by the use of a mixture composed of one part mammalian serum and three parts iron-poor medium (IPM) or iron-poor agar medium (IPAM). IPM contained the following ingredients in 1 L of double-distilled water: asparagin, 2.0 g; N-Z amine, 1.5 g; KH_2PO_4, 2.5 g; $MgSO_4$, 10.0 mg; $CaCl_2$, 0.5 mg; $ZnSO_4$, 0.1 mg; $CuSO_4$, 0.1 mg; and dextrose, 5.0 g. IPAM was made by adding 15.0 g agar per 1.0 L IPM. The growth of fastidious tubercle bacilli in these media was promoted by adding 2 mg bovine albumin (fraction V) per ml of the medium. Although the minute amount of iron present in IPAM (0.09 μg/ml) is sufficient to support microbial growth, the medium becomes microbiostatic when combined with mammalian serum. Transferrin (Tr) in serum binds the iron in the medium, and in the absence of ionic iron, this serum–medium mixture does not support the growth of various microorganisms. To eliminate the antibacterial activity of antibody-complement system, sera used in these experiments were heated at 56°C for 30 min to destroy the complement.

In most of our studies we have used an agar-plate diffusion test (*16, 17*) to study the effects of restricted amounts of iron on the growth of various bacteria and the molecular competition for iron between iron-binding substances of animal and microbial origin. The test is performed in plastic petri dishes (No. 3002, 60 × 15 mm; Falcon Plastics) filled with 12 ml of either growth-supporting IPAM or growth-inhibiting serum-IPAM mixture. A well, 10 mm in diameter, is made in the agar medium by placing a glass cylinder in the center of each plate. After removing the glass cylinders from the gelled media, the wells in growth-supporting medium are filled with 0.4 ml saline containing iron-binding proteins (such as Tr, lactoferrin, and ovotransferrin), and the wells in microbiostatic serum-agar medium are filled with bacterial siderophores. Thus, depending on the nature of the medium, wells are filled with substances which either prevent or facilitate the acquisition of iron by bacteria. After the diffusion of test materials from wells into the agar (12–15 hr), the glass cylinders are replaced, and the surface of the gelled medium is inoculated uniformly with 0.25 ml bacterial suspension containing 15,000 cells. The plates inoculated with slow-growing tubercle bacilli are incubated for about 3 weeks at 37°C and 70% humidity. Mycobactin, the hydrophobic siderophore of tubercle bacilli, does not leave the lipoidal bacterial wall and, therefore, does not neutralize tuberculostasis in zones containing Tr–iron complexes. The plates inoculated with fast-growing E. coli are incubated at 37°C and are examined periodically during the 48-hr incubation. Usually after 35 hr these plates show little or no serum-imposed microbiostasis, because enterochelin, produced at places of bacterial growth, diffuses rapidly into zones of microbiostasis and alleviates it by making iron available for iron-starved bacteria. After an appropriate period of incubation, the growth-inhibiting and growth-supporting activities of various iron-binding substances are determined by measuring the widths of growth-void and growth-filled zones around wells charged with iron-restricting and iron-donating substances, respectively.

The fate of tubercle bacilli or E. coli in human or bovine serum was determined not only by the use of the agar-plate diffusion test but also by experiments performed in serum-containing test tubes. In tube experiments, serum is inoculated with bacterial suspensions adjusted to contain a known number of bacteria in the inoculum. Infected sera are incubated at 37°C and, after various periods of time, bacterial numbers are determined by plating the samples on the growth-supporting IPAM. The number of developed colonies on agar medium shows the number of bacteria in the experimental tubes.

Effects of mammalian sera on avirulent and virulent bacteria were determined by using attenuated (BCG) and virulent ($H_{37}Rv$) strains of tubercle bacilli and avirulent (strain A) and virulent (strain C) strains of E. coli. All strains were obtained from the culture collection of the Department of Microbiology, Miami University. Bacterial virulence and the infection-promoting effect of iron were determined in Swiss–Webster mice. The degree of virulence of tubercle bacilli for iron-untreated and iron-treated mice has been shown previously (2). In this study, the pathogenicity of E. coli strains was tested by three experiments in which groups of 10 mice were injected intraperitoneally with 5×10^8 cells of strain A and strain C. Mice infected with the strain C died within two days whereas 80–90% of mice inoculated with strain A survived the infection. On the basis of these and several similar experiments (see

Table VIII), bacteria of the strain C were designated as virulent and those of strain A as avirulent.

Before the use in experiments, *E. coli* strains were cultured in trypticase soy broth for 18 hr at 37°C. After this growth period, bacteria were collected by centrifugation, bacterial pellets were resuspended in saline, and bacterial suspensions were adjusted to a desired density by the Klett spectrophotometer. Methods for the cultivation, collection, and experimental use of tubercle bacilli have been described previously (*9, 16*).

Microbiostasis in Mammalian Sera

Thirty years ago, Shade and Caroline described the antimicrobial iron-binding protein Tr in human plasma and suggested that this ligand withholds the metal from microbial invaders (*18*). Tr binds two atoms of ferric iron by tyrosyl and histidyl residues with a stability constant of around 10^{30} (*19*). The stability of Tr-iron complexes is influenced by pH while on the alkaline side of pH 7, complexes are quite stable, on the acidic side they dissociate. We have found that bacterial growth in serum becomes more prolific when its pH is progressively lowered from pH 7 to 5. At the physiological pH, Tr binds iron well and carries the metal to various cells of the body; 100 ml human serum contains 200–300 mg Tr and only 110 μg iron. Since 1 mg Tr can bind 1.25 μg iron, Tr is usually saturated only to one third of its total iron-binding capacity. Considering the loss of less than 0.1 mg iron per day in urine, the concentration of free iron in plasma is less than 0.01 μg per ml, and this is insufficient to support the growth of bacteria. Thus, in order to grow in animal blood, bacteria have to possess a way to obtain iron from Tr–iron complexes.

The correlation between the degree of bacterial inhibition in serum and the percentage of iron-free Tr was investigated by determinations of serum iron (SI), total iron-binding capacity (TIBC), and tuberculostasis in various mammalian sera. The results showed that most mammalian sera do not differ in TIBC values. Pronounced differences, however, were observed in SI values (Table I). The fate of tubercle bacilli in sera was determined by the degree of Tr saturation with iron; human and bovine sera with approximately 30% saturated Tr were strongly tuberculostatic, rabbit and mouse sera with approximately 60% saturated Tr were weakly tuberculostatic, and guinea pig serum with approximately 85% saturated Tr supported bacillary growth (*20*). These and similar findings demonstrated that the higher the percentage of iron-saturated Tr, the less inhibitory is the serum for the growth of tubercle bacilli.

The effect of Tr–iron interplay on the fate of tubercle bacilli has been investigated by adding ferric chloride and iron-free Tr in various

Table I.   Correlations between Levels of Iron-saturated
Tr and Degrees of Bacillary Growth
in Mammalian Sera (20)

Source of Serum	Number of Tests	Amt. of Iron (μg)[a] TIBC[b]	SI[c]	Tr Saturation[d] (%)	Tuberculostasis in Serum[e]
Human	10	327	97	30.0	present
Cow	4	490	191	39.0	present
Rabbit	8	317	204	64.3	limited
Mouse	10	382	230	60.2	limited
Guinea pig	20	323	273	84.4	absent

[a] Individual determinations of TIBC and SI fell within 10% variation of the mean value shown in the table.
[b] TIBC value shows the mean of the amount of iron present in 100 ml iron-saturated serum.
[c] SI value shows the mean of the amount of iron present in 100 ml untreated serum.
[d] Percentage of iron-saturated Tr in serum sample equals (SI/TIBC) × 100.
[e] Tuberculostasis was scored as "present" when bacillary growth was less than one generation, "limited" when it was less than five generations, and "absent" when it varied between five and 14 generations during a two-week incubation period.

Journal of Bacteriology

proportions to tuberculosatic human serum (21). These results showed that the addition of iron vitiated the tuberculostasis of serum; between 0.5 and 1.0 μg of iron neutralized the tuberculostatic activity in 1 ml of 1:4 diluted serum. Furthermore, the addition of Tr to iron-neutralized serum reconstituted tuberculostasis (Table II). The data showed quite clearly that the amounts of Tr which were effective in the reconstitution of tuberculostasis in iron-neutralized serum correlated closely with the known iron-binding capacity of this protein.

Table II.   Iron Neutralization of Tuberculostasis in a 1:4 Dilution of
Human Serum and the Reconstitution of Tuberculosis in
Iron-Neutralized Serum with Tranferrin (21)

Transferrin (mg/ml)	Generations in Serum with Iron[a] ($\mu g/ml$) 0.0	0.5	1.0	2.0	4.0	8.0	16.0
0.00	0.0	2.3	11.0	12.5	12.9	13.2	12.7
1.25	0.0	0.0	2.7	9.6	9.1	12.2	13.0
2.50	0.0	0.0	0.5	4.5	10.2	13.1	12.1
5.00	0.0	0.0	0.0	0.3	2.6	9.4	12.3
10.00	0.0	0.0	0.0	0.0	0.0	5.0	10.7
20.00	0.0	0.0	0.0	0.0	0.0	0.0	0.0

[a] Growth of BCG was determined after a two-week incubation period at 37°C. Iron was added as $FeCl_3 \cdot 6H_2O$ salt.

Journal of Infectious Diseases

Neutralization of Serum Microbiostasis

The growth-inhibitory activity of mammalian sera for tubercle bacilli and the neutralization of serum-tuberculostasis by iron or mycobactin were investigated by the agar-plate diffusion test. The tuberculostatic bovine serum–IPAM mixtures were inoculated with tubercle bacilli, and the wells were charged with various concentrations of iron or iron-free mycobactin. Results showed that with each increased concentration of iron or mycobactin in wells, an increased surface area covered by bacillary growth was observed around wells made in the growth-inhibitory serum–medium mixture (Figure 1). These findings demonstrated that

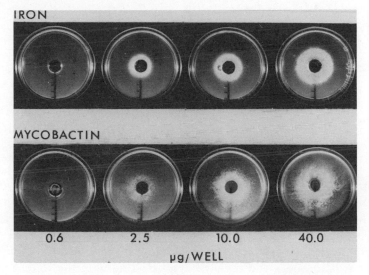

Infection and Immunity

Figure 1.   Three-week growth of tubercle bacilli around wells charged with various amounts of iron or mycobactin on micro-biostatic serum-IPAM mixture (16)

serum-imposed iron starvation of tubercle bacilli can be alleviated by adding exogenous iron or iron-binding mycobacterial product, mycobactin. The biochemical function of mycobactin in the growth of tubercle bacilli was determined by adding various amounts of mycobactin to tuberculostatic serum which also received 5 mg/ml of iron-free or iron-saturated Tr (22). Because the growth-promoting activity of mycobactin varied with the degree of saturation of Tr with iron (100%, 33%, 0%), it is safe to conclude that mycobactin is not acting as an essential growth factor (as has been believed since its discovery) but as the carrier of iron to bacillary cells (Figure 2). Mycobactin removes iron from Tr–iron complexes and supplies it to tubercle bacilli. Thus, mycobactin arms

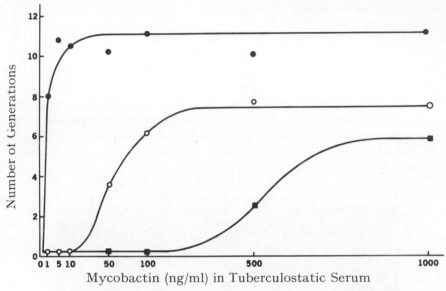

Figure 2. Two-week growth of tubercle bacilli in tuberculostatic bovine serum containing 5 mg Tr and different concentrations of mycobactin in ml of solution. (●) iron-saturated Tr; (■) iron-free Tr; (○) no Tr (22).

bacillary cells with a survival mechanism that plays an important role in host–tubercle bacillus interaction.

The stimulating effect of iron compounds on the growth of various fast-growing bacteria exposed to mammalian sera has received much attention in the last few years. This interest is illustrated by the appearance of several books and review papers dealing with the role of iron in the life of bacteria and in bacterial infections. Bullen and Rogers have shown that the inhibitory activity of rabbit serum for *E. coli* can be abolished by saturating Tr with iron (*23*). Iron and hemoglobin neutralized the antibacterial activity of human serum against a strain of *E. coli* isolated from a patient with pyelonephritis (*24*). The growth-supporting effect of iron and enterochelin for *E. coli* exposed to mammalian sera was shown quite clearly by the use of the agar-plate diffusion test in our laboratory. Typical results obtained in this study are presented in Figure 3. Wells in the growth-supporting agar medium were filled with 0.4 ml human serum or 0.4 ml saline containing 2.5 mg iron-free Tr. Wells in the growth-inhibitory serum–agar medium mixture were filled with 0.4 ml saline containing 20 μg iron (added in the form of ferric ammonium citrate) or 0.4 ml IPM in which *E. coli* cells were grown for 18 hr at 37°C (spent medium). The growth-inhibiting effects of Tr and the growth-promoting effects of iron or enterochelin in spent medium were determined by measuring the widths of bacterial inhibition or growth around charged wells at 15-, 25-, and 35-hr incubation periods.

The results presented in the upper part of the Figure 3 show that human serum and purified Tr inhibited the growth of *E. coli* (strain A). The examination of plates at various periods showed that microbiostatic areas around Tr-charged wells became progressively smaller and nearly disappeared at the 35-hr reading. The lower part of the Figure 3 shows that bacterial growth can be facilitated on the microbiostatic serum medium by adding iron or spent medium. Since unused medium during the 35-hr incubation period did not promote bacterial growth, the microbiostasis alleviating effect of spent medium was attributed to entero-

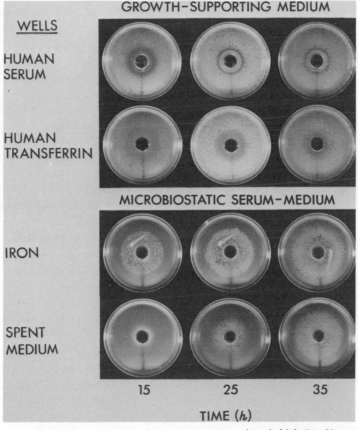

Journal of Infectious Diseases

Figure 3.   The inhibition of E. coli *around wells charged with human serum or Tr and the neutralization of microbiostasis on serum medium around wells charged with a solution of ferric ammonium citrate (20 μg iron/well) or spent medium at 15-, 25-, and 35-hr incubation period (25)*

chelin. In subsequent experiments, enterochelin isolated from spent media by Rogers' method (8) neutralized serum microbiostasis as effectively as the addition of iron. On prolonged incubation, areas covered by bacterial growth progressively increased and, at 35-hr examination, nearly all medium in each plate was covered by bacterial colonies.

The progressive increase in bacterial growth-covered areas around iron- or enterochelin-charged wells, as well as the progressive decrease of microbiostatic areas around Tr-charged wells, is attributed to the production of enterochelin by growing bacteria. The released enterochelin diffuses from the area of bacterial growth into the area of microbiostasis and provides iron of Tr–iron complexes for the use by bacterial cells. This process of the vitalization of inhibited bacteria at one place by enterochelin produced at some other place of the gelled medium may clearly illustrate the role of enterochelin in the promotion and spread of some bacterial infections in animal tissues. The phenomenon of vitalization of iron-starved bacteria has not been observed around iron- or mycobactin-charged wells (see Figure 1). In distinction to the water-soluble enterochelin, the hydrophobic mycobactin remains associated with the lipoidal cell wall of tubercle bacilli and does not vitalize cells exposed to the serum-imposed iron starvation. This may be the reason for the well defined tuberculous lesions in the body of infected animals. The water solubility of the majority of siderophores may favor not only the spread of bacterial infections, but it may also influence the state of antibacterial resistance of animals. Bacteria which reside in the intestinal tract may predispose the animal to an infection with homologous bacteria by producing diffusible siderophores.

In our further study we investigated effects of various numbers of virulent (strain C) and avirulent (strain A) *E. coli* on serum microbiostasis. Bovine serum was distributed in 1-ml quantities into small screw-topped tubes and was infected with various numbers of strain A or strain C bacteria. After 12- and 24-hr incubation at 37°C, samples of infected sera were plated on IPAM. Numbers of bacteria were determined by counting bacterial colonies on plates after 12-hr incubation. Results showed that bovine serum inhibited bacilli of avirulent strain A if the inoculated serum contained less than 12,000 bacilli per 1 ml of serum. Larger inocula of strain A multiplied in serum but at a slower rate than in broth medium (Table III). Bacteria of virulent strain C overcame the iron starvation in bovine serum even when serum was inoculated with as few as 100 cells/ml of serum. The results of this study showed that the rate of bacterial multiplication in bovine serum is determined partly by bacterial numbers in the inoculum. Small inocula of virulent and avirulent bacteria were more inhibited than large inocula. The growth of even minute inocula of virulent bacteria in bovine serum

Table III.   Effects of Different Bacteria Inocula on the Fate of
Avirulent (Strain A) and Virulent (Strain C)
Bacteria in Bovine Serum (25)

Bacterial Strain	Bacteria/0.1 ml Serum	Number of Generations[a] at (hr)	
		12	24
Strain A	10	0	0
	100	0	0
	600	1	2
	1,200	8	11
	6,000	10	13
Strain C	10	9	15
	100	11	16
	600	11	15
	1,200	14	15
	6,000	16	18

[a] Serum neutralized with iron (5 µg/ml) and nutrient broth contained 16 and 18 generations at 12 and 24 hr, respectively. Zero generation indicates the same or lower number of bacteria than in the inoculum.

Journal of Infectious Diseases

showed that these cells are much better equipped than avirulent cells to overcome the iron-starvation imposed by mammalian sera.

Effect of Iron Starvation on Bacterial Viability

The effect of iron starvation on the longevity of bacterial survival was tested by exposing strain A of *E. coli* to bovine serum for various periods (25). Plates of the agar-plate diffusion test containing microbiostatic serum-agar medium were uniformly inoculated with bacteria and then incubated at 4° and 37°C. At various periods during the incubation, wells in some plates were filled with either 0.4 ml saline containing 5 µg iron, 0.4 ml spent medium, or 0.4 ml iron-spent medium mixture. Subsequently, the plates with charged wells were incubated at 37°C for 18 hr.

Results showed that adding iron-providing materials to wells of plates which contained bacteria exposed to progressively longer iron starvation periods at 37°C vitalized fewer and fewer bacteria (Figure 4). Thus, the addition of iron or spent medium to wells of plates inoculated 24 hr previously vitalized only a few bacteria or was without effect. The addition of iron-providing materials to wells of 4°C-incubated plates promoted similar bacterial growth irrespective of the incubation period. The death of bacteria exposed to iron-starvation in bovine serum at 37°C, but not at 4°C, indicates that only metabolizing bacteria exposed to iron-starvation develop lethal changes which cannot be alleviated by subsequent addition of iron or iron-containing enterochelin. That bacteria cannot survive long without iron has been observed by Perry and

4 C-PLATES

37 C-PLATES

10   15   24   36

BACTERIAL SURVIVAL AT (h)

Journal of Infectious Diseases

Figure 4. The survival of E. coli on microbiostatic serum–agar medium around wells charged at 10, 15, 24, and 36 hr of incubation at 4°C and 37°C with a solution of ferric ammonium citrate (5 μg iron/well), spent medium, or iron-spent medium mixture (25)

Weinberg (26), who found that the bacterial death rate in synthetic media was accelerated by iron deprivation. The survival of serum-exposed bacteria incubated at 4°C makes it most unlikely that the iron-starvation per se is responsible for the killing of bacterial cells. Griffiths (5) found that the inhibition of E. coli by serum is accompanied by the appearance of abnormal bacterial phenylalanyl t-RNA. Such abnormal t-RNA was not found in iron-supplemented serum. It is possible, therefore, that permanent lesions develop in iron-starved bacteria which not only stop bacterial multiplication but eventually cause bacterial death.

Production of Bacterial Siderophores

On the basis of the production of iron-binding compounds, Neilands divided microorganisms into two types (27); the majority of microbes are autosequesteric and produce iron-binding compounds while a few species are anautosequesteric and cannot manufacture their own siderophores. Siderophores facilitate the acquisition of iron by microbes, and, therefore, they promote microbial growth in nature and in the animal host. In most cases, these iron-binding bacterial products were considered to be growth factors, and only recently has their iron-binding activity been considered in terms of microbial iron-transport mechanisms.

The production of siderophores can be affected by the level of iron in the bacterial environment and by temperature. The exposure of tubercle bacilli to an iron-poor medium stimulates the production of increased amounts of mycobactin (*11, 22*). Also, there is good evidence that enterochelin synthetase activity can be induced by exposing bacteria to media low in available iron (*28, 29*). Thus the accumulation of siderophores occurs only under imposed iron-limiting nutritional conditions. The effect of temperature on the growth of bacteria in the host has been of interest since Pasteur showed that by reducing the normally high body temperature of chickens, he was able to infect the cooled birds with anthrax bacillus. Hyperthermia has been used in treatments of various infectious diseases. Recent studies indicate that high temperatures suppress the production of siderophores. Garibaldi has shown that the production of enterobactin, a siderophore of *Salmonella typhimurium*, was decreased as the temperature of incubation was raised from 31° to 37°C (*30*). These bacteria would not multiply at temperatures over 40°C unless exogenous siderophore was added to the medium. The production of enterochelin by *E. coli* at various temperatures was tested by determining the amount of the siderophore in spent media (Table IV). The amount of enterochelin in spent medium of bacteria grown at 26°C was much larger than in spent medium of bacteria grown at 37°C; *E. coli* grown at 41°C produced little or no enterochelin. In view of these findings, fever, which develops in response to bacterial infections, might be considered as a defense mechanism because it limits the production of

Table IV. Bacterial Growth on Microbiostatic Serum–Medium Mixture around Wells Charged with Spent and Unused Media[a]

Wells Charged with	*Width (mm) of* E. coli *Growth around Wells at (hr)*			
	24	*30*	*40*	*48*
26 C-spent medium	9	13	18	20
33 C-spent medium	6	9	12	15
37 C-spent medium	5	8	12	15
41 C-spent medium	0	0	3	3
44 C-spent medium	0	0	3	3
Unused medium	0	0	0	1

[a] Spent media were prepared by growing *E. coli* at various temperatures for 48 hr in iron-poor medium. After the growth period, the cultures were adjusted to the same turbidity, bacteria were removed by filtration, and spent media were tested for the ability to alleviate the microbiostasis in bovine serum–agar medium mixture.

siderophores and, consequently, inhibits multiplication and spread of bacterial parasites.

The efficiency with which virulent bacteria invade and grow in fluids and tissues of animals suggests that they may differ from avirulent bacteria by the prolific production of siderophores. Recent experimental results showed that bacteria differ in the ability to acquire iron in the body of the host. The virulence and the ability of *Pseudomonas aeruginosa* to use iron in mice was significantly increased after 16 serial passages of the bacteria in these animals (*31*). Rogers' experiments suggested that there is a direct relationship between the ability to synthesize enterochelin and the virulence of different strains of *E. coli* for mice (*8*) since virulent bacteria exposed to iron limitation produced much more enterochelin than avirulent bacteria.

Determinations of mycobactin production by saprophytic and parasitic species of mycobacteria showed that saprophytic fast-growing mycobacteria produce more mycobactin than parasitic slow-growing mycobacteria (*11*). The extraction of virulent and avirulent strains of *M. tuberculosis* with surfactants showed that they contain similar amounts of mycobactin on their surfaces (*9*). Recently we have measured the production of enterochelin in IPM ($0.09 \, \mu g$ iron/ml) by virulent strain C and avirulent strain A of *E. coli* (*32*). Results showed that spent media of virulent and avirulent bacteria contain similar amounts of enterochelin. The possibility remains, however, that still smaller amounts of iron in synthetic medium may stimulate virulent cells to produce more enterochelin.

Acquired Hypoferremia in Vaccinated Animals

Low iron levels in human blood were observed by clinicians to be associated with infectious diseases. A decrease in plasma iron levels was induced by bacterial infections (*33, 34*) or by treatment with bacterial endotoxins (*35*). The treatment of animals with endotoxin-released mediator of hypoferremia protected them from lethal salmonellosis (*36*). Although the levels of hypoferremia or degrees of iron saturation of Tr were not examined in these studies, various experiments indicated that the fall in serum iron increased resistance to bacterial infections.

The relationship between hypoferremia and microbiostasis in sera of vaccinated or endotoxin-treated guinea pigs has been demonstrated by Kochan and co-workers (*20, 37*). This study originated from the finding of a difference in antimycobacterial activity of normal and immune sera (*38*). In contrast to the growth-supporting nature of normal guinea pig serum, serum of immune animals suppressed the growth of tubercle bacilli. This initial observation of acquired antimycobacterial activity of immune sera developed later into the study of the relationship between iron-free Tr and the tuberculostatic activity in sera of normal and BCG, *E. coli* lipopolysaccharide (LPS), or tuberculous cell wall (TCW)-vaccinated guinea pigs (20). The results showed that sera of animals vaccinated with live bacteria or their products gained antimicrobial activity in a direct proportion to the loss of iron and the decrease in iron saturation of Tr (Table V). The induced tuberculostasis in sera of vaccinated animals can be readily neutralized by adding exogenous iron. Since the hypoferremic response transforms guinea pig plasma from a bacterial growth-supporting medium to one that exerts microbiostasis, it should be considered to be a protective response of the animal body to parasitic invasion. Thus, the induced hypoferremia can be considered to be the acquired arm of nutritional immunity.

Table V. Iron Neutralization of Tuberculostasis in Sera of Animals Vaccinated with LPS, TCW, or Live BCG Bacilli (20)

Days after Treatments[a]	Tr Saturation (%)	Generations in Serum with Iron (μg/ml)[b]				
		0	1	2	4	8
LPS-1	17.5	0.0	0.0	0.4	8.0	9.2
2	42.0	0.0	10.6	11.3	10.7	12.2
3	60.2	0.0	11.8	11.9	9.5	11.1
5	74.7	8.7	12.0	11.4	10.7	11.3
10	93.8	12.5	11.6	12.0	12.0	11.3
TCW-1	26.5	0.0	5.1	9.9	9.5	9.5
2	42.1	0.5	10.2	10.3	10.0	10.6
3	59.3	1.2	9.7	9.3	10.7	11.0
5	66.3	0.7	7.3	12.0	11.3	12.3
14	78.8	11.8	12.6	12.1	11.9	12.7
BCG-3	86.2	10.5	11.5	11.9	11.0	11.2
7	79.4	9.1	10.6	11.3	12.1	11.7
14	75.0	4.3	9.5	10.2	12.3	11.1
21	75.3	0.4	9.1	10.3	9.9	10.7
28	68.6	0.0	9.7	10.2	10.9	11.3
Saline-1	85.6	11.5	11.8	11.6	11.5	12.1

[a] On day "0" animals were injected intraperitoneally with 0.05 mg LPS, 1 mg TCW preparation, or 1 mg BCG cells per 100 g body weight.
[b] Tests for the presence and neutralization of tuberculostasis were performed in a 1:4 dilution of serum samples. Bacillary fate determined after a 14-day incubation.

Journal of Bacteriology

In distinction to the development of generalized hypoferremia in LPS-treated animals, there is evidence for the development of localized hypoferremia at places of inflammatory reaction. It has been shown that polymorphonuclear cells release lactoferrin when stimulated by antigen–antibody complexes (39) or when destroyed in the acute inflammatory process (40). Lactoferrin is one of the major proteins present in cytoplasmic granules of polymorphonuclear cells (41, 42). In its iron-binding activity, lactoferrin is very similar to Tr. However, complexes of Tr and lactoferrin with iron are influenced differently with acidic pH; Tr–iron complexes dissociate around pH 5 whereas lactoferrin–iron complexes remain stable until pH is below 4 (43). At the acidic pH of inflammation, iron is removed from the loose iron–Tr complexes by lactoferrin of degranulated or destroyed leukocytes, and the stable lactoferrin–iron complexes are taken up by cells of reticuloendothelial system (40). Thus, the lactoferrin-induced depletion of iron in inflammatory or immunological processes may increase the resistance of the host against extracellular and intracellular parasites.

Virulence-Associated Acquisition of Iron

In our efforts to identify factors which permit virulent bacteria of *E. coli* to multiply in bovine serum (*see* Table III) we compared virulent and avirulent strains of *E. coli* as to their abilities to neutralize serum microbiostasis, their content of iron, and their amounts of cell-wall associated siderophores. Although we have shown that virulent strain C and avirulent strain A produce similar amounts of enterochelin, a possibility remained that virulent bacteria can neutralize microbiostasis in serum by rapid production of enterochelin and then multiply without much hindrance.

Evidence for the failure of extracellular enterochelin to participate in the initiation of growth of virulent bacteria in serum was obtained by examining sera in which virulent strain C had multiplied for various periods for their ability to support the growth of avirulent strain A. Heat-inactivated bovine serum was inoculated with strain C bacteria (1300 cells/ml) and incubated at 37°C for 12 hr. During the incubation period, samples of the spent serum were collected at 2, 4, 6, and 12 hr and tested for both the number of strain C bacteria and the growth-supporting quality for strain A bacteria (Table VI). Plating of spent serum samples showed that strain C bacteria multiplied in bovine serum. The tests of growth-supporting quality of the filtered spent sera showed that they inhibited avirulent bacteria. Only the 12-hr sample of spent serum contained enough enterochelin to promote the growth of avirulent bacteria. These results showed that serum in which virulent bacteria have multiplied remained microbiostatic for avirulent bacteria. Therefore, the growth of virulent bacteria in serum cannot be attributed to the neutralization of the serum by enterochelin but to iron-providing constituents of the virulent cell itself.

The growth of virulent bacteria in serum suggested the presence of a cell-associated factor which helps the cell to overcome the serum-imposed iron starvation. The most likely explanation of the ability of virulent bacteria to grow in serum is the presence of a large amount of stored intracellular iron. This possibility has been investigated by determining the amount of iron in ashed materials of virulent and avirulent cells. These determinations showed that the amount of iron in virulent and avirulent bacteria was similar. On an average, virulent bacteria contained 0.46 μg and avirulent bacteria 0.43 μg iron per 100 mg dried cells. This difference is too small to account for the ability of virulent cells to grow in serum.

Table VI. Growth of Avirulent Strain A and Virulent Strain C in Samples of Spent Serum Collected at Various Times during the Growth of Strain C (25)

Strain of E. coli	*Number of Generations in Spent Serum Collected at (hr)* [a]			
	2	*4*	*6*	*12*
Strain A	1.9	1.6	2.6	12.3
Strain C	16.8	16.0	15.9	15.6

[a] Determinations of bacterial growth in spent serum was done at 12 hr of incubation period at 37°C. Unused serum inoculated with strain C (1300 cells/ml) contained 2.4, 5.0, 7.6, and 15.9 generations after 2-,4-,6,- and 12-hr incubation periods, respectively. Unused serum was microbiostatic for bacteria of strain A.

Journal of Infectious Diseases

The presence and the amount of cell-wall associated siderophores on virulent and avirulent bacteria were determined by repeated extractions with saline and 0.05% Tween 80 solution in saline (25). The same amount of avirulent and virulent bacteria (250 mg wet-weight cell/ml of the extractant) was mixed for 1 hr in saline, and the first saline extract was collected by centrifugation and filtration. Bacteria were extracted two additional times with an appropriate quantity of fresh saline. In some experiments, twice saline-extracted bacteria were extracted with Tween 80, and Tween extract was tested for its ability to neutralize serum microbiostasis by the agar-plate diffusion test. Results showed that first and second saline extracts of either strain A or strain C promoted the growth of strain A bacteria around extract-charged wells made in microbiostatic serum–agar medium (Figure 5). Each extract of virulent bacteria exerted a somewhat stronger serum-neutralizing activity than the corresponding extract of avirulent bacteria. The third saline extract exerted little or no serum neutralization whereas the Tween extract of twice saline-extracted bacteria neutralized serum quite effectivly. Again, Tween extracts of virulent cells neutralized serum microbiostasis more effectively than the corresponding extracts of avirulent cells. The differential extractions suggested that cells of *E. coli* possess two iron-pro-

AVIRULENT STRAIN-A

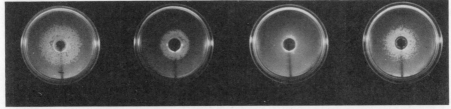

VIRULENT STRAIN-C

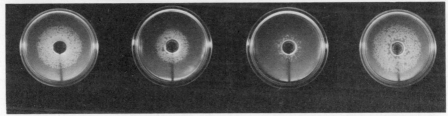

1st	2nd	3rd	3rd
SALINE EXTRACT	**SALINE EXTRACT**	**SALINE EXTRACT**	**TWEEN EXTRACT**

Journal of Infectious Diseases

Figure 5. The growth of strain A around wells charged with first, second, and third saline extracts of avirulent and virulent bacteria. The last entry in the table shows the bacterial growth around wells charged with Tween 80 (or serum) extract of twice saline-extracted bacteria (25).

viding materials—water-soluble (enterochelin) and surfactant-soluble substances.

The inhibitory activity of *E. coli*–LPS for tubercle bacilli (44) and the toxicity-neutralizing effect of iron for LPS (45) suggested that the iron-providing, surfactant-soluble substances on bacterial cells could be cell-wall associated LPS. Therefore, the iron-providing activity of *E. coli*–LPS (Difco) for serum-inhibited bacteria of strain A has been investigated by the agar-plate diffusion test (25). Saline containing various concentrations of untreated, dialyzed, or spent medium-treated LPS was added to wells made in microbiostatic serum–agar medium inoculated with strain A bacteria. Results showed that LPS of *E. coli* can provide iron to serum-inhibited bacteria. This activity cannot be attributed to the contamination of LPS with enterochelin, because the dialysis and the treatment of LPS with enterochelin esterase of spent medium did not decrease the serum-neutralizing activity of LPS (Table VII). Determination of LPS-content of virulent and avirulent cells by the Boivin method (46) showed that virulent cells contain 3.7 mg and avirulent cells

1.6 mg LPS per 100 mg dry-weight bacterial mass. The significantly higher amount of LPS on virulent cells seems to provide these cells with a stronger iron-acquiring activity. According to these findings, the cell-wall associated siderophores (enterochelin, LPS, or their combination) enable the bacteria to overcome the initial iron-starvation in serum whereas the released enterochelin helps the subsequent uninhibited growth and spread of bacteria in animal tissues.

The presented study showed that LPS of *E. coli* exerts a dual effect on the mechanism of the acquisition of iron by bacteria. When LPS is injected into animals, it induces the development of hypoferremia. Thus, sera of LPS-treated animals exert much stronger inhibitory activity for bacteria than sera of untreated animals. However, when LPS is introduced on bacterial cells, it helps them to acquire iron from Tr–iron complexes and thus promotes bacterial multiplication in mammalian sera. In view of this, the LPS-induced hypoferremia should be considered as a protective response of the host to the iron-providing activity of LPS.

Neutralization of Nutritional Immunity with Iron

The enhancing effect of iron on the progression of infection with *Yersinia pestis* was observed 20 years ago by Jackson and Burrows (47). Since that time many investigators have made similar observations. It has been reported that the administration of iron or iron-containing compounds such as hematin, hemin, or hemoglobin promoted infections with *E. coli* (48), *Clostridium welchii* (49), *Y. septica* (50, 51), *Listeria monocytogenes* (33), *Pseudomonas aeruginosu* (52, 53), *Salmonella typhimurium* (54), *Klebsiella pneumoniae* (54), *M. tuberculosis* (2), and *Neisseria gonorrhoeae* (10). Clinical observations suggest that diseases associated with raised serum iron concentrations predispose patients to microbial infections. The consequences of high iron saturation of Tr

Table VII. Bacterial Growth on Microbiostatic Serum–Agar Medium around Wells Charged with Untreated and Treated LPS Solutions

	Width (mm) of Growth around LPS-Charged Wells (mg/well)		
Treatment of LPS	2	1	0.5
No treatment	13	10	6
Dialysis	11	9	5
Incubated in spent medium[a]	12	11	8
Incubated in unused medium	14	10	8

[a] The incubation of LPS in spent medium was done at 37°C for 24 hr. Such incubation of enterochelin destroyed its bacterial growth-promoting activity in bovine serum.

in patients with sickle cell anemia (55), virus hepatitis (56), malaria (57), kwashiorkor (58), acute leukemia, or thalassemia major (59, 60) on susceptibility to microbial infections suggest that a close correlation exists between the hyperferremic state and the development of infectious diseases.

The potentiating effect of iron on bacterial virulence has been determined in mice injected intraperitoneally with strain A or strain C of *E. coli.* The infection dose of 5×10^8 bacilli was suspended in 1 ml saline containing 0.0 or 0.1 mg iron. Results showed that bacteria of strain C are four times more effective in killing mice than bacteria of strain A (Table VIII). Mice injected with the same number of heat-killed cells of either strain demonstrated no toxic reaction and remained alive. The addition of iron to the infection dose of strain A increased the

Table VIII. The Mortality of Mice Injected with Untreated, Iron-treated, and Heat-killed Cells of Strain A and Strain C of *E. coli* (25)

	Mortality[a]	
Treatment of Infection Dose	Strain A	Strain C
Untreated	6/27	22/27
Iron[b]	13/16	23/28
Heat (80°C for 1 hr)	0/10	0/10

[a] Number dead during two days/number injected.
[b] Each mouse received 0.1 mg iron added to the infection dose of 5×10^8 cells in the form of ferric ammonium citrate. Neither this quantity of iron in normal mice nor 0.567 mg corresponding ammonium citrate in infected mice increased mortality in a group of 10 mice.

Journal of Infectious Diseases

mortality from 20 to 80%; the mortality of strain C-infected, iron-treated, or untreated mice remained the same (80%). These results show that the virulence of a relatively avirulent strain A can be potentiated with iron and suggest that bacterial virulence can be related to the ability of bacterial cells to obtain iron in fluids and tissues of the infected host.

The frequent bacterial infections in hyperferremic patients or iron-treated animals suggest that elevated levels of iron either promote bacterial growth or neutralize natural defense mechanisms of the host. Experimental findings in several laboratories indicate that iron compounds do not interfere with antibacterial defenses of animal body. Studying the disease-promoting role of iron in the host–parasite relationship, Burrows found that iron compounds do not interfere with phagocytosis, intracellular digestion in phagocytic cells, or the production of antibodies (61). Similar findings were reported by Bullen and his associates who observed that iron did not interfere with phagocytosis and fixation of complement by antigen–antibody complexes (49). The blockage of the reticuloendothelial system in mice infected with attenuated plague bacilli did not stimulate infection whereas treatment of such mice with iron promoted bacterial growth and the development of the disease (*62*). There is also no evidence to suggest that iron impairs some other defense mechanisms of the host. Burrows observed that iron injected even in large quantities into animals infected with an avirulent strain of *Y. pestis* failed to establish an infection. However, much smaller amounts of iron enhanced the pathogenicity of virulent bacteria (*61*). If the activity of iron were to neutralize the defense mechanism of the host, iron-treated animals might be expected to become susceptible to infections with avirulent bacteria.

That the development of bacterial infections in iron-treated animals is caused by the alleviation of iron-starvation for infecting bacteria rather than by the neutralization of defense mechanisms by iron has been shown most clearly in siderophore-treated animals. Treatment of mice with enterochelin produced no observable effect in uninfected mice whereas in infected mice, *E. coli* grew logarithmically, and the animals died within 18 hr. This iron-binding compound facilitated the development of the overwhelming infection by being able to remove iron from Tr (*8*).

The resistance of newborn animals to intestinal infections with *E. coli* seems to be determined by the amount of lactoferrin in milk and of exogenous iron. Colostrum-deprived calves tend to develop infections with *E. coli* more frequently than do colostrum-fed animals (*64*). The evidence that milk actually suppresses the growth of *E. coli* was provided by experiments in which colostrum-fed guinea pigs were dosed orally with bacteria and hematin (*64*). Results showed that the presence of hematin enhanced the growth of *E. coli* in the small intestine by a factor of over 10,000. We have found that human and bovine colostrum inhibit the growth of *E. coli*. However, in distinction to strong antibacterial activity exerted by human milk collected at various times after the parturition, bovine milk is active only during the period of colostrum production; even five-times concentrated bovine milk does not exert any antibacterial activity. The antibacterial activity of colostrum and milk samples can be alleviated by adding iron or enterochelin (*25*). These findings may explain why the count of *E. coli* in the feces of bottle-fed babies tends to be much higher than in breast-fed babies (*65*). It seems that lactoferrin in milk may play a protective role in the resistance of newborn animals to enteritis caused by *E. coli*.

Literature Cited

1. Kochan, I., *Curr. Top. Microbiol. Immunol.* (1973) **60**, 1.
2. Kochan, I., in "Microbiology—1974," D. Schlessinger, Ed., p. 273, Amer. Soc. Microbiol., Washington, 1975.

3. Weinberg, E. D., *J. Am. Med. Assoc.* (1975) **231**, 39.
4. Weinberg, E. D., *J. Infect. Dis.* (1971) **124**, 401.
5. Bullen, J. J., Rogers, H. J., Griffith, E., in "Microbial Iron Metabolism," J. B. Neilands, Ed., p. 517, Academic, New York, 1974.
6. Sussman, M., in "Iron in Biochemistry and Medicine," A. Jacobs, M. Worwood, Eds., p. 649, Academic, New York, 1974.
7. Kochan, I., in "Microorganisms and Minerals," E. D. Weinberg, Ed., Marcel Dekker, Inc., New York, 1977.
8. Rogers, H. J., *Infect. Immunity* (1973) **7**, 445.
9. Golden, C. A., Kochan, I., Spriggs, D. R., *Infect. Immunity* (1974) **9**, 34.
10. Payne, S. M., Finkelstein, R. A., *Infect. Immunity* (1975) **12**, 1313.
11. Snow, G. A., *Bacteriol. Rev.* (1970) **34**, 99.
12. O'Brien, I. G., Gibson, F., *Biochim. Biophys. Acta* (1970) **215**, 393.
13. Pollack, J. R., Neilands, J. B., *Biochem. Biophys. Res. Commun.* (1970) **38**, 989.
14. Wawszkiewicz, E. J., Schneider, H. A., *Infect. Immunity* (1975) **11**, 69.
15. Wawszkiewicz, E. J., Schneider, H. A., Starcher, B., Pollack, J., Neilands, J. B., *Proc. Natl. Acad. Sci. U.S.A.* (1971) **68**, 2870.
16. Kochan, I., Cahall, D. L., Golden, C. A., *Infect. Immunity* (1971) **4**, 130.
17. Kochan, I., Golden, C. A., *Infect. Immunity* (1973) **8**, 388.
18. Shade, A. L., Caroline, L., *Science* (1946) **104**, 340.
19. Komatsu, S. K., Feeney, R. E., *Biochemistry* (1967) **6**, 1136.
20. Kochan, I., Golden, C. A., Bukovic, J. A., *J. Bacteriol.* (1969) **100**, 64.
21. Kochan, I., *J. Infect. Dis.* (1969) **119**, 11.
22. Kochan, I., Pellis, N. R., Golden, C. A., *Infect. Immunity* (1971) **3**, 553.
23. Bullen, J. J., Rogers, H. J., *Nature* (1969) **224**, 380.
24. Fletcher, J., *Immunology* (1971) **20**, 493.
25. Kochan, I., Kvach, J. T., Wiles, T. I., *J. Infect. Dis.* (1977), April.
26. Perry, R. D., Weinberg, E. D., *Microbios* (1973) **8**, 129.
27. Neilands, J. B., *Bacteriol. Rev.* (1957) **21**, 101.
28. Young, I. G., Gibson, F., *Biochim. Biophys. Acta* (1969) **177**, 401.
29. Bryce, G. F., Brot, N., *Arch. Biochem. Biophys.* (1971) **142**, 399.
30. Garibaldi, J. A., *J. Bacteriol.* (1972) **110**, 262.
31. Forsberg, C. M., Bullen, J. J., *J. Clin. Pathol.* (1972) **25**, 65.
32. Kochan, I., unpublished results.
33. Sword, C. P., *J. Bacteriol.* (1966) **92**, 536.
34. Pekarek, R. S., Bostian, K. A., Bartellony, P. J., Calia, F. M., Beisel, W. R., *Am. J. Med. Sci.* (1969) **258**, 14.
35. Baker, P. J., Wilson, J. B., *J. Bacteriol.* (1965) **90**, 903.
36. Kampschmidt, R. F., *Ann. Okla. Acad. Sci.* (1974) **4**, 62.
37. Kochan, I., Berendt, M., *J. Infect. Dis.* (1974) **129**, 696.
38. Kochan, I., Raffel, S., *J. Immunol.* (1960) **84**, 374.
39. Leffell, M. S., Spitznagel, J. K., *Infect. Immunity* (1975) **12**, 813.
40. Van Snick, J. L., Masson, P. L., Heremans, J. F., *J. Exp. Med.* (1974) **140**, 1068.
41. Masson, P. L., Heremans, J. F., Schonne, E., *J. Exp. Med.* (1969) **130**, 643.
42. Leffell, M. S., Spitznagel, J. K., *Infect. Immunity* (1972) **6**, 761.
43. Groves, M. L., *J. Am. Chem. Soc.* (1960) **82**, 3345.
44. Kochan, I., Golden, C. A., unpublished results.
45. Prigal, S. J., Herp., A., Gernstein, J., *J. Reticuloendothel. Soc.* (1973) **14**, 250.
46. Sutherland, I. W., in "Handbook of Experimental Immunology," D. M. Weir, Ed., p. 355, F. A. Davis Co., Philadelphia, 1967.
47. Jackson, S., Burrows, T. W., *Br. J. Exp. Pathol.* (1956) **37**, 577.
48. Bullen, J. J., Leigh, L. C., Rogers, H. J., *Immunology* (1968) **15**, 581.
49. Bullen, J. J., Cushnie, G. H., Rogers, H. J., *Immunology* (1967) **12**, 303.

50. Bullen, J. J., Wilson, A. B., Cushnie, G. H., Rogers, H. J., *Immunology* (1968) **14**, 889.
51. Wake, A., Morita, H., Yamamoto, M., *Jpn. J. Med. Sci. Biol.* (1972) **25**, 75.
52. Martin, C. M., Handl, J. H., Finland, M., *J. Infect. Dis.* (1963) **112**, 158.
53. Forsberg, C. M., Bullen, J. J., *J. Clin. Pathol.* (1972) **25**, 65.
54. Chandlee, G. C., Fukui, G. M., *Bacteriol. Proc.* (1965) 45.
55. Ringelhann, B., Konotey-Ahulu, F., Dodu, A. R. A., *J. Clin. Pathol.* (1970) **23**, 127.
56. Petersen, R. E., *J. Lab. Clin. Med.* (1952) **39**, 225.
57. Black, P. H., Kunz, I. J., Swartz, M. N., *N. Engl. J. Med.* (1960) **262**, 811, 864, 921.
58. McFarlane, H., Reddy, S., Adcock, K. J., Adeshina, H., Cooke, A. R., Akene, J., *Br. Med. J.* (1970) **4**, 268.
59. Caroline, L., Rosner, F., Kozinn, P. J., *Blood* (1969) **34**, 441.
60. Caroline, L., Kozinn, P. J., Feldman, F., Stiefel, F. H., Lichtman, H., *Ann. N.Y. Acad. Sci.* (1969) **165**, 148.
61. Burrows, T. W., *Curr. Top. Microbiol. Immunol.* (1963) **37**, 59.
62. Wake, A., Yamamoto, M., Morita, H., *Jpn. J. Med. Sci. Biol.* (1974) **27**, 229.
63. Ingram, P. L., Lowell, R., *Vet. Rec.* (1960) **72**, 1183.
64. Bullen, J. J., Rogers, H. J., Leigh, L., *Br. Med. J.* (1972) **1**, 69.
65. Bullen, C. L., Willis, A. T., *Br. Med. J.* (1971) **3**, 338.

RECEIVED July 26, 1976.

4

Iron Transporting System of Mitochondria

TORGEIR FLATMARK

Department of Biochemistry, University of Bergen, N-5000 Bergen, Norway

INGE ROMSLO

Laboratory of Clinical Biochemistry, University of Bergen,
N-5016 Haukeland Sykehus, Norway

Mammalian mitochondria have evolved a transport system to accumulate iron from the environment (cytosol) and are able to utilize Fe(III) of certain synthetic low molecular weight complexes. The process represents a unidirectional transport of Fe(II) across the inner membrane and is driven by energy derived from coupled respiration or ATP hydrolysis and by dissipation of an electric potential gradient of K^+. The uptake requires reducing equivalents, such as those supplied by the respiratory chain. The iron appears to be tightly bound to protein ligands although not detectable by EPR spectroscopy. The iron can be used for biosynthesis of heme and possibly for iron–sulfur centers within the mitochondria. The chemical nature of the cytosolic iron donor and regulatory mechanisms of the mitochondrial accumulation of iron are discussed.

All living organisms require iron and are faced with the problem of accumulating a sufficient amount of this metal from the environment. In mammals iron is taken up predominantly in the duodenum and jejunum (1) and is transported to body cells in extracellular fluid by the specific iron-binding protein transferrin (2). Taken up by individual cells, iron is either stored as ferritin and hemosiderin (3) or is used for biosynthetic purposes, notably synthesis of heme proteins (4, 5) and non-heme iron proteins (6). Thus, in the overall metabolism of iron in mammalian organisms several permeability barriers exist, and a detailed understanding of the transport mechanism of the metal across biological membranes is essential for an understanding of its metabolism in higher organisms. So far, however, the mechanism by which iron is taken up and transferred across biological membranes is poorly understood.

Within the past few years, there has been considerable progress in understanding the role played by the mitochondria in the cellular homeostasis of iron. Thus, erythroid cells devoid of mitochondria do not accumulate iron (7, 8), and inhibitors of the mitochondrial respiratory chain completely inhibit iron uptake (8) and heme biosynthesis (9) by reticulocytes. Furthermore, the enzyme ferrochelatase (protoheme ferrolyase, EC 4.99.1.1) which catalyzes the insertion of $Fe(II)$ into porphyrins, appears to be mainly a mitochondrial enzyme (10, 11, 12, 13, 14) confined to the inner membrane (15, 16, 17). Finally, the importance of mitochondria in the intracellular metabolism of iron is also evident from the fact that in disorders with deranged heme biosynthesis, the mitochondria are heavily loaded with iron (see "Mitochondrial Iron Pool," below). It would therefore be expected that mitochondria, of all mammalian cells, should be able to accumulate iron from the cytosol. From the permeability characteristics of the mitochondrial inner membrane (18) a specialized transport system analogous to that of the other multivalent cations (for review, see Ref. 19) may be expected. The relatively slow development of this field of study, however, mainly reflects the difficulties in studying the chemistry of iron.

Methodological Problems

In contrast to studies on the accumulation of Ca^{2+}, Sr^{2+}, Mg^{2+}, La^{3+}, and K^+ (19), studies on the accumulation of iron by isolated mitochondria have to take into account the extreme insolubility of $Fe(III)$ at neutral pH, the formation of iron complexes with high stability constants, and the redox activity of iron. Thus, in order to overcome these difficulties of working in the physiological pH range, it is necessary to use a complex of iron which is soluble at neutral pH.

Although there is some experimental evidence which points to a binding of iron ions by specific cytosolic proteins (see "Cytosolic Iron Donor," below), these proteins, with the exception of transferrin, are available only in minute quantities, and the nature and extent of iron–protein interactions are poorly understood. Therefore, a number of non-protein iron chelates have been studied as possible model donor complexes (Table I). Because of the high stability constants of, for example, the $Fe(II)/Fe(III)$–8-hydroxyquinoline and $Fe(III)$–ADP complexes (20), these iron–chelate complexes are unfavorable as iron donors, and in fact no energy-dependent uptake of iron has been detected using these complexes (21, 23).

We have found, however, that the ferric complex with sucrose is suitable for this purpose. Although the complex does not represent a homogenous molecular species, it is stable and highly soluble in water at neutral pH at a ratio of iron:sucrose < 1:40 and has a favorable over-

Table I.   Iron Complexes Used in the Study of Mitochondrial Iron Accumulation by Heart and Liver Mitochondria

Compound	Type of Accumulation		References
	Energy-dependent	Energy-independent	
Nonprotein chelates			
Fe(III)–ADP	—	+	15
Fe(III)–8-hydroxyquinoline	—	+	21
Fe(II)–8-hydroxyquinoline	—	+	21
Fe(III)–sucrose[a]	+[b]	+	23, 24, 27–32
Protein chelates			
Fe(III)–phosvitin[a, c]	—	+	50
Transferrin[d]	+[e]	+	53

[a] Polynuclear Fe(III) complexes.
[b] Approximately 7 nmol · 30 sec^{-1} · mg of protein^{-1}.
[c] From avian egg yolk.
[d] Partly saturated.
[e] *See* text.

all dissociation constant, which permits the complex to be dissociated by interaction with the mitochondrial membranes (Figure 1 A) (23). A high concentration of the ligand (sucrose) in the incubation medium contributes to this stability. Furthermore, the complex is amphoteric (24) and slightly negatively charged at neutral pH; the main species has a pI_4°c of 5.7. The ligand is metabolically inert (25), and the mitochondrial inner membrane is impermeable to the complex as well as to the ligand (26). In spite of some complicating factors related to its unique chemistry, the Fe(III)–sucrose complex has been successfully used in our laboratory to study the process of iron accumulation by mitochondria (23, 24, 27–32).

Energy-Dependent Accumulation of Iron by Isolated Rat Liver Mitochondria

Detailed kinetic measurements of the initial rates of iron uptake by rat liver mitochondria have been carried out in this laboratory during the last few years using ^{59}Fe(III)–sucrose as the donor complex (23, 24, 27–32). The general features given in Table II support the concept that iron, like other cations, is accumulated by an energy-dependent as well as an energy-independent mechanism, and these two processes have different time, pH, and temperature dependencies.

The energy-dependent uptake, which by definition is sensitive to uncoupler, is supported by endogenous respiration and low concentrations of ATP, and uncoupler-sensitive uptake is also increased by an electric potential gradient of K^+ (*see* Table I in Ref. *31*). The energized uptake is a rapid process with saturation kinetics. It is stimulated by K^+ and Mg^{2+} and is inhibited by complexing agents such as phosphate, carboxylates, and ADP as well as ATP at higher concentrations. It is sensitive to the same cations and SH-group reagents which inhibit Ca^{2+}

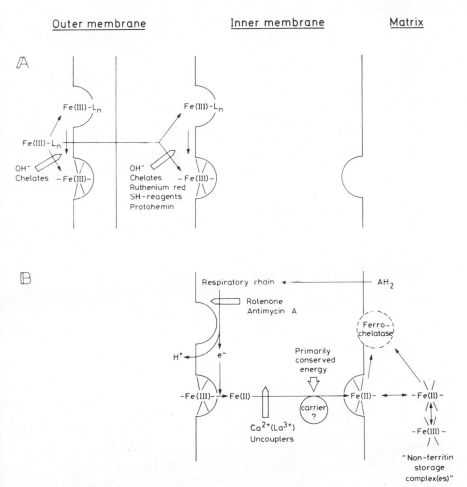

Figure 1. Schematic of binding, oxidation–reduction, and transport of iron by intact rat liver mitochondria.

A. Passive binding of Fe(III)–L_n complex and Fe(III) to ligands (acceptor sites) of the outer membrane and the outer phase of the inner membrane. Fe(III)–L_n represents a soluble chelate complex of Fe(III) with a sufficiently low stability constant, e.g., polynuclear Fe(III)–sucrose complex(es). B. reduction of membrane-bound Fe(III) to Fe(II) by the respiratory chain at the level of cytochrome c/cytochrome a and energy-dependent active transport of Fe(II) across the inner membrane. See text. This model is based on Refs. 23, 24, 27–32.

Table II.   Properties of the Iron Accumulation by Isolated Rat Liver Mitochondria Using $^{59}Fe(III)$–Sucrose as Iron Donor[a]

Properties	Energy-dependent	Energy-independent
Sensitive to uncoupler	+[b]	—
Supported by		
endogenous respiration	+	—
ATP (low conc.)	+	—
electric potential gradient of K^+[c]	+	—
Kinetics		
rate of accumulation		
(nmol · 30 sec^{-1} · mg prot^{-1})	7.6 ± 1.5	20–40
saturation	+	—
pH optimum	7.3	6.3
Tp optimum	25°C	> 30°C
K^+	stimulation	
Mg^{2+}	stimulation	stimulation
Pi, carboxylates, ADP	inhibition	inhibition
Inhibition by		
EDTA		+
TTFA[d]		—
La^{3+} or Ca^{2+}		—
ruthenium red	+	—
hexamine cobalt(III)chloride		—
PHMB[d]		(+)
NEM[d]		(+)

[a] From Refs. 23, 24, 27–32.
[b] By definition.
[c] Generated by suddenly releasing internal K^+ with valinomycin.
[d] TTFA, 1[thenoyl-(2′)]-3,3,3-trifluoracetone; PHMB, p-hydroxymercuribenzoate; NEM, n-ethylmaleimide.

uptake. Furthermore, Ca^{2+} inhibits competitively the energy-dependent accumulation of iron whereas it has no effect on the energy-independent accumulation (Figure 2). Rat liver mitochondria possess so-called high- and low-affinity binding sites of iron as previously described for Ca^{2+} (27), and the iron accumulated by the energy-dependent mechanism was recovered mainly in the matrix (~ 60% of the total) and the inner membrane (~ 30% of the total). This supports the view that the so-called high-affinity binding represents an iron transfer across the inner membrane (27, 28) (Figure 1 B, see also "Transport Form of Iron," below). On the other hand, most of the energy-independent iron accumulation is confined to the outer (~ 55%) and inner membrane (~ 35%) and largely represents a binding of the Fe(III)–sucrose complex(es) to these membranes (27, 28) (Figure 1 A). As expected the degree of energy-dependent uptake in state 1 (endogenous) respiration depends on the energy state (Figure 3), and the accumulation of Ca^{2+} and iron appears to operate at almost the same level of the mitochondrial energy potential (*31*). On the other hand, whereas the accumulation of Ca^{2+} is stimulated by adding exogenous substrates, that of iron is largely inhibited even in tightly coupled mitochondria. This inhibition of iron accumulation is caused by the high stability of the iron complexes wtih di- and tricarboxylic acids (*20*).

Depending on the metabolic state, Fe(III)–sucrose may stimulate as well as inhibit the respiration rate of isolated rat liver mitochondria (*32*). In state 1 respiration (i.e., in the absence of exogenous substrate or ADP) the addition of Fe(III)–sucrose lowers the rate of oxygen uptake. This inhibition of respiration is caused by a reduction of Fe(III) to Fe(II) prior to its transport across the inner membrane (*23, 32*) (*see* Figure 1 B and the text, below).

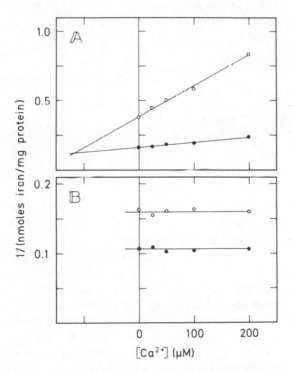

Biochimica et Biophysica Acta

Figure 2. Effect of Ca^{2+} on the energy-dependent (respiratory chain linked) (A) and the energy-independent (B) iron accumulation by rat liver mitochondria. The concentrations of iron were 250 μM (●) and 83 μM (○) (27).

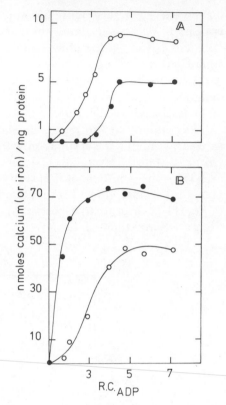

Biochimica et Biophysica Acta

Figure 3.   Effect of the degree of energy coupling, measured as respiratory control with ADP (R.C.$_{ADP}$), on the energy-dependent (respiratory chain linked) accumulation of iron (A) and calcium (B) in respiratory state 1 (in the absence of P$_i$,o) and state 4 (with 3.3 mM succinate as the substrate, ●). Rat liver mitochondria (31).

Transport Form of Iron

It has been generally assumed that iron is transported across biological membranes in the ferrous form and that ferric iron would have to be reduced before it can be used by the organism. Thus, based on nutritional studies it has long been recognized that Fe(II) is more effectively absorbed than Fe(III), and this has been attributed to differences in the thermodynamic and kinetic stability of the complexes and chelates formed by these cations (for review, *see* Ref. 2). The experimental proof of a transport in the ferrous form has, however, not been given until quite recently in studies of iron transport in isolated mitochondria (23) as well as in enterobacteria (33). In rat liver mitochondria we have found that Fe(III) donated from a metabolically inert water soluble complex of sucrose interacts with the respiratory chain at the level of cytochrome c (and possibly cytochrome a) (23, 32) (Figure 1 B), which has a oxidation–reduction potential of around $+250$ mV (34) and is localized to the outer phase of the mitochondrial inner membrane (35).

Thus, a primary event of the energy-linked uptake of iron appears to be Fe(III) binding to ligands on the outer phase of the inner membrane (Figure 1). These ligands have a unique microenvironment giving the metal a half-reduction potential which is high enough to establish a oxidation–reduction equilibrium with the respiratory chain. In contrast to the energy-dependent accumulation of Ca^{2+}, that of iron depends not only on the presence of an energized mitochondrial membrane (e.g., by ATP), but it is closely linked to the availability of reducing equivalents supported by the respiratory chain (Figure 4). Recent studies (36) have revealed that respiring mitochondria catalyze the insertion of Fe(II) into deuteroporphyrin. When Fe(III) is used as the metal donor, the rate of heme formation is largely determined by the rate at which reducing equivalents are supplied by the respiratory chain. The finding that the energized iron uptake revealed an absolute requirement for reducing equivalents indicates that iron crosses the inner membrane only in the ferrous form (23).

From a functional point of view it should be noted that Fe(II), reaching the inner compartment of the mitochondria by the energy-dependent uptake, is not immediately oxidized. This conclusion is sup-

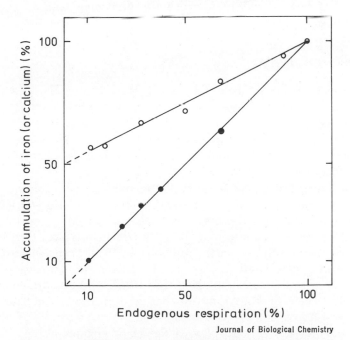

Figure 4. Effect of increasing degree of respiratory inhibition on the energy-dependent accumulation of iron (●) and calcium (○) by rat liver mitochondria energized with ATP (23)

ported by spectroscopic and oxygraphic experiments (32, 36) as well as by electron paramagnetic resonance (EPR) spectroscopy at 13.8°K of guinea pig heart and liver mitochondria (37). Thus, the signal generally observed at around $g = 4.3$ in isolated mitochondria and characteristic of Fe(III) in the high-spin state (see "Mitochondrial Iron Pool," below), does not change in intensity following energized accumulation of iron using Fe(III)–sucrose as the iron donor. This stability towards oxidation is essential for using iron in heme biosynthesis (16).

There is experimental evidence (23) that the energized iron accumulation represents a unidirectional flux of Fe(II) from the outer to the inner phase of the inner membrane and the matrix where it is tightly bound to ligands of a high molecular weight component ($> 10^6$ dalton) not yet characterized (see Figure 1 B). As for the molecular basis of the transport of Fe(II) across the mitochondrial inner membrane, little is known except that it has many features in common with that of Ca^{2+} (27, 31). Considerable evidence suggests that the transport of Ca^{2+} is mediated by a carrier (38), but the chemical nature of this carrier is still uncertain. Several protein factors have been isolated from mitochondria which can bind Ca^{2+} with high affinity, and an acidic glycoprotein has been found to increase the Ca^{2+} conductance of lecithin bilayers (39).

Cytosolic Iron Donor

Because of the low solubility of Fe(II) and Fe(III) in a medium corresponding to the cytoplasm of living cells, it is extremely improbable that ionized iron exists in the cytosol. Thus, iron is present either in metabolically active cell components or bound to intracellular ligands, and specific iron-binding molecules have evolved to maintain iron in soluble form useable by the cell.

It has been known for 40 years that mammalian cells store iron in the form of two complexes known as ferritin and hemosiderin (3). Whereas ferritin is a water-soluble protein containing about 20% iron by weight, hemosiderin represents insoluble granules in which typical ferritin molecules may be wholly or partly replaced by amorphous electron-dense iron deposits (40). Until recently ferritin and hemosiderin were considered solely as storage compounds of iron which could be drawn on when required. Recent findings indicate that ferritin also participates in the transport of iron from degraded red cells in the reticuloendothelial system to the hepatocytes (41). Our knowledge of other proteins and molecules in the cytosol which binds iron as well as the mechanism(s) involved in the cytosolic transport of iron in mammalian cells is as yet limited. Virtually nothing is known about the form(s) in which iron is delivered to the mitochondria in vivo. Although a NADH:(ferritin–Fe(III)$_n$) oxidoreductase system has been demonstrated in liver homogenates (*42*), its physiological significance in the rapid cytosolic transport of iron is still uncertain (*43, 44, 45, 46*). In the mucosal cells, ferritin may function as part of a stable iron pool in equilibrium with a more labile pool (*1*). Other possible donors have therefore been searched for.

It has been proposed that a low molecular weight phosphoprotein of phosvitin nature (*47*) which binds iron with high affinity may be involved in the cytosolic transport of iron in liver cells (*48, 49*), but no energized uptake of iron has been demonstrated such as in isolated rat liver mitochondria using a Fe(III)–phosvitin from avian egg yolk (mol wt 40,000) as the substrate (*50*). The biochemical significance of the cytosolic iron binding phosphoprotein is therefore still uncertain.

In mammals, iron is transported to cells of the body via the extracellular fluid tightly bound to transferrin as a mono- or diferric complex, and the mechanism of iron binding and release has been studied extensively in recent years (for review, *see* Ref. 2). Iron can be released from transferrin by cells such as erythroid cells (bone marrow cells and reticulocytes), but the mechanism by which these cells use iron from this carrier is still poorly understood. One of the hypotheses proposed is that the process of iron exchange between transferrin and erythropoietic cells involves at least four stages including a progressive uptake of transferrin molecules into the cytosol by surface endocytosis (for review, *see* Ref. 51), and at this stage iron becomes available to the cell. It has been proposed that mitochondria take an active part in the dissociation of the iron–transferrin complex (*8, 52*), but at least in vitro partly saturated transferrin functions as a poor iron donor for isolated rat liver mitochondria in a conventional incubation medium at neutral pH (*53*); an energized uptake of only 2–3 pmol · 15 min^{-1} · mg of protein^{-1} was measured as compared with 30–40 pmol · 15 min^{-1} · mg of protein^{-1} at pH 6.2. The low donor activity of transferrin around pH 7.0 as compared with its potency at lower pH primarily reflects the pH-dependence of the iron binding ability of transferrin (*51*). Thus, there is no conclusive evidence to support the recent proposal (*52*) that transferrin is the immediate donor of the mitochondrial iron in vivo.

A low molecular weight protein, different from metallothionine which reversibly binds iron with high affinity has been isolated from rabbit reticulocyte cytosol (*54, 55, 56*). Although very little is yet known about its physiological properties, the molecular weight is around 6000, and iron appears to be reversibly bound under physiological conditions. This protein may be able to mobilize iron from the plasma membrane and donate it for heme and ferritin biosynthesis (*56*), but no definitive physiological role for "siderochelin" has been established.

Cell Differentiation and Mitochondrial Iron Metabolism

The parenchymatous liver cells (hepatocytes) hold a key position in the overall metabolism of iron (57, 58, 59, 60), and since functionally intact liver mitochondria can be conveniently prepared at high yield, these mitochondria have been most extensively studied so far. The iron transporting system discussed above for liver mitochondria is present also in mitochondria from other tissues and animal species (Table III). Quantitatively, erythroid cells of the bone marrow play the most important role in the overall metabolism of iron (61), and it was therefore not unexpected to find that the energized uptake of iron by isolated reticulocyte mitochondria exceeds that of mitochondria isolated from, for example, liver, kidney, and heart (Table IV). Thus, a relationship appears to exist between the rate and extent of heme protein turnover in mitochondria isolated from different tissues and their energized iron accumulation (30). Thus, it is evident that cellular differentiation is expressed at the mitochondrial level by modulation of the activity of essential functions related to iron transport and heme biosynthesis.

Biochemical Significance of the Iron Transporting System of Mitochondria

The role of mitochondria in the regulation of cytoplasmic Ca^{2+} activity is well documented and now generally accepted (62, 63, 64, 65, 66). Recent studies have established important functions also in iron metabolism. In mammals, approximately 70% of the body iron is involved in the transport and storage of oxygen, and protoheme IX, the prosthetic group of hemoglobin and myoglobin, represents the most abundant organic iron-containing compound. Thus, the biosynthesis of protoheme IX no doubt quantitatively represents the most important pathway of the intracellular metabolism of iron (61). The role played by mitochondria is therefore very important since the final step in the

Table III. The Iron Content and Most Common Forms of Iron in Isolated Rat Liver Mitochondria[a]

$nmol \cdot mg \ of \ protein^{-1}$

Total iron	3.9 ± 0.9 $(n = 12)$
Heme proteins (cytochromes)[b]	1.38 –1.40
Iron–sulfur centers	not determined
Other non-heme iron[c]	1.47–1.53

[a] From Ref. 67.
[b] Determined from the reduced minus oxidized difference spectra.
[c] Iron which under mild reducing conditions (ascorbate + N,N,N',N'-tetramethyl-p-phenylenediamine) reacts with bathophenanthroline sulfonate in the presence of the Fe(II) ionophore X-537 A (Hoffman-La Roche Inc., Nutley, N.J.).

Table IV. Energy-Dependent Accumulation of Iron and Calcium
by Mitochondria Isolated from Reticulocytes
and Various Organs of Rabbit[a]

Mitochondria	Respiratory Control Values	Ion Accumulated (nmol/mg protein)		Ratio Fe:Ca
		Iron[a,b]	Calcium[c]	
Reticulocyte	1.6	18.3	20.1	0.910
Kidney	3.9	10.5	105.9	0.099
Liver	2.7	8.0	50.5	0.158
Heart	2.9	3.6	174.9	0.021
Spleen	1.6	2.2	34.6	0.042

[a] From Ref. *29*.
[b] Endogenous respiration.
[c] State 4 respiration.

biosynthesis of heme, the insertion of Fe(II) into protoporphyrin IX, appears to be mainly a mitochondrial enzyme (*10, 11, 12, 13, 14*) probably confined to the inner face of the inner membrane (*15, 16, 17*). The conclusions based on our in vitro studies are also supported by several observations in vivo. Thus, in reticulocytes the uptake of iron by mitochondria has been shown to depend on metabolic energy (*8*), and in disorders with deranged heme biosynthesis, the mitochondria are heavily loaded with iron (*see* below).

Mitochondrial Iron Pool and Its Regulation

When mammals ingest more iron than they need for biosynthetic purposes, this excess is stored in most tissues, but mainly in liver, spleen, and bone marrow. At physiological levels of iron in the tissues, the storage form is predominantly ferritin in the cytosol, but at higher concentrations of iron, hemosiderin is also formed and deposited in the same cell compartment (*19*). In addition to the cytosolic iron deposit, the mitochondria contain some non-heme iron different from the iron–sulfur proteins of the respiratory chain (*37, 67*). A fraction of this iron is detectable by EPR spectroscopy by its resonance at around $g = 4.3$ (*37*) characteristic of Fe(III) in the high-spin state and is accessible to reduction by substrates (*37*) and to reaction with iron chelators (*67*). Table III gives the distribution of iron in liver mitochondria in rat. According to iron and heme analyses there is more non-heme than heme iron, and the non-heme iron, which is not involved in the electron transport chain, accounts for a significant pool of the mitochondrial iron. Recent studies by Hanstein et al. (*68*) seem to indicate that the magnitude of this iron pool may vary with the total iron body content.

Iron loading of mitochondria is observed in sideroblastic anemia in both animals and man (for review, *see* Ref. *69*). The prototype of this group of anemias is pyridoxine deficiency, where the number of sidero-blasts in the bone marrow increases (70), and their mitochondria are heavily loaded with iron. Electron-dense iron particles have been found in the space between the mitochondrial cristae, and the mitochondria are morphologically distorted and swollen. Neither electron microscopic (71) nor immunological studies (72) support the conclusion that these particles represent ferritin as originally proposed (73). There is experimental evidence that the accumulation of mitochondrial iron results from an impaired heme biosynthetic pathway of, for example, the iron use (for review, see Ref. 69), as well as an increased iron uptake by erythroid cells from inhibited heme synthesis (74). Thus, the cytosolic hemin pool appears to be a natural negative feedback regulator of iron uptake (75) by inhibiting the dissociation of iron–transferrin complex in erythroid cells (76) as well as the energy-dependent iron accumulation (77).

However, during the most active phase of heme synthesis in maturing erythroid cells, there is a marked shift in the localization of cellular iron from a membrane bound pool to the cytosol (78). Moreover, as shown by Yoda and Israel (79), mitochondria incubated in cell sap or sucrose synthesize equivalent amounts of heme, but those in sucrose release only a small amount of heme to the surrounding medium. In other words, the release of heme from the mitochondria seems to depend on protein(s) in the suspending medium. Thus, the cytosol appears to facilitate the uptake of iron by the mitochondria as well as the release of heme from the mitochondria.

Literature Cited

1. Forth, W., Rummel, W., *Physiol. Rev.* (1973) **53**, 724.
2. Aisen, P., Brown, E. B., "Progress in Hematology," Vol. 10, p. 25, Grune and Stratton, New York, 1975.
3. Harrison, P. M., Hoare, R. J., Hoy, T. G., Macara, I. G., "Iron in Biochemistry and Medicine," p. 73, Academic, New York, 1974.
4. Nicholls, P., Elliott, W. B., "Iron in Biochemistry and Medicine," p. 221, Academic, New York, 1974.
5. Paul, K. -G., "The Enzymes," Vol. 3, p. 277, Academic, New York, 1960.
6. Beinert, H., "Iron-Sulfur Proteins," Vol. 1, p. 1, Academic, New York, 1973.
7. Jandl, J. H., Inman, J. K., Simmons, R. L., Allen, D., *J. Clin. Invest.* (1959) **38**, 161.
8. Morgan, E. H., Baker, E., *Biochim. Biophys. Acta* (1969) **184**, 422.
9. Morgan, E. H., *Biochim. Biophys. Acta* (1971) **244**, 103.
10. Barnes, R., Jones, M. S., Jones, O. T. G., Porra, R. J., *Biochem. J.* (1971) **124**, 633.
11. Labbe, R. F., Hubbard, N., *Biochim. Biophys. Acta* (1960) **41**, 185.
12. Lochhead, A. C., Goldberg, A., *Biochem. J.* (1961) **78**, 146.
13. Minakami, S., *J. Biochem. (Tokyo)* (1958) **45**, 833.
14. Nishida, G., Labbe, R. F., *Biochim. Biophys. Acta* (1959) **31**, 519.
15. Barnes, R., Conelly, J. L., Jones, O. T. G., *Biochem. J.* (1972) **128**, 1043.
16. Jones, M. S., Jones, O. T. G., *Biochem. J.* (1969) **113**, 507.
17. McKay, R., Druyan, R., Getz, G. S., Rabinowitz, M., *Biochem. J.* (1969) **114**, 455.

18. Klingenberg, M., *Essays Biochem.* (1970) **6**, 119.
19. Azzone, G. F., Massari, S., *Biochim. Biophys. Acta* (1973) **301**, 195.
20. Martell, A. E., "Stability Constants of Metal-Ion Complexes," p. 390, The Chemical Society, London, 1964.
21. Cederbaum, A. I., Wainio, W. W., *J. Biol. Chem.* (1972) **247**, 4593.
22. Stickland, E. H., Davis, B. C., *Biochim. Biophys. Acta* (1965) **104**, 596.
23. Flatmark, T., Romslo, I., *J. Biol. Chem.* (1975) **250**, 6433.
24. Romslo, I., Flatmark, T., *Biochim. Biophys. Acta* (1973) **305**, 29.
25. Pappius, H. M., Elliott, K. A. C., *Can. J. Biochem.* (1956) **34**, 1007.
26. Malamed, S., Recknagel, R. O., *J. Biol. Chem.* (1959) **234**, 3027.
27. Romslo, I., Flatmark, T., *Biochim. Biophys. Acta* (1973) **325**, 38.
28. *Ibid.* (1974) **347**, 160.
29. Romslo, I., *Biochim. Biophys. Acta* (1974) **357**, 34.
30. Romslo, I., *FEBS Lett.* (1974) **43**, 144.
31. Romslo, I., *Biochim. Biophys. Acta* (1975) **387**, 69.
32. Romslo, I., Flatmark, T., *Biochim. Biophys. Acta* (1975) **387**, 80.
33. Neilands, J. B., ADV. CHEM. SER. (1977) **162**, 3.
34. Flatmark, T., *J. Biol. Chem.* (1967) **242**, 2454.
35. Racker, E., *Essays Biochem.* (1970) **6**, 1.
36. Koller, M.-E., Romslo, I., Flatmark, T., *Biochim. Biophys. Acta* (1976) **449**, 480.
37. Flatmark, T., Beinert, H., unpublished data.
38. Carafoli, E., Sottocasa, G., *Dyn. Energy-transducing Membr. I.U.B. Symp. 1973* (1974) 455.
39. Carafoli, E., Prestipino, G. F., Ceccarelli, O., Conti, F., "Membrane Proteins in Transport and Phosphorylation," p. 85, North-Holland, Amsterdam, 1974.
40. Jacobs, A., Worwood, M., "Progress in Hematology," Vol. 10, p. 1, Grune and Stratton, New York, 1975.
41. Siimes, M. A., Dallman, P. R., *Br. J. Haematol.* (1974) **28**, 7.
42. Sirivech, S., Frieden, E., Osaki, S., *Biochem. J.* (1974) **143**, 311.
43. Hoy, T. G., Harrison, P. M., Shabbir, M., Macara, I. G., *Biochem. J.* (1974) **137**, 67.
44. Hoy, T. G., Harrison, P. M., Shabbir, M., *Biochem. J.* (1974) **139**, 603.
45. Primosigh, J. V., Thomas, E. D., *J. Clin. Invest.* (1968) **47**, 1473.
46. Zail, S. S., Charlton, R. W., Torrance, J. O., Bothwell, T. H., *J. Clin. Invest.* (1964) **43**, 670.
47. Pinna, L. A., Clari, G., Moret, V., *Biochim. Biophys. Acta* (1971) **236**, 270.
48. Clari, G., Pinna, L. A., Moret, V., Siliprandi, N., *FEBS Lett.* (1971) **17**, 300.
49. Donella, A., Pinna, L. A., Moret, V., *FEBS Lett.* (1972) **26**, 249.
50. Ulvik, R., Romslo, I., unpublished data.
51. Morgan, E. H., "Iron in Biochemistry and Medicine," p. 29, Academic, London, 1974.
52. Neuwirt, J., Borova, J. Poňka, P., "Proteins of Iron Storage and Transport in Biochemistry and Medicine," p. 161, North-Holland, Amsterdam, 1975.
53. Ulvik, R., Koller, M.-E., Prante, P. H., Romslo, I., *Scand. J. Clin. Lab. Invest.* (1976) **36**, 539.
54. Bates, G. W., Workman, E. F., *Fed. Proc.* (1975) **34**, 643.
55. Workman, E. F., Bates, G. W., *Biochem. Biophys. Res. Commun.* (1974) **58**, 787.
56. Workman, E. F., Bates, G. W., "Proteins of Iron Storage and Transport in Biochemistry and Medicine," p. 155, North-Holland, Amsterdam, 1975.
57. Levitt, M., Schacter, B. A., Zipursky, A., Israels, L. G., *J. Clin. Invest.* (1968) **47**, 1281.
58. Pollycove, M., Mortimer, R., *J. Clin. Invest.* (1961) **40**, 753.
59. Snyder, A. L., Schmid, R., *J. Lab. Clin. Med.* (1965) **65**, 817.

60. Theorell, H., Beznak, M., Bonnichsen, R., Paul, K. -G., Åkeson, Å., *Acta Chem. Scand.* (1951) **5**, 445.
61. Noyes, W. D., Hosain, F., Finch, C. A., *J. Lab. Clin. Med.* (1964) **64**, 574.
62. Borle, A. B., *Fed. Proc.* (1973) **32**, 1944.
63. Carafoli, E., *Biochimie* (1973) **55**, 755.
64. Carafoli, E., Tiozzo, R., Lugli, G., Crovetti, F., Kratzing, G., *J. Mol. Cell. Cardiol.* (1974) **6**, 361.
65. Lehninger, A. L., *Biochem. J.* (1970) **119**, 129.
66. Spencer, T., Bygrave, F. L., *Bioenergetics* (1973) **4**, 347.
67. Tangerås, A., Flatmark, T., unpublished data.
68. Hanstein, W. G. ,Sacks, P. V., Muller-Eberhard, U., *Biochem. Biophys. Res. Commun.* (1975) **67**, 1175.
69. Cartwright, G. E., Deiss, A., *N. Engl. J. Med.* (1975) **292**, 185.
70. Deiss, A., Kurth, D., Cartwright, G. E., *J. Clin. Invest.* (1966) **45**, 353.
71. Arstila, A. U., Bradford, W. D., Kinney, T. O., Trump, B. F., *Am. J. Pathol.* (1970) **58**, 419.
72. Romslo, I., Flatmark, T., unpublished data.
73. Bessis, M., Breton–Gorius, J., *C. R. Acad. Sci. (Paris)* (1957) **244**, 2846.
74. Poňka, P., Neuwirt, J., *N. Engl. J. Med.* (1975) **293**, 406.
75. Poňka, P., Neuwirt, J., *Br. J. Haematol.* (1974) **28**, 1.
76. Poňka, P., Neuwirt, J., Borová, J., *Enzyme* (1974) **17**, 91.
77. Koller, M.-E., Prante, P. H., Ulvik, R., Romslo, I., *Biochem. Biophys. Res. Commun.* (1976) **71**, 339.
78. Denton, M. J., Delves, H. T., Arnstein, H. R. V., *Biochem. Biophys. Res. Commun.* (1974) **61**, 8.
79. Yoda, B., Israels, L. G., *Can. J. Biochem.* (1972) **50**, 633.

RECEIVED July 26, 1976. This work was supported by the Norwegian Research Council for Science and the Humanities (grant No. C.11.14-3) and the Norwegian Cancer Society.

Genetically Controlled Chemical Factors Involved in Absorption and Transport of Iron by Plants

JOHN C. BROWN

U.S. Department of Agriculture, Agricultural Research Service,
Plant Stress Laboratory, Beltsville, Md. 20705

Plant use of iron depends on the plant's ability to respond chemically to iron stress. This response causes the roots to release H^+ and "reductants," to reduce Fe^{3+}, and to accumulate citrate, making iron available to the plant. Reduction sites are principally in the young lateral roots. Azide, arsenate, zinc, copper, and chelating agents may interfere with use of iron. Chemical reactions induced by iron stress affect nitrate reductase activity, use of iron from Fe^{3+} phosphate and Fe^{3+} chelate, and tolerance of plants to heavy metals. The iron stress–response mechanism is adaptive and genetically controlled, making it possible to tailor plants to grow under conditions of iron stress.

The amount of Fe^{3+} iron available in the aqueous solutions of most calcareous soils is insufficient for plant growth (1). Above pH 4.0, the Fe^{3+} activity in solution decreases a thousandfold for each unit increase in pH (2). Oertli and Jacobson (1) indicate that at pH 9, the saturation concentration of both cation forms of iron drops below 10^{-20} mol/l. in a solution in equilibrium with atmospheric oxygen. Yet an iron-deficient (chlorotic) and iron-sufficient (green) plant of the same species can grow side by side (Figure 1) in an alkaline environment (3). In such instances, the two cultivars usually differ only in their genetic ability to respond to iron stress. The chemical reactions induced by iron stress make iron available to the plant, and plants are classified iron-efficient if they respond to iron stress and iron-inefficient if they do not. Plants require a continuing supply of iron to maintain proper growth. Since iron stress occurs in many alkaline soils, iron-inefficient plants often become chlorotic and die. The chemical factors involved in iron absorption and transport in plants are described here by contrasting iron-inefficient and iron-efficient plants.

Iron Supply

Iron is most commonly supplied to plants by the seed, by the growth medium, or as a spray. The germinating seed usually contains sufficient iron to meet the plant's requirements in the seedling stage (4). Hyde et al. (5) suggested that phytoferritin was a form of iron stored for use

Figure 1.   Iron-efficient Hawkeye (left) and iron-ineffi-cient T203 soybean (right) grown together on an alka-line soil (pH 7.5). Only T203 soybean developed iron deficiency.

by the young seedling in cotyledons of pea. Ambler and Brown (6) showed that chemical factors that interfere with uptake of iron from the growth medium do not interfere with plant use of iron from the cotyledons. Iron added as a spray is not transported away from the application area. Thus, iron nutrition problems appear to be centered in the roots and on chemical factors that affect uptake and translocation of iron from the growth medium.

Role of the Rootstock in Iron Use

Reciprocal approach grafts of iron-inefficient on iron-efficient root-stocks showed that iron transport is controlled by the rootstock of tomato (Table I) (7) and of soybean (8). Although the iron concentration of iron-inefficient T3238fer tomato roots was similar to that of iron-efficient T3238FER tomato roots (Table I), less iron was transported to the T3238fer tops because T3238fer roots did not respond to iron stress. Likewise in soybeans, autoradiographs showed that ^{55}Fe was not transported to the new leaf that developed after the iron-efficient top was grafted to the iron-inefficient rootstock (8). The new leaf was chlorotic because the iron-inefficient rootstock could not supply ^{55}Fe to it.

Table I. Iron Concentration in Tops and Roots of Iron-inefficient T3238fer (t-fer) and Iron-efficient T3238FER (T-FER) Tomatoes as Affected by Tomato Rootstock (7)

| | Iron ($\mu g/g$) | | | |
| | Experiment (a) | | Experiment (b) | |
Approach Grafts	top	root	top	root
t-fer top on t-fer root	63	715	12	1160
t-fer top on T-FER root	124	340	192	618
T-FER top on t-fer root	43	229	17	338
T-FER top on T-FER root	121	692	130	740

Physiologia Plantarum

Plant Response to Iron Stress

Several products or biochemical reactions occur only in iron-efficient plants in response to iron stress:

(1) Hydrogen ions are released from the roots.

(2) Reducing compounds are released from the roots.

(3) Ferric iron is reduced at the roots.

(4) Organic acids (particularly citrate) are increased in roots.

Response to iron stress is adaptive and genetically controlled in corn (*Zea mays* L.) (9), soybeans (*Glycine max* (L.) Merr.) (10), and tomato (*Lycopersicon esculentum* Mill) (11).

Hydrogen Ion Release from Roots. In nutrient solutions with no added iron and nitrogen available only as NO_3–N, the first indication of iron stress is that the terminal leaves of iron-efficient T3238FER tomato develop incipient chlorosis. Almost simultaneously, hydrogen ions were released from their roots. This reduced the pH of 8 l. of nutrient from approximately 6.4 to 4.4 in 24 hr (iron stress) and increased pH from 4.4 to 6.6 within the next 24 hr (Figure 2), because iron repressed the

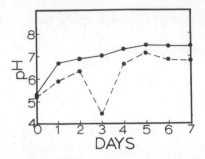

Figure 2. Plant-induced pH changes of nutrient solutions caused by differential iron-stress response of 25-day old T3238fer (———) and T328FER (– – – –) tomato plants placed in nutrient solutions lacking iron and containing only NO_3–N. The pH dropped at day 3 when the T3238FER plants developed iron stress.

iron stress–response mechanism when it was made available. Within this same time, the terminal leaves changed from green to slightly chlorotic and then became green again. The terminal leaves of the iron-inefficient T3238fer tomato also developed iron chlorosis but remained chlorotic throughout this period because it did not respond to iron stress.

Reducing Compound Release from Roots. In nutrient solutions with no iron and nitrogen as NH_4–N and NO_3–N, iron-efficient soybean (12) and tomato (7) released "reductants" from their roots in response to iron stress. The term "reductants" designates compounds released by roots that reduce Fe^{3+} to Fe^{2+} (Figure 3). "Reductant" is released into solution in greatest quantity when the pH of the nutrient is below 4.5. The amount of reductant released was determined spectrochemically in vitro using 2,4,6-tripyridyl-s-triazine (TPTZ) (13, 14, 15). The Fe^{2+} iron combines with TPTZ to form the color complex $Fe^{2+}(TPTZ)_2$ which in water conforms to Beer's law up to about 60 μmol Fe^{2+} (15). Chelating agents (in vitro) interfered with reduction of Fe^{3+} by the "reductants," but this interference was eliminated by increasing the concentration of "reductant" in solution (14).

Iron uptake by iron-inefficient soybeans was not increased when they were placed in nutrient solutions that contained "reductant" (14). This may mean that "reductants" in the external solution indicate a leaky root resulting from the release of hydrogen into the nutrient solution. More important may be the adaptive production of "reductants" inside the root or at the root surface that keeps iron in the more available Fe^{2+} form (13). We have concluded that iron absorption and transport is controlled inside the root, and iron uptake is greatest while the iron-stress–response mechanism is functioning.

The "reductant" maintained its reducing capacity even after being paper chromatographed, dried, and placed in a thin film of ferricyanide–ferrichloride solution (10 μmol) (14). The formation of three Prussian blue spots on the paper indicated that the roots released at least three different reducing compounds.

Over the past 20 yr, compounds have been identified and mechanisms established for microbial transport of iron. For example, Ito and

Neilands (*16*) suggested that excretion of metal-binding phenolic acids during iron deprivation of *Bacillus substilis* might combat iron deficiency. Walsh and Warren (*17*) indicated that the phenolic acids accumulated by iron-deficient cultures of *B. subtilis* do not seem to be involved in iron uptake but serve to solubilize the iron in the growth medium. Byers and Lankford (*18*) found that the addition of iron to growth cultures inhibited phenolic acid excretion by *B. subtilis.* Our findings with iron-efficient soybean and tomato agree with the above observations, except that we have not identified the "reductants."

Ferric Iron Reduction at the Root. Sites of Fe^{3+} reduction in iron-efficient soybean (*13, 19*) and tomato (*20*) are principally in the young lateral roots (Figure 4). Iron-inefficient T3238fer tomato showed practically no reduction in these roots. Reduction sites were determined by transferring iron-stressed T3238fer and T3238FER tomato to nutrient solutions containing FeHEDTA (iron-hydroxyethylenediaminetriacetic acid) and $K_3Fe(CN)_6$ (*20*). A blue precipitate, Prussian blue, appeared in the epidermal areas of the root where Fe^{3+} was reduced by the root.

Most of the Fe^{3+} was reduced outside the root in areas accessible to BPDS (bathophenanthrolinedisulfonate) (*20*). This was established

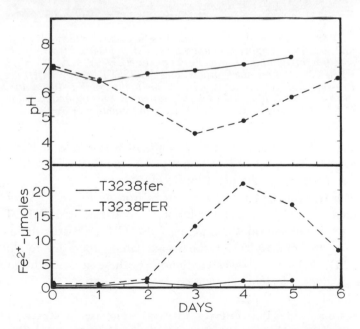

Figure 3.   Plant-induced pH changes of nutrient solutions (top) and release of "reductant" Fe^{3+} to Fe^{2+} (bottom) caused by differential iron stress response of 25-day old iron-stressed T3238fer and T3238FER tomato plants placed in nutrient solutions lacking iron and containing NH_4–N and NO_3–N

*Figure 4.   Roots of 25-day-old iron-stressed T3238FER (left)
and T3238fer (right) tomatoes after being placed in a nutrient
solution containing 5 mg Fe/l. as FeHEDTA and 33 mg/l.
$K_3Fe(CN)_6$.   The dark areas on T3238FER roots (left) are
Prussian blue precipitates formed when the roots reduced Fe^{3+}
to Fe^{2+}.   The blue precipitate indicates the reduction sites.
T3238fer roots (right) showed no reduction of Fe^{3+}.*

by adding BPDS to the nutrient solutions, 10% in excess of Fe^{3+}. As the Fe^{3+} was reduced, most of it was trapped in solution as $Fe^{2+}BPDS_3$ and was not transported to the plant top (*21*).

If the iron-efficient roots were given iron as FeHEDTA for 20 hr, then taken out of the nutrient solutions, rinsed free of FeHEDTA, and placed in nutrient solutions containing $K_3Fe(CN)_6$, Prussian blue formed throughout the protoxylem of the young lateral roots up to the metaxylem (*13*). Ferrous iron was continuous in these areas of the roots and in the regions of root elongation and maturation of the primary root (*13, 19*).

Organic Acid (Particularly Citrate) Increase in Roots. Organic acids keep iron mobile in external solution, and Rogers and Shive (*22*) suggested that they might function similarly inside the plant. Chlorotic plants usually contain more citric and malic acids than normal green plants (*23, 24, 25, 26*).

Schmid and Gerloff (27) reported a naturally occurring iron-complexing agent in tobacco xylem exudate that prevented iron precipitation and was a vehicle for transporting iron in the plant. Tiffin and Brown (28) showed that iron in xylem exudate is mostly in negatively charged forms. Iron stress, as it controls iron supply to the plant, may be the controlling factor affecting the appearance of citrate in xylem exudate (21, 29). The parallel between ion transported and citrate in the xylem exudate was striking. When iron increased, the citrate increased; a decrease in iron was paralleled by a decrease in citrate. This relationship held true whether iron stress was produced in the plant by limiting the iron supply or by using zinc, azide, or arsenate to induce iron stress (29). Trapping Fe^{2+} at the root as $Fe^{2+}BPDS_3$ decreased the iron supply to the plant and simultaneously decreased the citrate in stem exudate (21). Tiffin (30, 31, 32) identified ferric citrate in the xylem exudate of several plant species by using electrophoresis to follow the migration of the chelated iron. Sufficient citrate was always present to chelate the metal, and any excess migrated as an iron-free fraction behind the iron–citrate band.

Clark et al. (33) found that malic, acetic, and *trans*-aconitic acids were ineffective in moving ^{59}Fe electrophoretically in acetate, citrate, isocitrate, *trans*-aconitate, and malate buffers. Citric acid moved iron anodically whenever present on the electropherogram and successfully competed with the other acids for iron. Clark et al. (33) further showed that iron-efficient corn absorbed and transported more iron in the xylem exudate than iron-inefficient corn. But when ^{59}Fe was added in vitro to xylem exudate from the iron-inefficient corn, the ^{59}Fe moved as ^{59}Fe citrate, indicating that there was sufficient citric acid in the exudate to chelate the added ^{59}Fe. The iron-inefficient corn roots do not respond to iron stress and lack the mechanism to supply iron for movement into the root. The translocation of iron in the plant involves more than citrate chelation of iron in the root.

Mechanism of Iron Absorption and Transport

Iron-efficient and iron-inefficient plants can have several hundred μgFe/g of root, but the iron-inefficient plant may die from lack of iron in its tops. In contrast, iron-efficient plants respond to iron stress, and the root makes iron available for transport and use in tops. In a similar way, iron may remain in the nutrient solution as Fe^{3+} chelate or Fe^{3+} phosphate and not be transported to the plant top until it is made available for transport through chemical reactions induced by iron stress. These observations stress the importance of a plant being able to respond to iron stress. Iron is usually used in plant tops once it is made available for transport by the roots.

The individual reactions affected by iron stress can be considered as regulated biochemical pathways, although regulation by iron is not understood. The mechanism of iron absorption and transport involves the release of hydrogen ions by the root, which lowers the pH of the root zone. This favors Fe^{3+} solubility and reduction of Fe^{3+} to Fe^{2+}. "Reductants" are released by roots or accumulate in roots of plants that are under iron stress. These "reductants," along with Fe^{3+} reduction by the root, reduce Fe^{3+} to Fe^{2+}, and Fe^{2+} can enter the root. Ferrous iron has been detected throughout the protoxylem of the young lateral roots. The Fe^{2+} is probably kept reduced by the "reductant" in the root, and it may or may not have entered the root by a carrier mechanism. The root-absorbed Fe^{2+} is believed to be oxidized to Fe^{3+}, chelated by citrate, and transported in the metaxylem to the tops of the plant for use. We assume Fe^{2+} is oxidized as it enters the metaxylem because there is no measureable Fe^{2+} there (13), and Fe^{3+} citrate is transported in the xylem exudate (30, 31, 32).

Effect of Iron Stress–Response Mechanism on Other Biochemical Reactions

Induced iron stress alters biochemical reactions more in iron-efficient than in iron-inefficient plants, which makes the former more versatile than the latter. Listed below are some examples where this may occur.

Nitrate Reductase Activity. There are similarities between induced nitrate reductase activity and induced iron stress response. In both, biochemical reactions are induced, and a substrate is reduced; NO_3 to NO_2 by nitrate reductase and Fe^{3+} to Fe^{2+} by a reductant activated in response to iron stress. Chemical reactions induced by iron stress increased the use of iron, and simultaneously increased nitrate reductase activity in roots (Figure 5) and in tops of iron-efficient tomato. This induced nitrate reductase activity declined when iron was made available to the plants.

Use of Iron from Fe^{3+} Phosphate. Iron phosphate precipitate was used by iron-efficient soybeans when Fe^{3+} was reduced to Fe^{2+} (19). The Fe^{2+} was detected in solution as Fe^{2+} ferrozine [Fe^{2+}3-(2-pyridyl)-5,6-bis(4-phenylsulfonic acid)-1,2,4, trizine]. The iron-inefficient soybeans developed iron chlorosis because they did not reduce Fe^{3+} to Fe^{2+}, and they could not use the iron from Fe^{3+} phosphate.

Use of Iron from Fe^{3+}EDDHA (Iron-ethylenediamine di(o-hydroxyphenylacetic acid)). Iron-inefficient T3238fer tomato developed iron deficiency because it could not absorb iron from FeEDDHA (7). In contrast, the iron-efficient T3238FER tomato used iron from FeEDDHA because it could reduce Fe^{3+} to Fe^{2+}. Chaney et al. (34) showed that for

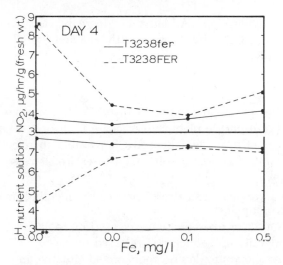

Figure 5. Nitrate reductase activity and nutrient solution pH for 28-day-old T3238fer and T3238FER tomatoes grown for 8 days at various levels of iron stress. In solutions lacking iron and where iron had been removed from the roots, iron stress developed in T3238FER tomato, and the pH of the nutrient solution decreased from 7.1 (day 2) to 4.35 (day 4). Nitrate reductase activity (NRA) increased in the roots from 2.8 (day 2) to 8.5 (day 4). No significant differences were noted between NRA of T3238fer and T3238FER roots when they did not respond to iron stress. *, significantly different at 1% level according to Duncan's multiple range test. **, some iron was removed from roots with 21μM NaEDDHA (ethylenediamine di(o-hydroxyphenylacetic acid)) before these plants were placed in the nutrient solution.

plants to use Fe^{3+} from several Fe^{3+} chelates, it was first necessary to reduce Fe^{3+} chelate to Fe^{2+} chelate. The latter usually has a much lower stability constant than the former.

Tolerance to Heavy Metals. In many plants, heavy metals induce iron stress. These metals seem to interfere with the iron stress–response mechanism (*13*) and in this way cause iron chlorosis to develop. Plants under these conditions will die unless they can respond to iron stress and make more iron available (*35*). Additional iron counteracts the effect of the heavy metals.

Zinc Stress Induction of Iron Uptake. For some unexplained reason, zinc stress may induce symptoms similar to, if not the same, as iron stress (*36*). In these plants, iron uptake may increase so much that iron toxicity symptoms develop. Added zinc decreases the absorption and transport of iron (13, 29) by decreasing the efficiency of the iron stress–response mechanisms.

Future Needs

Some of the chemical factors involved in the mechanism of iron absorption and transport in plants have been established. These reactions are genetically controlled, which makes it possible to select or develop iron-efficient plants for use in soil that causes iron stress. In addition we need to know:

(1) How and where iron stress changes the metabolism of the plant to accentuate iron uptake.

(2) The source of hydrogen ions released by roots.

(3) The identity of "reductants" released by roots.

(4) The reason young lateral roots are so effective in reducing Fe^{3+}.

(5) How Fe^{2+} moves in the root.

(6) Where Fe^{2+} is oxidized to Fe^{3+} and chelated by citrate.

(7) How heavy metals, arsenate, and azide inhibit chemical reactions induced by iron stress.

(8) How the chemical reactions induced by iron stress are related to nitrate nutrition and the activity of nitrate reductase.

(9) How iron from Fe^{3+} is used in leaves.

(10) How light affects iron use.

An understanding of the basic physiology, biochemistry, genetics, and nutrient element interactions involved in iron nutrition will contribute tremendously to our understanding of pertinent biological processes.

Literature Cited

1. Oertli, J. J., Jacobson, L., *Plant Physiol.* (1960) **35**, 683.
2. Lindsay, W. L., *Micronutrients Agri., Proc. Symp.* (1972) 341–357.
3. Brown, J. C., Ambler, J. E., Chaney, R. L., Foy, D. D., *Micronutrients Agri., Proc. Symp.* (1972) 389-418.
4. Brown, J. C., *Adv. Agron.* (1961) **13**, 329.
5. Hyde, B. B., Hodge, A. J., Kahn, A., Birnstiel, M. L., *J. Ultrastruct. Res.* (1963) **9**, 248.
6. Ambler, J. E., Brown, J. C., *Agron. J.* (1974) **66**, 476.
7. Brown, J. C., Chaney, R. L., Ambler, J. E., *Physiol. Plant.* (1971) **25**, 48.
8. Brown, J. C., Holmes, R. S., Tiffin, L. O., *Soil Sci.* (1958) **86**, 75.
9. Bell, W. D., Bogorad, L., McIlrath, W. J., *Bot. Gaz. Chicago* (1958) **120**, 36.
10. Weiss, M. G., *Genetics* (1943) **28**, 253.
11. Wann, E. V., Hills, W. A., *J. Hered.* (1973) **64**, 370.
12. Brown, J. C., Holmes, R. S., Tiffin, L. O., *Soil Sci.* (1961) **91**, 127.
13. Ambler, J. E., Brown, J. C., Gaugh, H. G., *Agron. J.* (1971) **63**, 95.
14. Brown, J. C., Ambler, J. E., *Agron. J.* (1973) **65**, 311.
15. Diehl, A., Smith, G. H., "The Iron Reagents," pp 41-56, G. Frederick Smith Chem. Co., Columbus, 1965.

16. Ito, T., Nielands, J. B., *J. Am. Chem. Soc.* (1958) **80**, 4645.
17. Walsh, B. L., Warren, R. A. J., *J. Bacteriol.* (1968) **95**, 360.
18. Byers, B. R., Lankford, C. E., *Biochim. Biophys. Acta* (1968) **165**, 563.
19. Brown, J. C., *Agron. J.* (1972) **64**, 240.
20. Brown, J. C., Ambler, J. E., *Physiol. Plant.* (1974) **31**, 221.
21. Brown, J. C., Chaney, R. L., *Plant Physiol.* (1971) **47**, 836.
22. Rogers, C. H., Shive, J. W., *Plant Physiol.* (1932) **7**, 227.
23. DeKock, P. C., Morrison, R. J., *Biochem. J.* (1958) **70**, 272.
24. Iljin, W. S., *Plant Soil* (1951) **3**, 239.
25. *Ibid.* (1952) **4**, 11.
26. Rhoades, W. A., Wallace, A., *Soil Sci.* (1960) **89**, 248.
27. Schmid, W. E., Gerloff, G. C., *Plant Physiol.* (1961) **36**, 226.
28. Tiffin, L. O., Brown, J. C., *Plant Physiol.* (1961) **36**, 710.
29. Brown, J. C., Tiffin, L. O., *Plant Physiol.* (1965) **40**, 395.
30. Tiffin, L. O., *Plant Physiol.* (1966) **41**, 510.
31. *Ibid.* (1966) **41**, 515.
32. *Ibid.* (1970) **45**, 280.
33. Clark, R. B., Tiffin, L. O., Brown, J. C., *Plant Physiol.* (1973) **52**, 147.
34. Chaney, R. L., Brown, J. C., Tiffin, L. O., *Plant Physiol.* (1972) **50**, 208.
35. Brown, J. C., Jones, W. E., *Commun. Soil Sci. Plant Anal.* (1975) **6**, 421.
36. Ambler, J. E., Brown, J. C., *Agron. J.* (1969) **61**, 41.

RECEIVED July 26, 1976.

6

Transport of Iron by Transferrin

PHILIP AISEN and ADELA LEIBMAN

Departments of Biophysics and Medicine, Albert Einstein College of Medicine, Bronx, NY 10461

Transferrin provides iron for the biosynthesis of hemoglobin and probably also for other essential iron proteins. Both active sites bind iron tightly enough to resist hydrolysis but reversibly so that the protein undergoes many cycles of iron transport. These sites do not exist unless a stereochemically suitable anion is available to stabilize them, probably acting as a bridging ligand between metal and protein. The initial event in iron delivery by transferrin to hemoglobin-synthesizing red blood cells is protein binding to specific receptors on the cell membrane surface. In the rabbit, this receptor appears to be a glycoprotein. After binding, iron may be released from transferrin to the cell by attack on the stabilizing anion. The protein is then returned to the circulation for another cycle of iron transport.

Under the conditions found in most biologic fluids, iron is readily oxidized by molecular oxygen to the ferric state. Ferric iron, however, is prone to hydrolyze, forming insoluble polynuclear ferric hydroxide complexes, even at a pH as low as 2 (*1, 2*). To maintain iron in soluble form utilizable for the synthesis of essential iron-bearing enzymes and proteins, specific iron-binding molecules have evolved in many organisms. In the microbial world these iron-binders are relatively simple and well characterized structures, with hydroxamate or phenolate ligands at their active sites (*3*). Their function is to mobilize iron from its inorganic environment and to present it in soluble form to the organism. Furthermore they may be expendable; iron appears to be released from some of them for metabolic use by degradation of their molecular structure (*4*).

In the vertebrate kingdom, where elaborate circulatory systems serve tissues with widely varying iron requirements, the transferrins, comprising a more complex and more subtle class of iron-binding molecules, have evolved for the transport of iron. Serum transferrin is the prototype and most extensively studied member of this class. Other members of the transferrin group include lactoferrin or lactotransferrin, a protein widely distributed in cells and fluids of external secretion, and conalbumin or ovo-transferrin, found in avian egg white (5). The functions of these latter proteins are still only poorly understood. They may serve in the transport of iron (6), and because of their avid affinities for iron, they may also offer a bacteriostatic activity for their hosts by sequestering the metal and making it unavailable for microbial growth and metabolism (7). By far the most important property of serum transferrin is that it is the source of iron for the biosynthesis of hemoglobin by the developing red blood cell (8). Our concern is with some of the molecular events in the binding of iron by transferrin and its release to the hemoglobin synthesizing apparatus of the immature red cell.

Fundamental Properties of Serum Transferrin

Transferrin is a glycoprotein of unremarkable amino acid composition with a molecular weight in the range 76,000–80,000 (9, 10). It consists of a single polypeptide chain on which are disposed two active metal-binding sites. Whether these sites are identical in structure and equivalent in function is still a lively and controversial question some 30 years after the discovery of transferrin (11). When considered as a hydrodynamic prolate ellipsoid of revolution, the protein is asymmetric, with an axial ratio of about 2.5- or 3-to-1 for the apoprotein, which decreases as iron is bound (12, 13). The interactions responsible for the metal-dependent shape change, and its physiologic significance, if any, remains unknown.

Perhaps the most striking chemical property of transferrin is that its iron-binding activity is exhibited only when a suitable anion is available for concomitant binding (14, 15, 16). Carbonate, or possibly bicarbonate (17), is the anion preferred by the protein, but when all hydrated forms of carbon dioxide are excluded from solution, a number of other anions may also activate the metal-binding sites. These include, but are not limited to, oxalate, malonate, thioglycollate, glycinate, EDTA, and possibly citrate (16, 18, 19). One anion is obligated for each metal ion bound. The physiology and chemistry of the anion binding function of transferrin is currently a major area of research interest.

The occurrence of two iron-binding sites on a single-chain protein has generated much speculation about the possibility of gene duplication and fusion during the biochemical evolution of transferrin (9). Although work based on chemical cleavage and peptide mapping was largely inconclusive, recent analysis of the amino acid sequence has revealed internal homologies which seem to support the gene duplication-fusion hypothesis (20). A search for a primitive, single-sited transferrin in lower vertebrates had not yet met with undisputed success, however (21, 22). The biologic advantage, if any, offered to the organism by a two-sited iron-carrying protein is still being debated (10).

The isoelectric point of transferrin depends on its content of metal, since the binding of each metal ion increases the net negative charge of the protein according to the equation (23, 24):

$$Fe(III) + H_3\text{-transferrin} + HCO_3^- \rightleftharpoons Fe\text{-transferrin-}HCO_3^- + 3H^+$$

(Whether the hydrated form of carbon dioxide actually bound to transferrin is HCO_3^- or CO_3^{2-} is still unsettled. In either case, however, the binding of iron increases the net negative charge of the protein and, correspondingly, its anionic mobility.) This makes possible the electrophoretic resolution of diferric-, monoferric-, and apotransferrin in mixtures containing these three species (25, 26). As determined by isoelectric focusing, their isoelectric points for human transferrin are 5.0, 5.3, and 5.6, respectively (27).

Carbohydrate Chains of Transferrin

The carbohydrate residues of human serum transferrin are disposed in two identical branched chains attached to asparaginyl residues and terminating in sialic acid residues, but recent studies of their primary structure by Spik et al. (28) disagree with the model proposed earlier by Jamieson et al. (29). One of these chains is located near the C-terminal half-cystine of the peptide chain while the other is on the inner portion of the protein chain (30). Enzymatic removal of the terminal sialic acid residues does not appear to affect the iron-binding or iron-donating properties of the protein (31). In contrast to other desialylated serum glycoproteins, which are very rapidly cleared from the circulation in the liver, the biologic half-life of homologous asiolotransferrin is only slightly decreased compared with that of the native protein (32, 33). Experiments with glycosidase-treated transferrin have failed to demonstrate unequivocally a functional role for the carbohydrate moiety in the interaction of the protein with its target cell, the reticulocyte (34).

Active Sites of Transferrin

Although Fe(III) is the most important, if not the only metal ion bound under physiologic circumstances, the two specific sites of transferrin will accommodate a variety of multivalent metal ions which have been useful as spectroscopic probes of the sites. Detailed understanding of the chemical mechanisms underlying the tight but physiologically reversible binding of iron by transferrin will probably have to await crystallographic studies, but some insight has already been obtained by various spectroscopic methods. In some of the earliest studies of the transferrins, it was noted that phenolic complexes of Cu(II) are yellow while those of Fe(III) may be red. Corresponding complexes of the transferrins have similar colors. On this basis, and because some of the protein binding groups have pKs greater than 10, Warner and Weber suggested that tyrosyl residues may be coordinated to the metal ions (23). This was subsequently corroborated for each of the transferrins by difference spectrophotometric titration of the metal-loaded and metal-free proteins, which revealed two or three more titratable tyrosine residues in the absence of specifically bound metal ions (35, 36). Proton magnetic resonance spectra of the iron complexes of transferrin and conalbumin, compared with the diamagnetic gallium complexes and the apoproteins, displayed changes in the aromatic regions presumably resulting from broadenings of tyrosyl proton resonances by the proximity of the paramagnetic ferric ion (37). A marked enhancement of the fluorescence of terbium ion when bound to transferrin was interpreted as resulting from energy transfer from a tyrosyl ligand (38). Finally, Gaber et al. identified four resonance-enhanced Raman bands in the Cu(II) and Fe(III) complexes of transferrin which resembled closely bands given by a model Fe(III) phenolate compound (39). Similar observations were also reported by Tomimatsu et al. (40).

Taken together, these spectroscopic studies provide compelling evidence of the involvement of tyrosyl residues at the active sites of transferrin. Chemical modification of tyrosyl residues, which impair or abolish transferrin's metal-binding properties, offers further support for this picture (41). The number of tyrosyl groups participating in metal ion binding is more difficult to determine, however. In spectrophotometric titration studies, three phenolic OH groups were thought to be coordinated to Fe(III) (35). However, other work based on uv-difference spectrophotometry indicates that only two such groups may be involved (42). It may be pertinent that two homologous regions in the primary structure of the transferrin molecule, which may therefore represent corresponding parts of the two active sites, each contain only two tyrosyl residues (20).

The first suggestion that nitrogen ligands may also be coordinated to specifically bound ferric ions came from hydrogen ion titration studies of transferrin (43). For each active site, two groups, presumably histidyl residues, were titrated near pH 6 in the apoprotein but not in the iron-bearing molecule. Chemical modification studies (which should be cautiously interpreted since it may be difficult to distinguish long range conformational effects of modification from more specific influences at

the active sites) and resonance Raman spectroscopy also implicate histidyl residues as metal ligands (39, 44). Direct demonstration of at least
one nitrogen ligand at each specific metal-binding site has been provided
by electron paramagnetic resonance (EPR) spectroscopy of the Cu(II)
complex of transferrin (36), but it is not clear whether this nitrogen is
on the imidazole ring of histidine. Although earlier (EPR) studies were
interpreted as showing the influence of two nitrogen ligands (24, 45), the
use of isotopically pure copper and spectral simulation by computer indicates that interaction of the Cu(II) electron spin with only one ^{14}N
nucleus is probably sufficient to account for the major features of the
spectrum (36).

In detailed studies of the frequency dependence of water proton
relaxation rates in solutions of Cu(II)–transferrin and Fe(III)–transferrin, a water molecule was thought to be close enough to the metal ion
(about 2 Å for the proton–Fe distance) to be considered a ligand (46, 47).

From these data, it appears that the ligand structure at the active
sites includes two or three tyrosyl residues, one or two nitrogen groups
(possibly from the imidazole rings of histidyl residues), and a water
molecule. Since the usual coordination number of iron is six, it would
seem possible to satisfy the requirements of Fe(III) with these ligands
alone. The remarkable feature in the specific binding of iron to transferrin, however, is the cooperation displayed between the metal and
anion binding functions of the protein; neither appears to bond strongly
in the absence of the other (14, 15, 19, 48). This interdependence of
metal and anion binding may indicate that the anion is also an obligatory
ligand of the metal ion, perhaps by serving as a bridging ligand between
metal and protein (17). Magnetic resonance studies of ^{13}C- and spin-
labeled anions specifically bound to transferrin support this model (see
"Spatial Relationships of the Anion and Metal Binding Sites," below
(49)).

Anion Binding and the Iron–Protein Bond

The dramatic role of the anion can perhaps best be appreciated
from simple quantitative considerations. In the absence of a suitable
anion, specific binding of iron to transferrin does not occur at all; the
effective binding constant is zero. At physiologic pH and bicarbonate
concentrations, however, the effective binding constant is about 5×10^{23} M^{-1} (24, 50). This means that in 1 L of blood plasma, in which the
transferrin is only about 30% saturated with iron, there will be less than
one free ferric ion or that a molecule of the ferric–transferrin complex
will spontaneously dissociate only about once in 10,000 years. Since iron
is readily removed from the transferrin molecule during its interaction
with the reticulocyte without disrupting protein structure (51, 52), a

specific physiologic mechanism to effect this removal must exist. Preliminary studies suggest that the anion may be at the heart of such a mechanism (53, 54, 55).

The extraordinarily tight, yet physiologically reversible, binding of iron by transferrin has clinical as well as biochemical implications. Like other heavy metals, iron may be toxic and is a frequent cause of accidental poisoning in children (56). Furthermore individuals with chronic hemolytic anemia such as thalassemia (Cooley's anemia), whose lives are sustained by freqeunt transfusions, may ultimately succumb to iron overload from the amount of iron necessarily imposed by each unit of blood administered. A mainstay in the treatment of such patients is desferrioxamine (57), a trihydroxamate of microbial origin (3). The effective equilibrium constant for the chelation of iron by desferrioxamine at the pH of blood, 7.4, is about 10^{26} M^{-1} (58). Since this is substantially greater than the corresponding constant for transferrin, and since pharmacologically attainable concentrations of desferrioxamine in blood are about 10 times greater than the concentration of transferrin-binding sites, it might be expected that the clinical use of desferrioxamine would deplete iron from transferrin. This, however, does not occur, at least in the early stages of treatment (59). In vitro experiments have also failed to demonstrate significant removal of iron from transferrin by desferrioxamine (60). A likely explanation for this seeming paradox is that there is a kinetic barrier to equilibrium imposed by the exceedingly low concentration of available ferric ion at the pH of these studies. It seemed worthwhile, therefore, to see whether the expected exchange of iron from transferrin to desferrioxamine could be facilitated by iron-complexing agents which could also fulfill the role of the anion in transferrin (61).

Desferrioxamine methane sulfonate binds 1 g atom of iron per mole forming a complex with an absorption maximum at 428 nm and a millimolar extinction coefficient of 2.77 (measured at pH 7.4). Transferrin has an absorption maximum at 466 nm with a millimolar extinction coefficient of 4.56. Accordingly, it is possible to monitor spectrophotometrically the exchange of iron between diferric transferrin and desferrioxamine. When only transferrin and desferrioxamine are present in the reaction mixture, a slow and incomplete transfer of iron from proteins to chelating agent is observed (Figure 1); after 140 min, 86% of the iron is still bound to transferrin. Addition of 2.9 $\times$ 10^{-3} M nitrilotriacetate, however, accelerates the exchange, so that at 140 min only 43% of the iron is complexed to transferrin. Increasing the concentration of nitrilotriacetate to 2.9 $\times$ 10^{-2} further enhances the transfer rate, with less than 10% of the iron then remaining bound to transferrin after 140 min. No significant binding of iron to nitrilotriacetate is observable in these

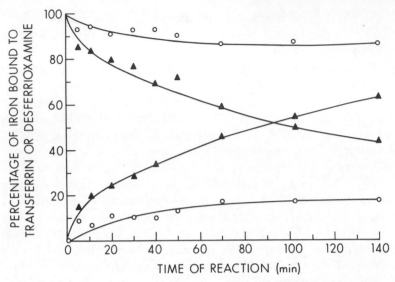

Figure 1. Exchange of iron between transferrin (3.75 × 10⁻⁵M) and desferrioxamine (3.74 × 10⁻³M) in 0.1M HEPES buffer, pH 7.4. The curves originating at 100% describe the iron bound to transferrin; the curves originating at 0% describe iron bound in ferrioxamine. (O—O) no addition; (▲—▲) in the presence of 2.9 × 10⁻³M NTA (61).

studies, since all the iron in the reaction mixtures was spectrophotometrically recovered in either transferrin or ferrioxamine. The nitrilotriacetate must act, therefore, as a mediating agent to bring about the thermodynamically predicted exchange of iron.

If, indeed, the coordination requirements of iron bound to transferrin are fully satisfied by ligands from the protein, water, and the stabilizing anion, a ligand exchange mechanism must be the means by which chelating anions such as citrate and nitrilotriacetate promote the exchange of iron between transferrin and desferrioxamine (61). The fact that these anions are also iron-complexing agents in their own right might be at the heart of such a mechanism, according to the following scheme:

(a) Fe–transferrin–bicarbonate + anion ⇌
$$\text{Fe–transferrin–anion + bicarbonate}$$

(b) Fe–transferrin–anion ⇌ transferrin + Fe–anion

(c) Fe–anion + desferrioxamine ⇌ ferrioxamine + anion

The concentrations of the ternary Fe–transferrin–anion species as well as the Fe–anion complex are too small at any instant during the experiment to be demonstrable spectroscopically (8), but they are sufficient to mediate the transfer of iron from transferrin to desferrioxamine until equilibrium is attained. Furthermore, a mechanism such as this may also underlie the exchange of iron between transferrin and other physiologic iron-accepting molecules (59). Such exchange mechanisms are well known in coordination chemistry and have been extensively studied by Margerum and his co-workers. For example, the transfer of Cu(II) from triglycine, tetraglycine, or tetraalanine to EDTA, a much stronger chelating agent than any of the oligopeptides, proceeds slowly unless a species to facilitate transfer, such as a simple amino acid, is present in the reaction mixture (62). As in chelate-mediated exchange of iron from transferrin to desferrioxamine, the postulated reaction sequence entails the formation of transient ternary complexes, in this case of Cu(II), oligopeptide, and amino acid. A ligand substitution reaction then produces a ternary complex of Cu(II), oligopeptide, and EDTA from which the oligopeptide is released to yield the final and thermodynamically stable product, Cu(II)–EDTA. Because hydrolysis prevents appreciable concentrations of free or aquated Fe(III) from occurring in biologic systems, it seems likely that ligand substitution sequences such as those postulated by Margerum are particularly important in the exchange of iron among sites of iron storage, transport, and utilization.

The exchangeable water molecule coordinated to the specifically bound ferric ion might be another site at which ligand exchange could be initiated. Since the anion is so critical in stabilizing the metal–protein bond, and since it too is exchangeable (65), it seems a more attractive target for attack by ligand substitution. Only further experimentation can choose between these possibilities.

Spatial Relationships of the Anion and Metal Binding Sites

In considering the structural basis for the striking interdependence of the iron- and anion-binding functions of transferrin, Schlabach and Bates treated three possible models (16):

(a) The anion binds only to the metal ion.

(b) The anion binds only to the protein, exerting its influence on metal binding by a conformational or allosteric effect.

(c) The anion binds to protein and to metal ion in an interlocking arrangement.

From a stereochemical analysis of binding anions, the interlocking-sites model is most consistent with available data. We have approached the question of the spatial relationship of metal ion and anion in human transferrin through electron and nuclear magnetic resonance spectroscopy (17, 49).

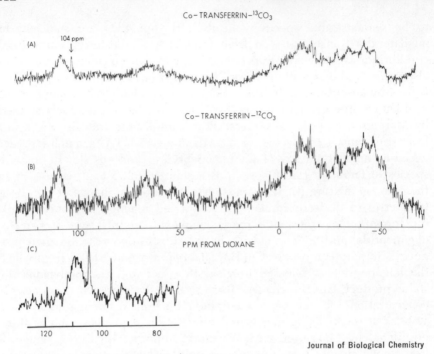

Figure 2. Undecoupled ^{13}C NMR Fourier transform spectra of Co(III)–transferrin–CO$_3$. (A) $^{13}CO_3$-labeled transferrin; (B) no label; (C) after adding $H^{13}CO_3^-$ to the preparation used for A. The line at 104 ppm is caused by $^{13}CO_3$ specifically bound to transferrin and is 14 Hz wide. The linewidth of free $H^{13}CO_3^-$ at 96 ppm is 7 Hz (32).

When no paramagnetic center is nearby, ^{13}C-labeled carbonate bound to transferrin should give rise to a sharp resonance line even in proton-undecoupled spectra. In the presence of a nearby paramagnetic metal ion, however, the ^{13}C-signal may be broadened. The extent of this broadening can provide a measure of the metal-^{13}C distance. Accordingly, the specific metal-binding sites of transferrin were loaded with paramagnetic Fe(III) or with diamagnetic Co(III) or Ga(III). In each case, the anion-binding requirement was met with $^{13}CO_3$; one carbonate being bound for each metal ion. A single, relatively sharp ^{13}C-NMR signal, shifted some 5 ppm downfield from free ^{13}C-bicarbonate in the same solution, was observed in the diamagnetic Co(III)–transferrin–$^{13}CO_3$ (Figure 2) and Ga(III)–transferrin–$^{13}CO_3$ complexes against the background of broad resonances from naturally abundant ^{13}C in the protein. By contrast, no signal for bound $^{13}CO_3$ could be located in the paramagnetic Fe(III)–transferrin–$^{13}CO_3$ complex within ± 400 ppm of its position in the spectra of the diamagnetic proteins (17).

The absence of a detectable $^{13}CO_3$ signal from the paramagnetic Fe(III) complex of transferrin probably reflects broadening caused by interaction of the nuclear spin with the electronic spin of the metal ion. From the Solomon–Bloembergen equation for T_2, the transverse relaxation time of the ^{13}C nucleus, it is possible to estimate the maximum distance that must separate the two spins for the resonance of the specifically bound $^{13}CO_3$ to become indistinguishable from noise in the spectrum. This distance has an upper bound of 9 Å but might in fact be much smaller (*17*).

This result is interesting but perhaps not very satisfying, since it puts an upper limit on the metal–anion distance without unequivocally revealing whether the anion is directly coordinated to the metal ion. To get at this question, we attached a nitroxyl spin-labeled anion to transferrin to determine whether its electron paramagnetic resonance signal was broadened by interaction with specifically bound Fe(III) (*49*). The mechanism of such broadening would be similar to that in the NMR experiments, but now there is an electron–electron interaction rather than an electron–nuclear interaction as before.

Since malonate (Structure **I**) is able to fulfill the role of the anion in transferrin, it seemed reasonable to see whether spin-labeled derivatives of malonate could serve as probes of the active sites. Two such spin-labled derivatives were prepared and tentatively identified as having structures **II** (*N*-4-(2,2,6,6-tetramethylpiperidin-1-oxyl)malonamide) and **III** (*N*-4-(2,2,6,6-tetramethylpiperidin-1-oxyl)malonate). Similar results were obtained with each (Figure 3). Upon mixing Fe(III), transferrin, and **II** at low pH, and then raising the pH to near-neutrality with CO_2-free ammonia, the characteristic orange–red color of the ternary Fe–transferrin–anion complex is promptly displayed. However, the anticipated EPR signal of the nitroxide spin-label is not observed, presumably because it is broadened beyond detectability by its proximity

$^-O_2CCH_2CO_2{}^-$

I

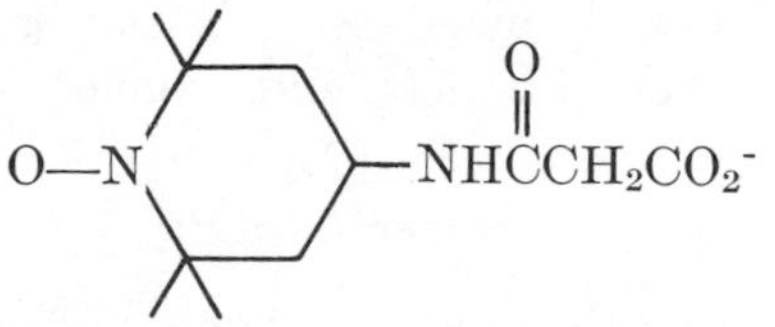

$O-N$ (ring) $-NH\overset{O}{\overset{\|}{C}}CH_2CO_2{}^-$

II

$O-N$ (ring) $-O\overset{O}{\overset{\|}{C}}CH_2CO_2{}^-$

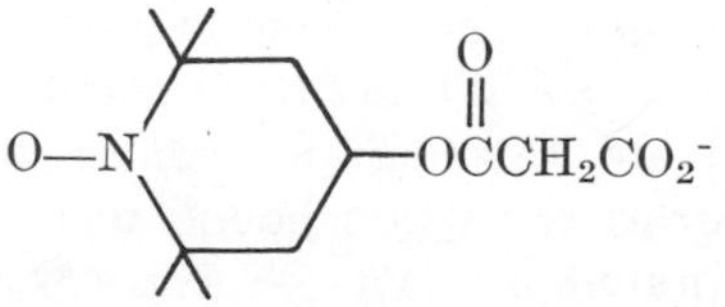

III

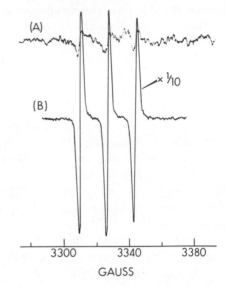

Proteins of Iron Storage and Transport

Figure 3. (A) EPR spectrum at 9.34 GHz Fe(III)–transferrin–II complex at room temperature. The 5 × 10⁻⁴M transferrin solution was about 90% saturated with Fe(III) and nitroxyl-labeled malonate (compound II). (B) After the preparation was made 5 × 10⁻³M in bicarbonate (49).

to the large magnetic moment of Fe(III). Addition of bicarbonate to the preparation caused an intense nitroxide free radical signal as the malonate derivative is displaced from the protein by the more tightly bound carbonate to become free in solution.

Calculations of the metal-nitroxyl distance, using either Leigh's analysis (63) or the Solomon–Bloembergen equation, indicate that the metal and the radical are separated by about 11 Å (49). Since space-filling models show that the distance from the nitroxyl nitrogen atom to either or both carboxyl oxygen atoms can vary from 6 to 12 Å, it seems likely that the spin-labeled malonate is a ligand of the iron atom in transferrin. Although our results must be considered preliminary and in need of confirmation and elaboration, they do offer evidence that the anion may stabilize the metal–protein bond in transferrin by acting as a bridging ligand between metal and protein.

Interaction of Transferrin with Reticulocytes

Probably the most intriguing and important properties of transferrin are those involved in its physiological role as the source of iron for the biosynthesis of hemoglobin by the immature red blood cell. About 30 mg of iron are incorporated into hemoglobin synthesized by the normal adult bone marrow each day, or about 10 times the amount of non-hemoglobin iron in the circulation at any time. Transferrin is the shuttle for this traffic. Since the half-life in serum iron bound to transferrin is 1–2 hr while the half-life of the protein is about 7–8 days (51), the protein must be conserved during its interaction with the immature red cell and recycled as an iron-carrier many times before it meets its biologic fate. How this occurs still eludes detailed understanding, despite numerous studies.

Although younger nucleated red cells in the marrow may have more avid iron requirements, the reticulocyte, because of its accessibility in peripheral blood of animals made anemic by bleeding or by the administration of acetylphenylhydrazine, has been most frequently used in an in vitro model for the hemoglobin-synthesizing, and therefore iron-demanding, red blood cell. A system has also been used in which conalbumin (ovotransferrin) functions as an iron source for the nucleated red cells of the developing chick embryo (6), but whether this is the physiologic role of that protein is still unclear.

The uptake of iron from transferrin by reticulocytes is a time-, temperature-, and energy-dependent process in which integrity of both protein and cells is required (64, 65). Synthetic iron chelates, once thought to be effective iron donors (66), appear to depend on membrane-bound transferrin as an intermediate agent; cells depleted of the protein by preincubation and washing no longer accept iron from such complexes (67). When such cells are reincubated with transferrin, their capacity to accept iron initially bound to synthetic chelators is largely restored.

Perhaps the most dramatic indication of the indispensable role of transferrin as an iron donor, however, is provided by an experiment of nature. Individuals afflicted with atransferrinemia, a genetic inability to synthesize transferrin, suffer from a paradoxical co-existence of iron deficiency anemia and generalized iron overload (68, 69). Without transferrin, neither the delivery of iron to hemoglobin-synthesizing cells nor its mobilization from stores is successfully regulated.

Transferrin Receptor of the Reticulocyte

The existence of specific receptors for transferrin on the reticulocyte membrane was implied by the work of Jandl and associates, who observed that trypsin virtually abolished the ability of reticulocytes to take up iron from transferrin without affecting other metabolic functions of the cells (8). Whether the effect of the enzyme was to cleave the receptor from the cell membrane or to degrade it proteolytically was not clear. Neuraminidase treatment also depressed iron uptake by reticulocytes, but to a much lesser degree than trypsin and only at much higher concentrations than needed to abolish the hemagglutinating effects of influenza virus. Subsequent work from Morgan's laboratory has confirmed these results and has shown further that binding of transferrin to the receptor protects it from proteolytic enzymes (70).

The development of techniques for solubilizing membrane proteins with detergents has stimulated a flurry of interest in the transferrin receptor of the reticulocyte (71, 76). Unhappily, however, the results of studies from different laboratories, using different methodologies and different species, are not entirely concordant, and the reasons for this are not clear. Our own efforts have focused on the transferrin receptor of the rabbit reticulocyte, using Triton X-100 as a membrane-solubilizing agent.

Isolation of the Transferrin Receptor of the Rabbit Reticulocyte

Reticulocytes were obtained from rabbits made anemic by repeated bleeding. Reticulocyte counts were generally 20–30% of total red cells. Standard methods were followed for preparing and ^{125}I-labeling of rabbit transferrin and its incubation with reticulocytes (54). Reticulocyte membranes were prepared by the method of Dodge (77) and solubilized for

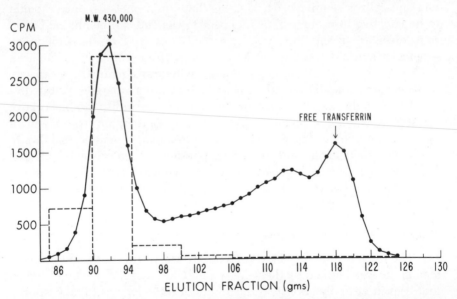

Figure 4. Chromatogram of Triton X-100 extract of membranes from reticulocytes incubated with ^{125}I-labeled transferrin. A 2.5 × 80-cm column of LKB AcA 22 gel was used. (· · ·) indicates the relative transferrin-binding activity of stromal fractions from reticulocytes not incubated with transferrin.

1 hr at 4° in 5-milliosmolar phosphate buffer, pH 7.4, containing 1% Triton X-100 (72, 77). The same buffer was used for column chromatography. Electrophoresis of membrane fractions in SDS gel was carried out according to Fairbanks et al. (78).

When reticulocytes were incubated with ^{125}I-labeled transferrin until a steady state in cell-bound radioactivity was achieved, a labeled membrane fraction could be demonstrated in reticulocyte ghosts solubilized with Triton X-100 (Figure 4). This fraction was well resolved from free transferrin and had an apparent molecular weight near 430,000. We presume it represents the solubilized transferrin–receptor complex. Additional evidence to support this presumption is obtained from a study of the transferrin-binding activity of membrane fractions from reticulocytes not previously incubated with transferrin. Such fractions were isolated chromatographically, incubated with ^{125}I-labeled transferrin, and again chromatographed on a column of Sephadex G-200. Since free transferrin is well resolved from larger molecular weight complexes of transferrin which migrate near the excluded volume of this column (75), the amount of radioactivity recovered in the early fractions from the Sephadex column is a measure of the transferrin-binding activity present. In Figure 4 is it evident that the transferrin-binding activity of reticulocyte membranes resides in a fraction with a molecular weight around the 400,000 range. Fractions from transferrin-free reticulocyte membranes, corresponding to the shoulder on the high molecular weight side of the transferrin peak in the chromatogram of transferrin-saturated membranes, did not display transferrin-binding activity. Furthermore, SDS-gel electrophoresis of the shoulder showed only a single protein band corresponding to free transferrin. It is possible, therefore, that this shoulder represents transferrin aggregates (presumably dimers) or an artifact of chromatography, rather than a second transferrin–receptor complex.

In order to characterize further the transferrin-bearing high molecular weight fraction of transferrin-saturated membranes, electrophoresis in SDS gels was carried out after removing the Triton X-100 (79) (Figure 5). Three prominent and distinct bands giving a protein stain are identifiable in the gels, with apparent molecular weights of about 165,000, 95,000, and 75,000, respectively. Over 85% of the total ^{125}I counts subjected to electrophoresis is present in the band corresponding to molecular weight 75,000. Since transferrin is a single-chain glycoprotein of molecular weight 77,000, this band is almost certainly transferrin-dissociated from its complex with the receptor. In agreement with other studies (52), no transferrin binding was demonstrated in membranes of mature erythrocytes.

We believe these results are consistent with the hypothesis that the transferrin receptor of the rabbit reticulocyte is a protein with a complex subunit structure and a molecular weight near 350,000, so that the receptor–transferrin complex has an aggregate molecular weight of approximately 430,000, as measured by gel-filtration chromatography. A possible structure for the receptor itself, then, might consist of two subunits of molecular weight 95,000 and one of 165,000, for a total molecular weight of 355,000.

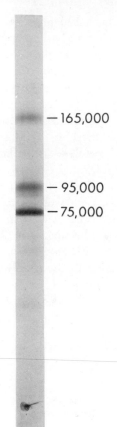

- 165,000

- 95,000
- 75,000

*Figure 5.   SDS gel electrophoresis
of the high molecular weight peak
in the chromatogram of Figure 4.
Over 85% of the ^{125}I activity is
present in the band of molecular
weight 75,000.*

Our experiments corroborate the findings of Garrett, Garrett, and
Archdeacon (71), the first to report solubilization with detergents of a
macromolecular fraction containing iron and iodine from the membranes
of reticulocytes incubated with transferrin doubly labeled with ^{125}I and
^{59}Fe. The labeled fractions were found in or near the void volume of a
Sephadex G-200 column, indicating that the molecular weight of the
active components probably exceeded 200,000. Similar results were obtained by von Bockxmeer, Hemmaplardh, and Morgan (75). They found
a transferrin complex in detergent-solubilized reticulocyte ghosts which
eluted just after the void volume of a Sephadex G-200 column and estimated its molecular weight to be 225,000. Since the reliability of molecular weight estimation by gel filtration falls off at the limits of the operating range of the supporting medium, it seems likely that the complex
identified by these workers is the same as ours which also elutes at or
just after the void volume of a Sephadex G-200 column. The complex
elutes within the operating range of the Ultragel column, and so we
believe the molecular weight estimation of 430,000 to be more reliable.

An essential caveat must be observed, however; the molecular weight estimation measures the size of the detergent–protein complex, and different proteins bind different amounts of detergent. All empirical methods, accordingly, are fraught with possible error, and the results reported or reviewed now should be looked upon as preliminary and indicative, rather than definitive.

In other studies, Fielding and Speyer, using Triton X-100 to solubilize reticulocyte stroma and gel chromatography with Sepharose 6B as the support medium to fractionate the stromal constituents, identified a protein of apparent molecular weight near 150,000 as the primary transferrin receptor of the human reticulocyte (72, 73). By contrast, Witt and Woodworth, using photo-affinity labeled conalbumin, found the conalbumin receptor molecule of the chick embryo red cell to have a molecular weight of about 35,000 (76). Finally, Sly, Grohlich, and Bezkorovainy found evidence for an intracellular transferrin receptor in the rabbit reticulocyte with a molecular weight in the 15–20,000, range, as well as a membrane receptor with an apparent molecular weight near 120,000 (74). Whether the seeming discrepancies among the studies cited reflect differing techniques and species or simply mirror the complexities of the initial events in the transferrin–reticulocyte interaction remains a challenging problem for continuing study.

History of Transferrin after Binding to Its Receptor

Observing that 90% of ^{131}I-labeled transferrin bound to reticulocytes remains associated with ghosts after hemolysis, Jandl and Katz suggested that the transferrin–reticulocyte interaction might be restricted to the surface of the cell, where the transfer of iron from protein to cell was thought to occur (80). Other investigators have also argued that the cell membrane was the locus of iron exchange (81). Several studies have called this model into question, however. Electron microscopic autoradiography showed decay tracks inside reticulocytes incubated with ^{125}I-labeled transferrin (82). When reticulocytes were incubated with ferritin-conjugated transferrin (83) or with transferrin and ferritin-conjugated antibodies to transferrin (84), electron micrographs revealed many of the ferritin markers within pinocytotic vesicles or, to a lesser degree, free in the cytoplasm. No ferritin was seen within cells incubated with ferritin alone or ferritin conjugated to nonspecific proteins. Finally, agents such as colchicine and the vinca alkaloids, which are thought to inhibit microtubular function and the intracellular movement of pinocytotic vesicles, depressed the uptake and release of transferrin by reticulocytes (85). It appears, therefore, that during the sojourn of transferrin on the reticulocyte, the protein translocates from the surface membrane of the cell to its interior. Where iron exchange occurs, however, remains uncertain, as does the fate of the specific transferrin receptor during endocytosis and exocytosis.

The dwell time of a transferrin molecule with the reticulocyte may be only a minute or two (52, 80) or possibly as long as 10 min (86), after which the protein is released for another cycle of iron transport. The affinity of reticulocyte receptors for apotransferrin appears less than for iron-loaded molecules (80). This property may facilitate the release of the protein from the receptor after it has donated its iron and ensures that the apoprotein will not impede the delivery of iron to reticulocytes by competing with iron-bearing molecules for available receptors on the cell surface.

Mechanism and Regulation of Iron Release from Transferrin to Reticulocytes

It is tempting to speculate about a physiologic role for the anion-binding function of transferrin, since a stable iron–protein bond in transferrin does not seem to occur in the absence of a stereochemically suitable, synergistic anion while an extremely tight bond is found in the ternary iron–transferrin–carbonate complex. The carbonate anion is exchangeable with bicarbonate free in solution (87) and therefore may be sufficiently labile for attack by the reticulocyte. Such an attack might be the initial event in the removal of iron from transferrin by the reticulocyte. Furthermore, the protein itself is conserved during its interaction with the reticulocyte, so that the iron-removing mechanism must not alter the primary structure of the protein. A disruption of specifically bound carbonate meets this requirement.

Early experimental evidence in support of the hypothesis that an attack on the anion is at the heart of the iron-exchange mechanism (53) was soon corroborated by work from several laboratories (54, 55, 88). Replacing carbonate with oxalate at the specific anion-binding site of transferrin results in a relatively stable ternary Fe(III)–transferrin–oxalate complex. Over the time course of many hours or even days the oxalate complex slowly reverts to the physiologic Fe(III)–transferrin–carbonate form, but since in vitro studies seldom require more than an hour or two, the biologic properties of the oxalate complex can be tested.

A comparison was made of the ability of the oxalate and carbonate complexes to donate iron to rabbit reticulocytes (Figure 6). After incubation for 1 hr, 5.0 μg of iron were taken up from the carbonate complex by 1 ml of reticulocytes while only 1.75 μg of iron was removed from the oxalate complex. As the oxalate complex was prepared in a physiologic buffer with a relatively high bicarbonate content, about 20% of its iron was actually bound in the carbonate form of transferrin. This may well have been responsible for some of the observed uptake of iron in this experiment, so that the true uptake from Fe(III)–transferrin–oxalate is even lower than indicated in Figure 5. The defect in the iron-donating ability of the oxalate complex lies in the removal of iron from it and not in its binding to transferrin receptors, since the uptake of each form of the protein by reticulocytes is about the same (54). Still another indication that the anion-binding site of transferrin may be involved in the

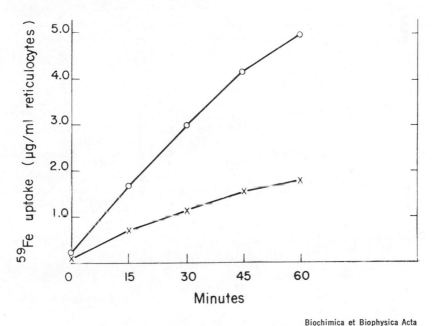

Figure 6.   Uptake of iron by reticulocytes from ^{59}Fe–transferrin–carbonate (○—○) and ^{59}Fe–transferrin–oxalate (×—×) (54)

mechanism of iron release comes from studies showing that metabolic inhibitors of iron uptake invariably affect the release of iron and carbonate from the protein by the reticulocyte in parallel (88). The carbonate- and iron-binding functions of transferrin seem to be tightly coupled physiologically as well as chemically.

The specific reactions involved in the physiologic release of iron from transferrin have not been elucidated. An enzymatic mechanism has been postulated (55), but without direct evidence. Recently, Ponka and Neuwirt have found that cyclic AMP may be involved, possibly by activating a protein kinase (89). If so, it may be worthwhile to consider whether the specifically bound carbonate itself is the site of a phosphorylation reaction, perhaps entailing its removal as carbamyl phosphate. It is also possible that a ligand exchange mechanism, similar to that postulated for the chelate-mediated transfer of iron from transferrin to desferrioxamine, is operative or that the carbonate is released by a protonic attack (54).

Other mechanisms are also involved in the reticulocyte-mediated release of iron from transferrin, and its subsequent incorporation into heme. Intact oxidation–reduction pathways are essential and seem to operate at a step following the attack on the anion (90). Heme exerts an inhibitory influence on iron transfer from protein to cell, seemingly by interfering at an early stage in the release of the metal from transferrin (89). Non-transferrin carriers for iron have also been identified in the reticulocyte and may participate in its incorporation into heme (81).

Literature Cited

1. Sylva, R. N., *Rev. Pure Appl. Chem.* (1972) **22**, 115.
2. Spiro, T. G., Saltman, P., in "Iron Biochemistry and Medicine," A. Jacobs and M. Worwood, Eds., pp. 1–28, Academic, New York, 1974.
3. Neilands, J. B., in "Inorganic Biochemistry," G. Eichhorn, Ed., pp. 167–202, Elsevier, Amsterdam, 1973.
4. Rosenberg, H., Young, I. G., in "Microbial Iron Metabolism, A Comprehensive Treatise," J. B. Neilands, Ed., pp. 1167–1182, Academic, New York, 1974.
5. Feeney, R. E., Komatsu, S. K., *Struct. Bonding* (1966) **1**, 149–206.
6. Williams, S. C., Woodworth, R. C., *J. Biol. Chem.* (1973) **248**, 5848–5853.
7. Masson, P. L., Heremans, J. F., Schonne, E., *J. Exp. Med.* (1969) **130**, 643–658.
8. Jandl, J. H., Inman, J. K., Simmons, R. L., Allen, D. W., *J. Clin. Invest.* (1959) **38**, 161–185.
9. Greene, F. C., Feeney, R. E., *Biochemistry* (1968) **7**, 1366.
10. Mann, K. G., Fish, W. W., Cox, A. C., Tanford, C., *Biochemistry* (1970) **9**, 1348–1354.
11. Aisen, P., Brown, E. B., *Prog. Hematol.* (1975) **9**, 25–56.
12. Bezkorovainy, A., Rafelson, M. E., Jr., *Arch. Biochem. Biophys.* (1964) **107**, 302–304.
13. Rosseneu-Motreff, M. W., Soetewey, F., Lamote, R., Peeters, H., *Biopolymers* (1967) **10**, 1039–1048.
14. Aisen, P., Aasa, R., Malmstrom, B. G., Vanngard, T., *J. Biol. Chem.* (1967) **242**, 2484–2490.
15. Price, E. M., Gibson, J. F., *Biochem. Biophys. Res. Commun.* (1972) **46**, 646–651.
16. Schlabach, M. R., Bates, G. W., *J. Biol. Chem.* (1975) **250**, 2182–2188.
17. Harris, D. C., Gray, G. A., Aisen, P., *J. Biol. Chem.* (1974) **249**, 5261–5264.
18. Aisen, P., Leibman, A., Pinkowitz, R. A., in "Protein-Metal Interactions, Advances in Experimental Biology and Medicine," M. Friedman, Ed., Vol. 48, pp. 125–140, Plenum, New York, 1974.
19. Woodworth, R. C., Virkaitis, L. M., Woodbury, R. G., Fava, R. A., in "Proteins of Iron Storage and Transport in Biology and Medicine," R. R. Crichton, Ed., pp. 39–50, North Holland, Amsterdam, 1975.
20. MacGillivray, R. T. A., Brew, K., *Science* (1975) **190**, 1306–1307.
21. Palmour, R. M., Sutton, H. E., *Biochemistry* (1971) **10**, 4026–4032.
22. Aisen, P., Leibman, A., Sia, C.-L., *Biochemistry* (1972) **11**, 3461–3464.

23. Warner, R.C., Weber, I., *J. Am. Chem. Soc.* (1953) **75**, 5094–5101.
24. Aasa, R., Malmstrom, B. G., Saltman, P., Vanngard, T., *Biochim. Biophys. Acta* (1963) **75**, 203–222.
25. Aisen, P., Leibman, A., Reich, H. A., *J. Biol. Chem.* (1966) **241**, 1666–1671.
26. Wenn, R. V., Williams, J., *Biochem. J.* (1968) **108**, 69–74.
27. Harris, D. C., Aisen, P., *Biochemistry* (1975) **14**, 262–268.
28. Spik, G., Bayard, B., Strecker, G., Bouquelet, S., Montreuil, J., *FEBS Lett.* (1975) **50**, 296–299.
29. Jamieson, G. A., Jett, M., De Bernardo, S. L., *J. Biol. Chem.* (1971) **246**, 3686–3693.
30. Montreuil, J., Spik, G., in "Proteins of Iron Storage and Transport in Biology and Medicine," R. R. Crichton, Ed., pp. 27–38, North-Holland, Amsterdam, 1975.
31. Morgan, E. H., Marsaglia, G., Giblett, E. R., Finch, C. A., *J. Lab. Clin. Med.* (1967) **69**, 370–381.
32. Morell, A. G., Gregoriadis, G., Scheinberg, I. H., Hickman, J., Ashwell, G., *J. Biol. Chem.* (1971) **246**, 1461–1467.
33. Regoeczi, E., Hatton, M. W. C., Wong, K.-L., *Can. J. Biochem.* (1974) **52**, 155–161.
34. Kornfeld, S., *Biochemistry* (1968) **7**, 945–954.
35. Wishnia, A., Weber, I., Warner, R. C., *J. Am. Chem. Soc.* (1961) **83**, 2071–2080.
36. Aasa, R., Aisen, P., *J. Biol. Chem.* (1968) **243**, 2399–2404.
37. Woodworth, R. C., Morallee, K. G., Williams, R. J. P., *Biochemistry* (1970) **9**, 839–842.
38. Luk, C. K., *Biochemistry* (1971) **10**, 2838–2843.
39. Gaber, B. P., Miskowski, V., Spiro, T. G., *J. Am. Chem. Soc.* (1974) **96**, 6868–6873.
40. Tomimatsu, Y., Kint, S., Scherer, J. R., *Biochem. Biophys. Res. Commun.* (1973) **54**, 1067–1074.
41. Komatsu, S. K., Feeney, R. E., *Biochemistry* (1967) **6**, 1136–1141.
42. Tan, A. T., Woodworth, R. C., *Biochemistry* (1969) **8**, 3711–3716.
43. Hazen, E. E., Jr., "A Titration Study of Transferrin," Ph.D. Thesis, Harvard University, 1962.
44. Bezkorovainy, A., Grohlich, D., *Biochem. J.* (1971) **123**, 125–126.
45. Windle, J. J., Wiersema, A. K., Clark, J. R., Feeney, R. E., *Biochemistry* (1963) **2**, 1341–1345.
46. Gaber, B. P., Schillinger, W. E., Koenig, S. H., Aisen, P., *J. Biol. Chem.* (1970) **245**, 4251.
47. Koenig, S. H., Schillinger, W. E., *J. Biol. Chem.* (1969) **244**, 6520–6526.
48. Bates, G. W., Schlabach, M. R., *J. Biol. Chem.* (1975) **250**, 2177–2181.
49. Harris, D. C., Aisen, P., in "Proteins of Iron Storage and Transport," R. R. Crichton, Ed., pp. 59–66, North-Holland, Amsterdam, 1975.
50. Aisen, P., Leibman, A., *Biochem. Biophys. Res. Commun.* (1968) **32**, 220–226.
51. Katz, J. H., *J. Clin. Invest.* (1961) **40**, 2143–2152.
52. Katz, J. H., Jandl, J. H., in "Iron Metabolism," F. Gross, Ed., pp. 103–117, Springer-Verlag, Berlin, 1964.
53. Aisen, P., *Mt. Sinai J. Med. N.Y.* (1970) **3**, 213–222.
54. Aisen, P., Leibman, A., *Biochim. Biophys. Acta* (1973) **304**, 797–804.
55. Egyed, A., *Biochim. Biophys. Acta* (1973) **304**, 805–813.
56. Fairbanks, V. F., Fahey, J. L., Beutler, E., "Clinical Disorders of Iron Metabolism," pp. 359–374, Grune and Stratton, New York, 1971.
57. Modell, C. B., Beck, J., *Ann. N.Y. Acad. Sci.* (1974) **232**, 201–210.
58. Schubert, J., *Iron Metab. Int. Symp.* (1964) 466–494.
59. Hallberg, L., Hedenberg, L., *Scand. J. Haematol.* (1965) **2**, 67–69.

60. Keberle, H., *Ann. N.Y. Acad. Sci.* (1964) **119**, 758–768.
61. Pollack, S., Aisen, P., Lasky, F., Vanderhoff, G., *Br. J. Haematol.* (1976) **34**, 231–235.
62. Dukes, G. R., Margerum, D. W., *Inorg. Chem.* (1972) **11**, 2952.
63. Leigh, J. S., Jr., *J. Chem. Phys.* (1970) **52**, 2608–2612.
64. Schade, A. L., *Behringwerk-Mitt.* (1961) **39**, 3–23.
65. Morgan, E. H., Baker, E., *Biochim. Biophys. Acta* (1974) **363**, 240–248.
66. Princiotto, J. V., Rubin, M., Shashasty, G. C., Zapolski, E. J., *J. Clin. Invest.* (1964) **43**, 825–833.
67. Hemmaplardh, D., Morgan, E. H., *Biochim. Biophys. Acta* (1974) **373**, 84–99.
68. Heilmeyer, L., *Iron Metab. Int. Symp.* (1964) 201–213.
69. Goya, N., Miyazaki, S., Kodate, S., Ushio, B., *Blood* (1972) **40**, 239–245.
70. Hemmaplardh, D., Morgan, E. H., *Biochim. Biophys. Acta* (1976) **426**, 385–398.
71. Garrett, N. E., Garrett, R. J., Archdeacon, J. W., *Biochem. Biophys. Res. Commun.* (1973) **52**, 466–474.
72. Fielding, J., Speyer, B. E., *Biochim. Biophys. Acta* (1974) **363**, 387–396.
73. Fielding, J., Speyer, B. E., in "Proteins of Iron Storage and Transport in Biochemistry and Medicine," R. R. Crichton, Ed., pp. 121–126, North Holland, Amsterdam, 1975.
74. Sly, D. A., Grohlich, D., Bezkorovainy, A., in "Proteins of Iron Storage and Transport in Biochemistry and Medicine," R. R. Crichton, Ed., pp. 141–145, North Holland, Amsterdam, 1975.
75. Van Bockxmeer, F., Hemmaplardh, D., Morgan, E. H., in "Proteins of Iron Storage and Transport in Biochemistry and Medicine," R. R. Crichton, Ed., pp. 111–119, North-Holland, Amsterdam, 1975.
76. Witt, D. P., Woodworth, R. C., in "Proteins of Iron Storage and Transport in Biochemistry and Medicine," R. R. Crichton, Ed., pp. 133–140, North-Holland, Amsterdam, 1975.
77. Dodge, J. T., Mitchell, C., Hanahan, D. J., *Arch. Biochem. Biophys.* (1963) **110**, 119–130.
78. Fairbanks, G., Steck, L., Wallach, D. F. H., *Biochemistry* (1971) **10**, 2606–2617.
79. Holloway, P. W., *Anal. Biochem.* (1973) **53**, 309–312.
80. Jandl, J. H., Katz, J. H., *J. Clin. Invest.* (1963) **42**, 314–326.
81. Workman, E. F., Jr., Bates, G. W., in "Proteins of Iron Storage and Transport in Biochemistry and Medicine," R. R. Crichton, Ed., pp. 155–160, North-Holland, Amsterdam, 1975.
82. Morgan, E. H., Appleton, T. C., *Nature* (1969) **223**, 1371–1372.
83. Appleton, T. C., Morgan, E. H., Baker, E., *Regul. Erythropoiesis Haemoglobin Synth. Proc. Int. Symp.* (1971).
84. Sullivan, A. L., Grasso, J. A., Weintraub, L. R., *Blood* (1976) **47**, 133–143.
85. Hemmaplardh, D., Kailis, S. G., Morgan, E. H., *Br. J. Haematol.* (1974) **28**, 53–65.
86. Baker, E., Morgan, E. H., *Biochemistry* (1969) **8**, 2954–2958.
87. Aisen, P., Leibman, A., Pinkowitz, R. A., Pollack, S., *Biochemistry* (1973) **12**, 3679–3684.
88. Schulman, H. M., Martinez-Medellin, J., Sidloi, R., *Biochim. Biophys. Acta* (1974) **343**, 529–534.
89. Ponka, P., Neuwirt, J., in "Proteins of Iron Storage and Transport," R. R. Crichton, Ed., pp. 147–154, North-Holland, Amsterdam, 1975.
90. Egyed, A., *Arch. Biochim. Biophys. Acad. Sci. Hung.* (1974) **9**, 43–52.

RECEIVED July 26, 1976.

Biological Electron Transport and Copper-Containing Biomolecules

Nonadiabatic Electron Transfer in Oxidation–Reduction Reactions

HENRY TAUBE

Department of Chemistry, Stanford University, Stanford, Calif. 94305

Several criteria are applied in searching for evidence of nonadiabatic electron transfer in oxidation–reducton reactions. There is direct evidence for a nonadiabatic factor only for some self-exchange reactions of $Mn(CNR)_6^{2+,+}$ and $Fe(Rphen)_3^{3+,2+}$ where R is a bulky group, although the factor also may affect rates for some f-electron couples. Refinement of the criteria may show that the nonadiabatic factor decreases the rates in other cases. One approach is to study the nonadiabatic regime by using electron transfer from Ru(II) to Co(III) in an intramolecular mode. Observations on such systems and on the intensities of the intervalence bands in Ru(II)–Ru(III) mixed valence complexes offer clues to the extent of coupling by the bridging groups.

The measurement of the rate and determination of the rate law *(1, 2)* for the self-exchange process $Fe_{aq}^{3+,2+}$ mark an important stage in the study of electron transfer in oxidation–reduction reactions. Dodson's results attracted a great deal of attention to the field, stimulating other experimental work and, by providing some definite data at a critical time, also stimulating discussion of the mechanism of the electron transfer process (*see,* for example, discussion reported in Ref. *3*). In retrospect, the development of the subject could as well have been based on studies of orthodox oxidation–reduction reactions of simple chemistry. However, until some specific proposals about the mechanism of electron transfer had been made, there was little incentive for measuring the rates of the ordinary reactions. The self-exchange reactions came to the fore not only because they were carried by the momentum of interest which then prevailed in applying artificial radioactivity to problems in chemistry but also because the element of symmetry simplifies the understanding of the observations.

With the demonstration that an activated complex for the $Fe_{aq}^{3+, 2+}$ self-exchange contains one of each of the reactants (1) and that there are other activated complexes (2) which contain, in addition, anions such as OH^- or X^-, productive discussion of the electron transfer processes in terms of molecular models began. Quite early (4), attention was directed to the energy barrier to electron transfer which is imposed by the Franck–Condon restriction. But when a molecular model for the activated complex for electron transfer is proposed, a distance of approach for the reactant ions needs to be specified, and at once the question of the "conductivity" of the matter intervening between oxidant and reductant arises. In the discussions of the large amount of experimental data which has been accumulated during the past two decades, more attention has probably been devoted to this aspect of the electron transfer process than to the Franck–Condon barrier. This is somewhat ironical in view of the conclusion that will be reached that in few of the systems studied until now are differences in ligand "conductivity" useful in understanding differences in reaction rates. This applies also to systems in which oxidation–reduction involves electron transfer over large distances, where the temptation to ascribe rate differences to the conductivity of ligands has been particularly difficult to resist. The first intimations of remote electron transfer in artificial systems were reported in 1955 (5), but convincing evidence for remote attack was not provided until 1966 (6, 7). In biochemical systems, electron transfer over many bond lengths is a particularly appealing idea, and it may turn out to be important as well. Electron transport over large distances in proteins was proposed many years ago by Szent–Györgi (8, 9), and he may well have been preceded by others who suggested the possibility seriously. In fact, strong evidence for electron transfer over large distances has been obtained by DeVault and Change (10) in excited states of systems related to the photosynthetic cycle.

One purpose of this paper is to examine the evidence that the rates of oxidation–reduction reactions are related to the conductivity of the medium separating the oxidant and reductant. This survey will then describe experiments now in progress to investigate systematically the nonadiabatic regime in oxidation–reduction reactions. First the relationship between what has loosely been referred to as the conductivity of the medium and the title term, "nonadiabatic," should be defined.

Figure 1 shows the double-well potential which is often used to represent the electron transfer act in oxidation–reduction. The special case of a self-exchange process was chosen for simplicity. (For a full discussion of the issues being discussed in relation to the potential energy diagram, see Ref. 11. Ref. 13 gives a more compact treatment.) The implications of the diagram may not be immediately obvious, and the

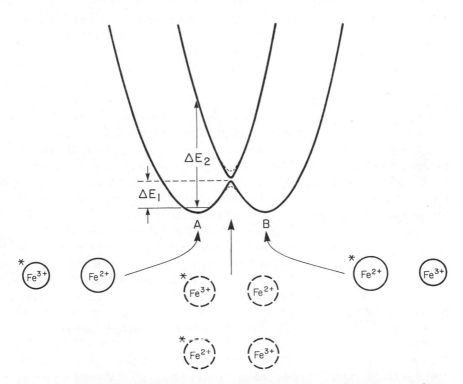

Figure 1. Potential energy as a function of reaction coordinate for a self-exchange reaction. ΔE_1, energy barrier for thermal electron transfer (weak coupling); ΔE_2, energy of an intervalence transition which is possible for the system.

important ones will therefore be explained by reference to a specific system. If the minimum on the left at A is taken to represent the system $^*Fe(H_2O)_6^{3+} + Fe(H_2O)_6^{2+}$, minimum B represents the final product of the electron transfer act, namely $^*Fe(H_2O)_6^{2+} + Fe(H_2O)_6^{3+}$. The reaction coordinate is a combination of nuclear motions which results in the movement of the electron from $Fe(H_2O)_6^{2+}$ to $^*Fe(H_2O)_6^{3+}$. For present purposes, it is arbitrarily assumed to be made up of the breathing frequencies for the water molecules in the first coordination spheres of the two reactant ions. The reaction coordinate taken in the direction to the right of A then represents the motion of the ligands away from $^*Fe^{3+}$ in phase with those attached to Fe^{2+} moving toward it, the separation of the ions remaining fixed. At a point defined by the intersection of the two curves, the coordination spheres about both metal ions are the same, and the energy of the system arrived at from A is the same as that arrived at from B. At this point along the reaction coordinate, the condition imposed by the Franck–Condon restriction is met, but if the ions are far apart so that the interaction between the orbitals is weak (very small energy gap at the cross-over point), the probability of passing from the state $*Fe(H_2O)_6{}^{3+} + Fe(H_2O)_6{}^{2+}$ to $*Fe(H_2O)_6{}^{2+} + Fe(H_2O)_6{}^{3+}$ will depend on tunnelling probability (hence a relation to the "conductance" of the medium). (If tunnelling probability is interpreted as the probability that the system, once it has the requisite energy, will pass on to products, it has a direct relation to the probability of adiabatic transfer. Marcus (11) has pointed out that when tunnelling probability is calculated in the usual way (see for example Ref. 12), the simple direct relationship is lost. According to Marcus, the result of this kind of calculation is a rough measure of the energy gap at the crossover point, but this is only one of the factors which affect the probability of adiabatic tranfer.) In this, the nonadiabatic regime, the rate at a fixed distance of separation will be governed both by the Franck–Condon barrier and a transition probability. When the interaction between the orbitals increases, as expected when the ions approach, the energy gap at the crossover point increases. When this gap becomes sufficiently large—and according to theory (13) approximately 0.5 kcal suffices—the system will remain on the lower curve as it traverses the energy barrier separating A and B (adiabatic transfer). In this limit, not considering the energy involved in bringing the reactants to a suitable distance, the reaction rate is governed solely by the Franck–Condon barrier. In this regime, improving the interaction between the orbitals will not lead to a rate increase, unless the delocalization becomes great enough to lower the energy maximum significantly. The only thing that is learned from reaction rates about tunnelling probability in the adiabatic regime is that it is sufficiently high. The nonadiabatic reactions on the other hand are particularly interesting because by studying them, the transition probability as a function of distance and the properties of the medium can be studied. This relationship has not yet been examined for chemical reactions.

A number of criteria can be used to gage whether reactions are adiabatic or nonadiabatic, and they will be considered in turn. One criterion rests on the applicability of the Marcus (11) correlation of the rates of cross reactions with the self-exchange reactions of the couples involved. It seems unlikely that the correlation would be successful if tunnelling probability, which is exepcted to be a sensitive function of the dimensions of the barrier, were an important factor in determining the rates. This barrier is not that shown in Figure 1; it is the barrier which would be mapped out by considering the potential energy of the system as a function of distance when the electron is moved from the reducing agent through the intervening medium to the oxidizing agent. With the exception of reactions involving the Co(III)–Co(II) couple, where it has been tested in the simple inorganic systems, the Marcus correlation works reasonably well. Reasonable agreement is all that can be expected since the work required to assemble the reactants in the precursor complex cannot properly be allowed for. Even when the reactant couples are of the same charge type, idiosyncracies in interaction, perhaps because of changes in solvation, may become a factor. This is particularly likely to happen when the reaction partners approach closely on reaction. Factors such as the hydrophobic/hydrophilic nature of the reactants and, for large reactants, even their shapes can lead to idiosyncracies in the work required to assemble the precursor complexes. In addition, the measurements being considered are often made under varying conditions, and all are subject to experimental error. As a consequence, it is difficult to use the Marcus correlation to reveal small but real effects arising from non-adiabaticity. Reactions involving the Co(III)–Co(II) couple, for which there is a large spin change, might provide examples of nonadiabatic electron transfer (nonmixing of states of different multiplicity). However, in the usual interpretation of reactions in which Co(III) is reduced, the system circumvents a nonadiabatic transition in the activated complex by undergoing a spin change prior to or following the activated complex.

Table I.   Reactions of Ru(II) Complexes with Fe^{3+} aq at 25°

Reactant	Medium	k_{11} $(M^{-1} sec^{-1})$ [a]	k_{obs} $(M^{-1} sec^{-1})$	k_{22} $(calc)$ [b,c]
$Ru(NH_3)_6^{2+}$	$0.1M$ $HClO_4$	4.3×10^3 [d]	3.4×10^5	1.3×10^{-2}
$Ru(NH_3)_5isn^{2+}$ [e]	$1M$ HO_3SCF_3	4.3×10^5 [f]	2.6×10^4	0.6×10^{-2}
$Ru(NH_3)_4bipy^{2+}$ [e]	$1M$ $HClO_4$	2.1×10^6 [g]	7.2×10^3	0.8×10^{-2}

[a] Self-exchange rate for Ru(III)–Ru(II) couples.
[b] Calculated self-exchange rate for $Fe^{3+,2+}$ aq.
[c] K_{eq} determined for conditions specified by cyclic voltammetry (Ru couples) and potentiometric titration [$Ru(NH_3)_6^{3+,2+}$ and $Fe^{3+,2+}$ aq].
[d] Ref. 15.
[e] isn ≡ isonicotinamide; bipy ≡ 2,2′-bipyridine; nic ≡ nicotinic acid; phen ≡ 1,10-phenanthroline.
[f] From [$Ru(NH_3)_5isn$]$^{3+}$ + [$Ru(NH_3)_5nic$]$^{2+}$.
[g] From [$Ru(NH_3)_4phen$]$^{3+}$ + [$Ru(NH_3)_4bipy$]$^{2+}$.

In some recent experiments (14) on rates of oxidation by Fe^{3+} of a number of related Ru(II) species, the rates of the self-exchange reactions for Ru(III)–Ru(II) couples and the equilibrium data were measured as much as possible under uniform conditions. In determining self-exchange rates, a series of reactions of the type $Ru(NH_3)_5L^{2+} + Ru(NH_3)_5L'^{3+}$ were studied in which L and L′ are pyridines which differ in one substituent in the 3 or 4 position. No rate differences ascribable to differences in L and L′ were observed, apart from the effect on driving force. The result of these studies are summarized in Table I.

The rate of the $Fe_{aq}^{3+,2+}$ self-exchange as calculated from the data is several hundred times slower than that measured (ca. $4 M^{-1} sec^{-1}$ for the conditions of the experiments) (2). The discrepancy was noted also in the early work (15) on the $Ru(NH_3)_6^{3+,2+}$ self-exchange, but in view of the difficulty attending these measurements, not much significance was attached to it. A similar discrepancy has been noted for the $Fe_{aq}^{3+,2+}$ and a series of ferricinium–ferrocene cross reactions (16). The discrepancies can hardly be considered proof that there is a nonadiabatic contribution somewhere in the process. Perhaps they better illustrate the point that at the present level of refinement in calculating the work of bringing the reactants together and with existing data, the application of the Marcus relation is not very useful in revealing small effects resulting from nonadiabatic behavior.

A second criterion which can be applied in searching for the effects of nonadiabaticity is based on charge trapping by the orientation of solvent molecules. In illustrating the significance of Figure 1, it was assumed that the trapping of the charge is effected by the molecules in the first coordination sphere. The theoretical treatments of the electron transfer process in solution (17, 18) show that for species of ordinary size, charge trapping by the solvent makes an important contribution to the energy barrier. Observations on the change in energy of the inter-valence transition in μ-4,4'-bipyridinebis(pentaammineruthenium) in the mixed ([2, 3]) valence state bear directly on this point (19, 20) and lend support to the theoretical treatments (21). The results obtained for these systems suggest that in the ions of the type specified, about half of the overall energy barrier is ascribable to charge trapping by the solvent. Therefore if a series of complexes is studied in which the polar groups attached to the metal ions are held constant but the ligands are made more bulky by adding saturated hydrocarbon matter, the reaction rate should increase with the bulkiness of the groups if the reactions are adiabatic for the series. Implicit in this conclusion is the premise that the hydrocarbon matter is less effective at trapping the charge than is the solvent, and this seems reasonable for reactions in water. If the rate decreases on increasing reactant bulk in the manner specified, it is reasonable to assume that the rate decrease is caused by a reduction in electron transfer probability.

No systematic studies involving net chemical change have been reported which demonstrate nonadiabatic behavior, but two studies of self-exchange show effects which are expected if the probability of barrier penetration is becoming a rate-determining factor. The specific rates ($\times 10^{-4} M^{-1} sec^{-1}$) of self-exchanges as measured by NMR line broadening in ^{55}Mn for $Mn(CNR)_6^{2+,+}$ are 64 for $R = Et$ and 4.0 for $R = tert$-butyl at 7° in acetonitrile (22). For $Fe(phen)_3^{3+,2+}$, $Fe(3,4,5,8$-me-phen)$_3^{3+,2+}$, $Fe(4,7$-phenylphen)$_3^{3+,2+}$, and $Fe(4,7$-cyclohexylphen)$_3^{3+,2+}$,

specific rates ($\times 10^{-6} M^{-1} sec^{-1}$) in acetonitrile at 25° are 6 ± 0.6, 16.9 ±
1.2, 8.0 ± 0.8, and 0.41 ± 0.04, respectively, were measured by Chan
and Wahl. In the above, phen represents o-phenanthroline, and the
numerals specify the positions of substitution by CH_3, C_6H_5, and C_6H_{11}.
In both systems, the rate declines with bulky substituents. The result
recorded for the second set of systems—a small increase in bulk actually
causing a rate increase—is particularly interesting. Taken at face value,
the observation implies that self-exchange for $Fe(phen)_3^{3+,2+}$ is well within
the adiabatic regime, and it bolsters the conclusion that with the bulkiest
saturated group, nonadiabatic effects are being felt.

A low transition probability in the activated complex will be reflected in a decrease in entropy of activation, so that in principle, the
entropy of activation can reveal nonadiabatic electron transfer. Unfortunately, when reactions are studied in the bimolecular mode, and particularly when both the reacting species are charged, the entropy changes
associated with forming the precursor complex are large, and it is impossible to separate this contribution from the overall entropy change. Some
values of $\Delta S^{\ddagger}$ for systems of similar charge type but featuring differing
electronic structures are shown in Table II. This selection was made

**Table II.   Entropies of Activation as Functions of
Electronic Structure of Reactants**

Reaction	k at 25° ($M^{-1} sec^{-1}$)	$\Delta H^{\ddagger}$ (kcal mol^{-1})	$\Delta S^{\ddagger}$ (cal mol^{-1} deg^{-1})	I^a	Ref.
$(NH_3)_5CoOAc^{2+} + Cr^{2+}$	0.35	8.2	−33	1.0	24
$(NH_3)_5RuOAc^{2+} + Cr^{2+}$	2.7×10^4	1.0	−34	0.1	25
$(NH_3)_5CoNH_3^{3+} + Cr^{2+}$	8.8×10^{-5}	14.7	−30	0.4	26
$(NH_3)_5CoOH_2^{3+} + Eu^{2+}$	0.074	9.3	−33	0.4	27
$(NH_3)_5CoOH_2^{3+} + V^{2+}$	0.53	8.2	−32	1.0	28
$[(NH_3)_5CoO_2C—C—CH_3]^{2+} + V^{2+}$ $\overset{\|}{O}$	10.2	11.6	−15	1.0	29

^{a} Ionic strength.

because it seems reasonable that the transition probability will be sensitive to something as basic as orbital symmetry, and if it were indeed
partly rate determining, this would be revealed in the entropies of
activation.

The systems include an oxidizing agent, Co(III), which accepts an
electron in an orbital of σ symmetry; Ru(III), where the acceptor orbital

has π symmetry; and reducing agents which lose electrons from σd orbitals (Cr^{2+}), πd orbitals (V^{2+}), and f orbitals (Eu^{2+}). No significant differences in entropy of activation for the same charge type are observed, except for the last entry. The reaction of V^{2+} with the pyruvatopentaamminecobalt(III) is one of a large class in which substitution on $V(H_2O)_6{}^{2+}$ appears to be rate determining. In all the others, the precursor complex, whether inner-sphere or outer-sphere, is in equilibrium with the reactants, and $\Delta S^{\ddagger}$ overall reflects a contribution from the formation of the precursor complex as well as from the electron transfer act itself.

Even though a calculation of rates of electron transfer using only molecular parameters for the reactants and the dielectric properties of the solvent is beyond the competence of current theory, theory is nonetheless useful in rate comparisons and can reveal unusual behavior. For reasons which have already been stated, self-exchange rates are more amenable to theoretical analysis than are the rates of cross reactions. In the following, the $Fe^{3+,2+}$ self-exchange rate is taken as a reference value for other couples of the $3+, 2+$ charge type. According to Sutin (13), the increase in rate for the $Fe(phen)_3{}^{3+,2+}$ self-exchange compared with $Fe^{3+,2+}$ can be understood on the basis of the larger size of the former ions, which decreases the contribution by the solvent to the Franck–Condon barrier and of the reduced inner-sphere reorganization energy which arises from back-bonding interaction between Fe^{2+} and the π acid ligand. The calculated rate increase of 10^7 is consistent with the measurement of the self-exchange rate in acetonitrile (the self-exchange rates are apparently larger in water.) The self-exchange rate for $Ru(NH_3)_6{}^{3+,2+}$, ca. 10^3, seems reasonable in relation to that of $Fe(H_2O)_6{}^{3+,2+}$ because the change in bond distance with change in oxidation state is less for the ruthenium than it is for iron. A slower rate for $Cr(H_2O)_6{}^{3+,2+}$ self-exchange compared with $Fe(H_2O)_6{}^{3+,2+}$ is expected, because $Cr(H_2O)_6{}^{3+}$ absorbs an anti-bonding electron, and a large distortion attends the reduction, but it is not certain if this can account for a rate reduction by a factor (30, 31) in excess of 10^5. The $Eu^{3+,2+}$ self-exchange rate ($k < 3 \times 10^{-5} M^{-1} sec^{-1}$) seems more clearcut. Owing to the larger size of the ions, both the inner-sphere and solvent reorganization energies must be less than for $Fe(H_2O)_6{}^{3+,2+}$, yet the rate of self-exchange is so slow that it has not been measured for the aquo ions although the rate was measured for the path involving Cl^- (32). It is possible that the $Eu^{3+,2+}$ couple provides an example of nonadiabatic transfer. Since the f orbitals are buried in the kernel of the ion, it is reasonable that the Eu^{2+}–Eu^{3+} couple, of all those considered, would show nonadiabatic behavior. This argument was presented also in a review article (33). The chloride ion may affect the rate by stabilizing an inner-sphere activated complex, thus decreasing the barrier penetration distance.

In this connection, several groups of workers have had difficulty in getting reasonable and/or reproducible measurements with the f-electron couples in reactions with each other. Sullivan and Thompson observed (*34*) in the Eu(II)–Np(IV) system with the reactants in approximately equivalent concentrations that the rate of reaction is first order in Eu(II) but zero order in Np(IV). In separate studies of the U^{3+}–Eu^{3+} reaction (*35*) by Templeton and Nicolini, the rates were irreproducible and did not conform to second-order kinetics. The reaction Yb^{2+} and Eu^{3+} was found by Christensen also to be intractable (*35*). Lavallee and Newton (*36*) found that the U(III)–Np(IV) rates were not reproducible and that the data did not follow simple second-order kinetics. The researchers in all these cases are experienced, and great care was taken with purity of the materials. As a final example of unexplained irreproducibility in a seemingly simple system is the interaction of Cr^{2+} with $Ru(NH_3)_6^{3+}$ in the presence of Cl^-. Neither Armor nor Armstrong (*37*), who worked with the system independently, was able to get reproducible or reasonable behavior in the ratio of $Cr(H_2O)_6^{3+}$/$Cr(H_2O)_5Cl^{2+}$ produced as a function of Cl^-. Nonadiabatic processes may be unusually sensitive to changes in the environment, and the anomalies mentioned may be traceable to a nonadiabatic contribution to the rate, but of course cannot with certainty be taken as diagnostic of adiabatic effects. Clearly, the systems are worth further study, but it is not at all obvious what measures will bring the systems under control.

Remaining still to be considered are the oxidation–reduction reactions in which electron transfer occurs over many bond lengths. These are best dealt with as a function of electronic structure type. The most thoroughly studied among them are reactions of Cr^{2+} with pentaammine–cobalt(III) complexes of ligands having conjugated bond systems (σ donor, σ acceptor). Efforts have been made to relate the rates to properties such as the mobile bond order of the conjugated ligands. If such a relation existed, it would at once imply nonadiabatic transfer. But it now appears (*38, 39, 40*) that most of the reactions of Cr^{2+} with Co(III) complexes in which there is remote attack proceed by a stepwise mechanism in which Cr^{2+} transfers an electron to the ligand, and the resulting radical then reacts with Co(III) or with Cr(III):

$$Co^{III}[L' \ldots L] + Cr^{2+} \rightleftharpoons Co^{III}[L' \ldots L]^- Cr^{(III)}$$

$$Co^{III}[L' \ldots L]^- Cr^{(III)} \rightarrow Co(II) + [L' \ldots L]Cr^{(III)}$$

This conclusion was strongly supported by Nordmeyer's (*7*) results and has been amply borne out by further studies done chiefly by E. S. Gould and co-workers (*41*) (*see*, for example, Ref. *41* which refers to earlier work). In these systems, a stepwise mechanism may be forced on the system because of the symmetry mismatch between the redox active metal ion orbitals and the carrier orbital on the ligand, which in all likelihood is a low-lying π^* orbital. Two Franck–Condon barriers— Cr^{2+} to ligand and ligand radical to Co(III)—affect the rates. There is no reason to believe that tunnelling probability at either of these junctions is rate determining and certainly not that electron mobility in the ligand is rate determining. An important property of the ligand which makes remote electron transfer possible in these systems is the accessibility of the π^* orbital. Other things being equal, ready reducibility of the ligand is important in determining whether remote attack will occur. For example, it is probable that terephthalatopentaamminecobalt(III) is not reduced by remote attack (39) while p-formylbenzoatopentaamminecobalt(III) certainly is (42). Numerous other comparisons can be found in the extensive and instructive contributions which have been made to this subject by Gould (39, 41). When the orbital on the oxidizing agent has π symmetry as in a pentaammineruthenium(III) complex, and the complex is reduced by Cr(II), there is apparently only one Franck– Condon barrier to surmount—Cr(II) to ligand—the electron being trans- ferred to an orbital that is delocalized over metal ion and ligand. In these systems, transfer from Cr(II) to (ligand + metal) is rate deter- mining, and again there is no reason to invoke tunnelling probability as a reaction barrier. Systems in which the reducing agent loses a π electron and the oxidizing agent gains a π electron will be dealt with below. For completeness, it must be mentioned that in these systems when the bridging ligands are conjugated, it is likely that the electron transfer reactions are adiabatic.

To sum up the survey of the past work on oxidation–reduction reac- tions: the only experimental results obtained thus far which strongly indicate nonadiabatic effects are some obtained by Matteson and Bailey (22) and by Chan and Wahl (23) for self-exchange reactions. In addi- tion, it is very likely that such effects are significant also for reactions of f electron redox agents, particularly on reaction with one another. Marcus has advocated consistently the position that nonadiabatic effects are relatively unimportant for the ordinary oxidation–reduction reactions which have been studied. But many experimentalists, including myself, have been much slower to arrive at it. Further work may show that a nonadiabatic factor is significant in many other processes, but at the present level of development of the subject, there are not many cases where it needs to be invoked.

Before describing experiments designed to study the nonadiabatic regime for electron transfer in oxidation–reduction reactions systemati- cally, some systems in which electron tunnelling, in the sense that it determines rate of electron transfer, is a factor should be mentioned. Zamaraev and co-workers (43, 44) and others (45) have developed a strong case for electron transfer by tunnelling in condensed systems which have undergone radiation damage. Miller (46) also added evidence for tunnelling in studying the reaction:

[biphenyl]⁻ + triphenylethylene = biphenyl + [triphenylethylene]⁻

in rigid ethanol at −77°. The overall behavior in these systems is complicated because the reactants are at varying distances. However in the systems described below the oxidant and reductant are at fixed relative positions. They represent a great simplification particularly when compared with the second-order redox processes which have been studied. By having oxidant and reductant assembled in a single molecule, intramolecular electron transfer rates can be measured and many ambiguities attending the interpretation of the second-order processes are avoided. The systems currently being studied involving metal ions do have precedent in purely organic ones, for example, in the paracyclophane radical anions studied by Weissmann (47) and radical anions derived from the 4,4'-nitrobiphenyl studied by Harriman and Maki (48). Although in principle the results in these systems are in the direct line of our interest, in practice they go only a small way toward yielding the kind of information which is desired, namely rates as a function of a wide range in the structure of the bridging groups spanning the redox centers and as a function of temperature (as mentioned earlier, values of $\Delta S^\ddagger$ can be particularly significant). For the organic systems, rates were measured by the ESR technique, and this is competent only in a rather narrow range of specific rates. Even if other approaches for the organic systems were available, there would still be important reasons for studying metal ions as the redox reactants, because the rates will be sensitive to orbital symmetry, to differences in energy of orbitals, and perhaps to orbital overlap. Only by studying metal-ion containing systems are we likely to understand the many redox reactions of metal complexes which have been studied in the bimolecular mode.

The advantage of studying intramolecular rather than intermolecular electron transfer was appreciated quite early (49), but only rather recently have reports of the results of such measurements been appearing for metal-to-metal electron transfer. Gaswick and Haim (50) reported first-order rates for electron transfer in the outer-sphere complex $Co(NH_3)_5H_2O^{3+} \cdot Fe(CN)_6^{4-}$, Cannon and Gardiner (51) for the inner-sphere complex formed between $[(NH_3)_5CoO_2CCH_2N(CH_2CO_2H)_2]^{2+}$ and Fe^{2+}, and Hurst and Lane (52) for $[(NH_3)_5Ru$ fumarate$]^{2+}$ with Cu^+ bound to the double bond of the ligand. But since in each case mentioned the oxidizing–reducing agent bond is labile, the positions which the metal ions have relative to each other in the activated complex are not defined. The ambiguity referred to is much reduced in systems now under study. Isied and Taube (53) devised a strategy for measuring intramolecular electron transfer in complexes of the class $Co^{III}L'$ . . . $L Ru^{II}$, a strategy which should be applicable to more complex systems, including biological ones. By the proper choice of polar groups for the bifunctional bridging ligand L' . . . L, a substitution inert combination $Co^{III}L'$ . . . $L Ru^{III}$ is first formed. Many reducing agents react with Ru(III) (a π electron acceptor) more rapidly than with Co(III) (a σ electron acceptor), and the system can be "cocked" for intramolecular electron transfer by adding a reducing agent such as $Ru(NH_3)_6{}^{2+}$. For the complexes thus far studied, Ru(II) is bound to the bridging ligand by a heterocyclic nitrogen and aquation of this linkage has a half-life in excess of a month (54). Moreover, Ru(II) in combination with a pyridine-like ligand shows a very strong $\pi^* \leftarrow \pi d$ absorption, which can be used to follow the reaction. The cobalt–ligand bond is disrupted on net electron transfer from Ru(II) to Co(III) because the electron accepted by the cobalt center is antibonding, and the complex is labilized for substitution on absorbing an electron. Haim and co-workers (55, 56) have analogous studies under way but with $Fe(CN)_6{}^{3+}$ in place of Ru(II). Comparison of results for the two kinds of systems should be useful. Similarities can be expected, but it is likely that differences will be especially instructive. Such differences arise, for example, from the differences in charge type, in the reducing power of the metals, and in their back-bonding capability.

In current work (57) being done by Seidel, a series of complexes related to and including those originally studied by Isied is being investigated. The measurements are now being extended to rates as a function of temperature, and to eliminating the ambiguity left in the original work for the complexes showing very rapid rates. This ambiguity arises from the fact that $SO_4{}^{2-}$ is included in the coordination sphere, and if electron transfer is rapid enough, the reducing agent is a monosulfato complex of Ru(II) rather than a monoaquo complex as is the case for slower electron transfer rates. In the meantime, measurements have been made on complexes in which 4,4'-bipyridine and related bifunctional ligands serve as bridging groups. These measurements include the rates of intramolecular electron transfer as a function of temperature (58). For the same bridging ligands, it is possible to prepare the mixed valence complexes of the type $[(NH_3)_5Ru^{II}L \ . \ . \ . \ LRu^{III}(NH_3)_5]^{5+}$. When the coupling between the metal centers is strong enough, the mixed valence species feature a band in the nir. According to Hush (21), this absorption band is related to the thermal electron transfer process in the mixed valence species, at least for systems in which the valence delocalization is not too great. The band energy measures the Franck–Condon barrier for optical electron transfer, (ΔE_2 in Figure 1), and if the energy of the system as a function of the reaction coordinate for electron transfer is known, the Franck–Condon barrier for the thermal process can be calculated. The oscillator strength for the intervalence transition is related to the extent of valence delocalization and is therefore a measure of energy gap at the crossover point in Figure 1. Although the electron transfer process for $Ru(II) \rightarrow Co(III)$ is different from that for $Ru(II) \rightarrow Ru(III)$ [$Co(III)$ and $Ru(III)$ accept the reducing electrons into orbitals of σ and π symmetry respectively; in the reduction of $Co(III)$, but not of $Ru(III)$, bonds are broken] it nevertheless seems of interest to compare the thermal rates for the $Ru(II)$–$Co(III)$ system with the properties of the nir absorption band for $Ru(II)$–$Ru(III)$. It would be desirable to measure the rates of electron transfer for the latter systems directly, but until now, no reports of such measurements have been published. Results of some experiments designed to measure the rate of electron transfer by studying the dielectric constant as a function of frequency in rigid media are reported in Ref. *60.*

The data which have been obtained for the complexes bridged by 4,4′-bipyridine and related molecules are summarized in Table III. Referring first to the results on net electron transfer, perhaps the most notable among them is the small range in rate for what seem to be drastic changes in electronic coupling between the rings. This suggests strongly that the reactions lie very much toward the adiabatic regime and supports the position that a small coupling ensures adiabatic transfer. Leaving out of consideration the last system which will be discussed presently, the values of $\Delta H^{\ddagger}$ are much alike for the different systems, showing that the Franck–Condon barrier is rather constant within the series. This is expected because the immediate environment around each metal atom remains the same and, on a more empirical basis, is expected because the reduction potential for the $Ru(III)$–$Ru(II)$ couple remains rather constant. For the above arguments to be applicable, the proviso needs to be made that the metal–metal distance remain constant within the series, otherwise the barrier to electron transfer contributed by the solvent will not be constant. This distance, it must be admitted, is not quite constant.

Another striking result which has emerged from the study is that the large negative entropies which are characteristic of reactions of the $3+, 2+$ charge type when studied in the bimolecular mode have all but disappeared. The entropy of activation for the slowest reaction among the first five is more negative than that observed for the most rapid, and this perhaps indicates that the effect of nonadiabatic transfer is being felt for the dipyridyl methane system.

Table III.   Intramolecular Electron Transfer Rates the Intervalence Transitions

$L \ldots L$	$k \times 10^3$ $(M^{-1} sec^{-1})^{a,\, b}$	$\Delta H^{\ddagger}$ $(kcal\ mol^{-1})$
N◯—◯N	44	20.0 ± 0.3
N◯—C=C—◯N (H, H)	19	20.2 ± 0.3
N◯—◯N (CH₃, CH₃)	5.5	20.0 ± 0.3
N◯—S—◯N	5.0	19.7 ± 0.6
N◯—CH₂—◯N	2.1	18.4 ± 0.5
N◯—CH₂—CH₂—◯N	1.2	18.6 ± 1.0

[a] At 25°; ionic strength, 0.4.
[b] Values of $\Delta H^{\ddagger}$ and $\Delta S^{\ddagger}$ obtained by Fischer, Tom, and Taube (58) recalculated by Konrad Rieder. Error estimates according to the method of J. L. Cramer (59). The only substantial difference from those published is for the dipyridylethane where $\Delta H^{\ddagger} = 19.5 \pm 0.3$ and $\Delta S^{\ddagger} = -6.5 \pm 1.5$.

In the systems being described, a decrease in rate by a factor of about 20 is registered when a CH_2 group is inserted between the rings of 4,4'-bipyridine, but this factor does not necessarily apply in other cases. Thus, comparing isonicotinate with the same ligand with a CH_2 inserted between the ring and the carboxyl group, the rate decreases by a factor of only 1.5 (57). The small change is not expected on the basis of experiments done in the bimolecular mode. Whether this means that the interpretation of the bimolecular processes is badly confused by the complication of the stability of the precursor complex or whether the difference is a consequence of a difference in mechanism is not yet clear. (In the bimolecular processes Cr^{2+}, a σ electron donor is often used.

in Co(III)L . . . LRu(II) and Properties of
for Ru(III)L . . . LRu(II)

$\Delta S^{\ddagger}$ (cal mol^{-1} deg^{-1})	$E_f(V)$ [e]	λ max (nm) [d]	ϵ(M^{-1} cm^{-1})	Ref.
2.6 ± 1.2	0.35	1050	400	58
1.3 ± 1.0	0.33	960	400	58
-1.6 ± 0.6	0.34	860	90	58
-2.9 ± 2.1	0.36	855	70	58
-9.0 ± 1.7	—	825	~ 5	61
-9.6 ± 3.2	0.31	—	< 10	58

[e] Reduction potential for the 3+,2+ couple referred to NHE as obtained by cyclic voltammetry for the protonated complexes, [(NH$_3$)$_5$RuL . . . LH]$^{4+, +3}$.
[d] For the intervalence transition in the Ru(III)–Ru(II) complex, measured in D$_2$O.

The reduction of [(NH$_3$)$_5$CoO$_2$CH$_2$⟨○⟩N]$^{2+}$ apparently has not been studied. The general experience with reduction of Co(III) by Cr^{2+} however is that when CH$_2$ is inserted into a conjugated system, the reaction by remote attack is much less rapid than it is by the outer-sphere path, and this latter path is usually much slower than that involving remote attack, when this can occur.) Our limited capacity to predict the results of the experiments on intramolecular electron transfers is a strong motivation for continuing the work.

The last system dealt with in the table differs from the others in that the bridging group is flexible. The molecule can adopt a configuration in which the metal ions approach quite closely, and it is likely that in this case there is direct electron transfer from Ru(II) to Co(III), i.e., electron transfer does not depend on electronic coupling being provided by the bridging group. The fact that the reaction rate is more sensitive to ionic strength in this than in the other systems in which this parameter is varied (58) supports the idea that the configuration suitable for electron transfer is not the equilibrium one.

The results in Table III show that there is a rough parallel between the rates of net electron transfer from Ru(II) to Co(III) and the intensity of the intervalence band observed for the Ru(II)–Ru(III) mixed valence complex. Any such parallelism—more exactly a parallelism between $\Delta S^\ddagger$ and the intensities—would imply that a nonadiabatic factor does affect the rates of electron transfer. Unfortunately, it is not possible to examine this relationship in the sensitive region where the reactions are strongly nonadiabatic. The nir bands are very broad, and is difficult to measure extinction coefficients that are less than about $5\ M^{-1}\ cm^{-1}$.

An important application of the work on mixed valence complexes in which the metal ions are weakly but not too weakly coupled is in resolving the Franck–Condon barrier to electron transfer into the separate contributions from the inner-sphere and the solvent. Some work directed to this goal has been reported (62, 63). Because of the nature of the dependence of the solvent barrier on distance, those systems in which the separation of the metal centers is large will be particularly instructive. When this separation is sufficiently large, it may be possible to estimate the radius of the dielectrically saturated region about each ion so as to compare it with the radius of the coordination sphere itself.

The systems devoted to studying intramolecular electron transfer which have been described have not gone far toward mapping out the characteristics of the nonadiabatic regime, but it seems likely that they can be elaborated to make a systematic study possible. Success is by no means a certainty, however. Thus in the systems with a —CH$_2$— or —S— interrupting the conjugation, the electron may now leak from one π system to another. Placing a second saturated link in the chain for a rigid system may cause such a profound decrease in the rate that other processes, such as reduction of Co(III) in one center by Ru(II) in another, may intervene. But if the half-life is not increased beyond several days by further decoupling, first-order rates can still be measured. Synthetic work is in progress (61) to dispose pyridine groups on the cyclopropane ring (trans) and on the cyclohexane ring and to use the resulting molecules as bridging groups. Even apart from achieving strong decoupling, the results both of net electron transfer and of the intensities of the intervalence transition are useful in learning about weak interactions between separate electronic systems.

The strategy that has been used for the Co(III) . . . Ru(II) systems can be adapted also to include molecules of biological interest. In fact, for them an additional factor—geographical isolation of one ion in the interior of a structure while the second reactant is ligated to the exterior—can be used to discriminate between the centers in cocking the system for electron transfer. By using pulse radiolysis, the preferential oxidation or reduction of one center over another can be done on the time scale of a fraction of a microsecond, and rapid rates of electron transfer can be measured. It will also be of interest to attach ruthenium ammines to molecules of biological interest and to study the interaction between the ions in the mixed valence state using spectrophotometric and other techniques.

The question of facile electron transfer over many bond distances in biological systems appears still to be open. Just how strong the coupling is between metal ions suitably disposed with respect to the hydrogen bond system of a protein or as mediated by $\pi-\pi$ interactions is difficult to calculate. It may turn out to be strong enough to allow even adiabatic transfer over distances of the order of 20 Å.

Acknowledgment

The work described herein originating in my laboratories has been supported by the NSF and NIH, and the financial assistance by these agencies is gratefully acknowledged. I also want to express my appreciation to my collaborators, not only those who have been acknowledged in the references but also the many who have made indirect contributions to the advances which have been described.

Literature Cited

1. Dodson, R. W., *J. Am. Chem. Soc.* (1950) **72**, 3315.
2. Silverman, J., Dodson, R. W., *J. Phys. Chem.* (1952) **56**, 846.
3. *J. Phys. Chem.* (1952) **56**, 829–846.
4. Libby, W. F., *J. Phys. Chem.* (1952) **56**, 863.
5. Taube, H., *J. Am. Chem. Soc.* (1955) **77**, 4481.
6. Nordmeyer, F., Taube, H., *J. Am. Chem. Soc.* (1966) **88**, 4295.
7. *Ibid.* (1968) **90**, 1162.
8. Szent-Györgi, A., *Science* (1941) **93**, 609.
9. Schmitt, M., *Z. Naturforsch.* (1947) **2b**, 98.
10. DeVault, D., Chance, B., *Biophys. J.* (1966) **6**, 825.
11. Marcus, R. A., *Ann. Rev. Phys. Chem.* (1964) **15**, 155.
12. Zwolinski, B. J., Marcus, R. J., Eyring, H., *Chem. Rev.* (1957) **55**, 157.
13. Sutin, N., "Inorganic Biochemistry," G. Eichorn, Ed., Chapter 19, Elsevier, 1973.
14. Abe, M., Brown, G., Krentzien, H., unpublished data, 1976.
15. Meyer, T. J., Taube, H., *Inorg. Chem.* (1968) **7**, 2369.
16. Pladziewicz, J. R., Espenson, J. H., *J. Am. Chem. Soc.* (1973) **95**, 56.
17. Marcus, R. A., *J. Chem. Phys.* (1956) **24**, 979.
18. Hush, N. S., *Trans. Faraday Soc.* (1961) **57**, 557.

19. Tom, G., Creutz, C., Taube, H., *J. Am. Chem. Soc.* (1974) **96**, 7827.
20. Tom, G., Ph.D. Thesis, Stanford University, 1974.
21. Hush, N. S., *Prog. Inorg. Chem.* (1967) **8**, 391.
22. Matteson, D. S., Bailey, R. A., *J. Am. Chem. Soc.* (1969) **91**, 1975.
23. Chan, Man-Sheung, Ph.D. Thesis, Washington University, St. Louis, 1974.
24. Barrett, M. B., Swinehart, J. H., Taube, H., *Inorg. Chem.* (1971) **10**, 1983.
25. Stritar, J. A., Taube, H., *Inorg. Chem.* (1969) **8**, 2281.
26. Zwickel, A., Taube, H., *J. Am. Chem. Soc.* (1961) **83**, 793.
27. Doyle, J., Sykes, A. G., *J. Chem. Soc.* (1968) 2836.
28. Dodel, P., Taube, H., *Z. Phys. Chem. (Frankfurt)* (1965) **44**, 92.
29. Price, H. J., Taube, H., *Inorg. Chem.* (1968) **7**, 1.
30. Anderson, A., Bonner, N. A., *J. Am. Chem. Soc.* (1954) **76**, 3826.
31. Deutsch, E., Taube, H., *Inorg. Chem.* (1960) **7**, 1532.
32. Meier, D. J., Garner, C. S., *J. Phys. Chem.* (1952) **56**, 853.
33. Taube, H., "Advances in Inorganic Chemistry and Radiochemistry," Vol. 1, p. 47, Academic, New York.
34. Sullivan, J. C., private communication, 1976.
35. Espenson, J. H., private communication, 1976.
36. Newton, T. W., private communication, 1976.
37. Armor, J. N., Armstrong, R. A., unpublished data, 1970, 1975.
38. Gould E. S., Taube, H., *J. Am. Chem. Soc.* (1964) **86**, 1318.
39. Gould, E. S., *J. Am. Chem. Soc.* (1965) **87**, 4730.
40. Taube, H., Gould, E. S., *Acc. Chem. Res.* (1969) **2**, 321.
41. Martin, A. H., Gould, E. S., *Inorg. Chem.* (1975) **14**, 873.
42. Zanella, A., Taube, H., *J. Am. Chem. Soc.* (1972) **94**, 6403.
43. Zamaraev, K. I., Khairutidov, R. F., Mikhailov, A. I., Goldanskii, V. I., *Dokl. Akad. Nauk, SSSR* (1971) **199**, 640.
44. Zamaraev, K. I., Khairutidov, R. F., *Chem. Phys.* (1974) **4** (2), 181.
45. Kroh, J., Stradowski, C., *Int. J. Radiat. Phys. Chem.* (1973) **5**, 243.
46. Miller, J. R., *Science* (1975) **189**, 221.
47. Weissman, S., *J. Am. Chem. Soc.* (1960) **80**, 6462.
48. Harriman, J. E., Maki, A. H., *J. Chem. Phys.* (1963) **39**, 778.
49. Taube, H., Myers, H., *J. Am. Chem. Soc.* (1954) **76**, 2103.
50. Gaswick, D., Haim, A., *J. Am. Chem. Soc.* (1971) **93**, 7347.
51. Cannon, R. D., Gardiner, J., *J. Am. Chem. Soc.* (1970) **92**, 3800.
52. Hurst, J. R., Lane, R. H., *J. Am. Chem. Soc.* (1970) **95**, 1703.
53. Isied, S., Taube, H. (1973) **95**, 8198.
54. Shepherd, R. E., Taube, H., *Inorg. Chem.* (1973) **12**, 1392.
55. Gaswick, D., Haim, A., *J. Am. Chem. Soc.* (1974) **96**, 7845.
56. Jwo, J.-J., Haim, A., "Abstracts of Papers," Centennial Meeting, ACS, April 1976, Inorg. 15.
57. Seidel, S., unpublished data, 1975-76.
58. Fischer, H., Tom, G. M., Taube, H., *J. Am. Chem. Soc.* (1976) **98**, 5512.
59. Cramer, J. L., Ph.D. Thesis, University of North Carolina, p. 217, 1975.
60. Richman, R. M., Drago, R. S., "Abstracts of Papers," 170th National Meeting, ACS, August 1975, Inorg. 29.
61. Rieder, Konrad, unpublished data, 1976.
62. Tom, G. M., Creutz, C., Taube, H., *J. Am. Chem. Soc.* (1974) **96**, 7827.
63. Callahan, R. C., Brown, G. M., Meyer, T. J., *J. Am. Chem. Soc.* (1974) **96**, 7829.

RECEIVED July 26, 1976.

Structure and Electron Transfer Reactions of Blue Copper Proteins

HARRY B. GRAY, CATHERINE L. COYLE, DAVID M. DOOLEY,
PAULA J. GRUNTHANER, JEFFREY W. HARE, ROBERT A.
HOLWERDA, JAMES V. McARDLE, DAVID R. McMILLIN,
JILL RAWLINGS, ROBERT C. ROSENBERG, N. SAILASUTA,
EDWARD I. SOLOMON, P. J. STEPHENS, SCOT WHERLAND,
and JAMES A. WURZBACH

Arthur Amos Noyes Laboratory of Chemical Physics,
California Institute of Technology, Pasadena, CA 91125

Complete assignments of the electronic spectra of stella-cyanin, plastocyanin, and azurin have been made. Bands attributable to d–d transitions have been located in the near-infrared region for the first time, and their positions are consistent with a distorted tetrahedral geometry for the blue copper center. The kinetics of the electron transfer reactions of stellacyanin, azurin, and plastocyanin with $Fe(EDTA)^{2-}$ and $Co(phen)_3^{3+}$ have been studied. Kinetic parameters indicate that reduction of azurin and plasto-cyanin by $Fe(EDTA)^{2-}$ occurs by long distance transfer to a buried blue copper center. However, the pathway for oxidation involves substantial protein rearrangement, thereby allowing contact of $Co(phen)_3^{3+}$ with the copper ligands. In contrast, the blue copper center of stellacyanin is equally accessible in solution to redox agents.

There has been much recent interest in the mechanism of electron transfer involving blue copper proteins and nonphysiological as well as physiological redox agents (1). Questions such as how far apart the two redox centers are at the instant of electron transfer, whether there are multiple pathways, whether there is a high degree of specificity, and, if so, what is its origin, are being asked. Experiments in our laboratory and elsewhere have provided enough information on the spectroscopic properties and the kinetics of electron transfer reactions of blue copper proteins to make serious structural and mechanistic discussions possible. The blue copper proteins which we discuss here are bean plastocyanin (*Phaseolus vulgaris;* mol wt 10,700 (2); E 350 mV (3)), azurin (*Pseudo-*

monas aeruginosa; mol wt 13,900 (*4*); E 330 mV (*5*)), and stellacyanin (*Rhus vernicifera;* mol wt 20,000 (*6*); E 184 mV (*7*)). All these proteins are thought to have electron transfer functions, although a physiological partner is not known for stellacyanin. Plastocyanin accepts an electron from cytochrome *f* in the photosynthetic chain of green leaves, and azurin is believed to be the physiological oxidant for ferrocytochrome *c* 551 in *Pseudomonas aeruginosa* and related bacteria. The blue copper proteins as a class are particularly attractive for systematic kinetic studies, as basically the same chromophore spans a fairly large range of reduction potentials. The potential that is adopted apparently depends on finer details of the blue copper environment, allowing an opportunity for correlations of electron transfer reactivity with such structural differences.

Structural Considerations

X-ray crystal structure analysis has not been completed to date for any blue (or type 1) copper protein. The probability that the copper coordination environment is highly unusual, however, has long been recognized, as a result of various spectroscopic studies (*8, 9*). A typical blue copper protein is characterized by an intense electronic absorption band system, which peaks at about 600 nm ($\epsilon \geqslant 2 \times 10^3$), as well as by an extremely small $A_{||}$ EPR spectral parameter. Neither of these spectral properties has been duplicated satisfactorily in low molecular weight Cu(II) complexes. As square planar Cu(II) centers, in particular, exhibit optical and EPR spectra that are much different from those observed for blue proteins, most models have featured geometries based on tetrahedral or five coordination. Two different explanations of the intense 600-nm absorption have been proposed. One treats the band as arising from one or more allowed *d–d* transitions in a non-centrosymmetric center, and the other attributes the strong absorption to a charge transfer process, probably of the ligand-to-metal (LMCT) type (*10, 11*).

Spectroscopic studies of Co(II) derivatives of stellacyanin, plastocyanin, and azurin have established that the charge transfer interpretation is preferred (*10, 11*). Intense bands ($\epsilon \geqslant 2 \times 10^3$) that appear to be analogous to the 600-nm system of blue proteins are observed between 300 and 350 nm in the Co(II) derivatives. The shift in band position of about 16 kK [Cu(II) << Co(II)] accords well with expectation for an LMCT transition. The visible and near-infrared absorption, CD, and MCD spectra of Co(II) derivatives of stellacyanin, plastocyanin, and azurin have been interpreted (*12*) successfully in terms of the *d–d* transitions expected for distorted tetrahedral metal centers (Table I). Average ligand field parameters are the same for all three Co(II) proteins ($Dq = 490$, $B = 730$ cm^{-1}), which strongly suggests that the donor atom set of a blue site does not vary from case to case, even though the potential does. Interestingly, the splitting of the $^4T_1$ state is smallest for the Co(II) derivative of stellacyanin, which is the protein whose native form exhibits the lowest potential.

Table I. Energies of the Ligand Field States
in Co(II) Protein Derivatives[a]

Protein	$^4T_2$	$^4T_1$ (^4F)		Avg.	$^4T_1$ (^4P)		Avg.
Stellacyanin	5000	6800	9700	8250	15,500	18,500	17,000
Azurin	[b]	6600	10,150	8375	15,700	19,200	17,400
Plastocyanin	[b]	6900	10,000	8450	15,200	19,700	17,400

[a] Energies in cm^{-1}; ground state (zero) is $^4A_2$ (*12*).
[b] Not observed.

New absorptions and CD spectral features attributable to d–d transitions have been observed in the near-infrared region for stellacyanin, plastocyanin, and azurin (Table II). These d–d transitions have been analyzed (*13*) successfully in terms of a slightly flattened tetrahedral geometry for blue copper centers. In such a geometry the electronic energies (W) of the ligand field states of Cu(II) increase according to $^2B_2 < {}^2E < {}^2B_1 < {}^2A_1$ (*14*). The value of the angle (β) between the z-axis and the metal–ligand bonds ($\beta = 90°$ at the D_{4h} limit) for stellacyanin was fixed at 60° in order to give reasonable values for $\Delta W(^2A_1 - {}^2B_2)$ in the tetrahedral and square planar limiting geometries. Taking $\Delta W(^2E - {}^2B_2) = 5250$ and $\Delta W(^2B_1 - {}^2B_2) = 8750$ cm^{-1} ($Ds = 765$, $Dt = 444$ cm^{-1}), values of $\Delta W(^2A_1 - {}^2B_2)$ are calculated to be 22,800 and 6915 cm^{-1} at the D_{4h} and T_d ($\beta = 54.74°$) limits, respectively. The energy separation of 22,800 cm^{-1} is correctly denoted $\Delta W(^2A_{1g} - {}^2B_{1g})$, as $^2A_1$ and $^2B_2$ become $^2A_{1g}$ and $^2B_{1g}$, respectively, in D_{4h}. The other D_{4h} transitions, $^2B_{1g} \to {}^2E_g$ and $^2B_{1g} \to {}^2B_{2g}$, are predicted at 24,890 and 15,550 cm^{-1}, respectively. All the calculated D_{4h} values should be reduced by about 20%, allowing for the slight increase (~ 0.1 Å) in metal–ligand bond lengths that is expected to accompany a change from tetrahedral to square planar coordination. The adjusted values for the D_{4h} d–d transition energies are in reasonable agreement with the observed band positions in square planar Cu(II) complexes containing nitrogen-donor ligands (*14*). In sharp contrast, ligand field parameters derived from β values a few degrees below or above 60° give absurd energy differences in the D_{4h} limit. For example, $\Delta W(^2A_{1g} - {}^2B_{1g})$ is calculated to be 58,800 cm^{-1} from $\beta = 57°$ parameters ($Ds = 2240$, $Dt = 505$ cm^{-1}) and 5050 cm^{-1} from $\beta = 70°$ ones ($Ds = 77$, $Dt = 320$ cm^{-1}).

The derived ligand field parameters ($\beta = 60°$, $Ds = 765$, $Dt = 444$ cm^{-1}) predict $^2A_1$ to be 11,540 cm^{-1} above the $^2B_2$ ground state for stellacyanin. An absorption band and CD maximum are observed near this

Table II.	Absorption and CD Spectral	
Protein	$\bar{\nu}(cm^{-1})$	$\Delta\epsilon(M^{-1}\ cm^{-1})$
Stellacyanin	5250	0.45
	8750	−0.35
	11,470	2.4
	13,040	−5.0
	16,580	3.6
	17,840	0.75
	22,570	−7.35
Plastocyanin	5500	0.125
	10,300	−0.165
	1·1,940	~ 1.5
	13,540	−3.78
	16,600	4.08
	18,140	0.4
	21,540	−1.32
	23,640	1.26
Azurin	10,300	−0.475
	12,940	−3.6
	15,940	3.96
	17,770	0.72
	20,840	−1.11

[a] From Ref. *13*. Absorption for each blue protein are taken from a Gaussian resolution of the 35 K near-infrared and visible absorption spectrum.

energy. The resolved absorption band has a molar extinction coefficient of 565, which is approximately five times larger than that observed for the maximum at 8750 cm^{-1}. It is expected that the $^2B_2 \rightarrow {}^2A_1$ transition should be more intense, as it is electric-dipole allowed, whereas $^2B_2 \rightarrow {}^2B_1$ is not.

Excited d-level energies for plastocyanin and azurin are very similar. Taking $\Delta W(^2E - {}^2B_2) = 5500$, $\Delta W(^2B_1 - {}^2B_2) = 10,300$ cm^{-1}, and $\beta = 60°$ for both proteins, Ds and Dt are calculated to be 764 and 508 cm^{-1}, respectively, and the $^2B_2 \rightarrow {}^2A_1$ transition is predicted to be at 12,520 cm^{-1}. Plastocyanin exhibits a Gaussian-resolved peak at 11,940 cm^{-1} (Table II), which may be attributed to $^2B_2 \rightarrow {}^2A_1$. In azurin, the relatively intense band at ~ 13,000 cm^{-1} probably overlaps extensively with absorption owing to $^2B_2 \rightarrow {}^2A_1$.

Research aimed at identifying the ligands comprising the flattened tetrahedral blue copper center has been particularly intense in the case of plastocyanin. Direct evidence for a sulfur ligand has come from x-ray photoelectron spectral (XPS) experiments on bean plastocyanin, where a large shift of the S2p core energy of the single cysteine (Cys-85) residue in the protein upon metal incorporation (164.5, apo; 169.8, native; 168.8 eV, Co(II) derivative) was observed (*15*). The two histidines in spinach plastocyanin exhibit pK values below 5 in NMR titration experiments,

Data for Blue Copper Proteins[a]

$\epsilon (M^{-1}\ cm^{-1})$	γ	Assignment
~ 100	~ 0.018	$^2B_2 \rightarrow {}^2E$
~ 100	~ 0.014	$^2B_2 \rightarrow {}^2B_1$
565	0.017	$^2B_2 \rightarrow {}^2A_1$
341	0.059	$\pi S \rightarrow d_{x^2-y^2}$
3549	0.0041	$\sigma S \rightarrow d_{x^2-y^2}$ [b]
1542	0.0019	
942	0.031	$\pi N^* \rightarrow d_{x^2-y^2}$
~ 100	0.005	$^2B_2 \rightarrow {}^2E$
200	0.0033	$^2B_2 \rightarrow {}^2B_1$
1162	0.0052	$^2B_2 \rightarrow {}^2A_1$
1289	0.012	$\pi S \rightarrow d_{x^2-y^2}$
4364	0.0037	$\sigma S \rightarrow d_{x^2-y^2}$ [b]
1163	0.0014	
300	0.018	$\pi N^* \rightarrow d_{x^2-y^2}$ [b]
~ 100	~ 0.050	
82	0.023	$^2B_2 \rightarrow {}^2B_1$
686	0.021	$\pi S \rightarrow d_{x^2-y^2}$
3798	0.0042	$\sigma S \rightarrow d_{x^2-y^2}$ [b]
504	0.0057	
185	0.024	$\pi N^* \rightarrow d_{x^2-y^2}$

[b] Not assigned.

suggesting that they are coordinated to copper (*16*). It is reasonable to assume, therefore, that the analogous two residues in the bean protein, His-38 and His-88, are also ligands (*13*). The fourth ligand in the proposed donor set for bean plastocyanin has been identified in extensive infrared spectral studies (*17*). These experiments have revealed that a short section of α-helix in apoplastocyanin is strongly perturbed upon metal [Cu(II) or Co(II)] incorporation, thereby implicating a backbone peptide nitrogen or oxygen as a ligand. The preference of copper for nitrogen donors, as well as evidence from charge transfer spectra (vide infra), favors coordination by a deprotonated peptide nitrogen (N*). Consideration of the bean plastocyanin sequence places the α-helix, and therefore the backbone peptide nitrogen, a few residues above His-38 (*14*).

A flattened tetrahedral CuN_2N^*S model is also reasonable for azurin. Recent XPS experiments have shown (*18*) that a sulfur is bound to copper in azurin, and the electronic spectroscopic properties of both Cu(II) and Co(II) forms of the protein are closely similar to those of analogous bean plastocyanin derivatives. It is probable that the near-tetrahedral binding site in each blue protein is rather rigid, as ligand-field stabilization factors strongly favor a square planar geometrical structure for four-coordinate Cu(II). The site-structure rigidity built by the protein must overcome the electronic stabilization energy associated with a Jahn–Teller distortion toward square planar Cu(II) geometry, thereby contributing to the relative instability of the oxidized state of the system. Thus the relatively high reduction potentials of blue copper proteins may be attributable, at least in part, to electronic factors associated with the rigidly constrained, flattened tetrahedral CuN_2N^*S site structure.

Strong evidence for cysteine sulfur coordination in stellacyanin has been obtained (18) in XPS experiments. Thioether coordination is ruled out in this case, as the protein does not possess any methionine (6). It is probable that the other ligands are similar, but not necessarily identical, to those of bean plastocyanin.

An interpretation of the intense absorption bands observed at about 13,000, 16,000, and 22,000 cm^{-1} in blue copper proteins has been presented (13). The analysis is based on the assumption that the energies of the highest occupied ligand orbitals in a CuN_2N^*S unit decrease according to $\pi S > \sigma S > \pi N^* > \pi N$. Charge transfer excited states derived from transitions of the type $\pi \rightarrow d_{x^2-y^2}$ are both electric- and magnetic-dipole allowed. The transition $\sigma S \rightarrow d_{x^2-y^2}$, on the other hand, is only electric-dipole allowed. The rotational strength (R) of a CD band is related to the electric dipole ($\vec{\mu}_{el}$) and the magnetic dipole ($\vec{\mu}_{mag}$) moments by:

$$R = \text{Im}[\psi_i^* \vec{\mu}_{el} \psi_f d\tau \cdot \int \psi_i^* \vec{\mu}_{mag} \psi_f d\tau] \qquad (1)$$

where ψ_i and ψ_f are the initial and final states, respectively. The rotational strength may also be determined from the experimental quantity $\Delta\epsilon$, or $\epsilon_l - \epsilon_r$, according to Equation 2:

$$R = 22.9 \times 10^{-40} \int (\Delta\epsilon/v) dv \qquad (2)$$

Further, the dipole strength (D) is related to the molar extinction coefficient ϵ by:

$$D = 91.8 \times 10^{-40} \int (\epsilon/v) dv \qquad (3)$$

and $4R/D \approx \gamma$, the Kuhn anisotropy factor. Moscowitz has approximated (19) the integral $\int (\epsilon/v) dv$ as $\ln 2 \sqrt{\pi} \epsilon^\circ \delta^\circ / v^\circ$, where ϵ° is the maximum value of ϵ, δ° is the half-width at half-maximum, and v° is the frequency of ϵ°. Assuming that δ° and v° are the same for corresponding absorption and CD bands, γ may be calculated from Equation 4:

$$\gamma = |\Delta\epsilon/\epsilon| \qquad (4)$$

Bands associated with magnetic-dipole allowed, π charge transfer transitions are expected to have much larger γ values than those attributable to $\sigma S \to d_{x^2-y^2}$, as the intensity-giving mechanism in the latter case is purely electric dipole in origin. Values of γ, $\Delta\epsilon$, and ϵ for the electronic spectral features of stellacyanin, plastocyanin, and azurin are listed in Table II. The results clearly indicate that the bands at $\sim 13{,}000$ and $\sim 22{,}000$ cm^{-1} be assigned to π charge transfer, as they have relatively large values of γ. Specific assignments are $\pi S \to d_{x^2-y^2}$ at 13,000 cm^{-1} and $\pi N^* \to d_{x^2-y^2}$ at 22,000 cm^{-1}. The 16,000-cm^{-1} band in each protein exhibits a γ value well below 0.005 and is attributed to a $\sigma S \to d_{x^2-y^2}$ charge transfer transition.

Electron Transfer Reactions

An important point to be made at the outset is that blue copper centers appear to be built to minimize the inner sphere reorganization associated with electron transfer, Cu(II) $\overset{+e^-}{\underset{-e^-}{\rightleftharpoons}}$ Cu(I), as a distorted tetrahedral (or a related three- or five-coordinate) geometry would be an acceptable ground state for Cu(I) derivatives. Facile outer-sphere electron transfer is to be expected, therefore, to and from such copper centers. Furthermore, certain blue copper sites may be substantially buried in hydrophobic protein interiors and are therefore relatively inaccessible to solvent and other small molecules (1). Evidence will be presented in this section that access to these sites varies considerably from protein to protein.

The results of kinetic studies of the reduction of stellacyanin, plastocyanin, and azurin by Fe(EDTA)$^{2-}$ are summarized in Table III (20, 21). The order of cross reaction rate constants (k_{12} values) is stellacyanin > plastocyanin > azurin, which is surprising, as considerations based on driving force alone would predict stellacyanin to be the least reactive of

Table III. Rate Constants for the Reduction of Blue Copper Proteins by Fe(EDTA)$^{2-}$ (25°)

Protein	k_{12}(M^{-1} sec^{-1})	k_a(M^{-1} sec^{-1})	k_b(M^{-1} sec^{-1})	pK
Stellacyanin	4.3×10^{5a}	5.7×10^{5b}	5.1×10^{5b}	6.4[b]
Plastocyanin	8.2×10^{4c}	5.5×10^{4d}	3.1×10^{4d}	6.1[d]
Azurin	1.3×10^{3c}	2.4×10^{3d}	1.0×10^{3d}	6.4[d]

[a] $\mu = 0.5M$, pH 7 (20).
[b] $\mu = 0.5M$ (21).
[c] $\mu = 0.2M$, pH 7 (20).
[d] $\mu = 0.2M$ (21).

the three. Each reaction shows a small pH dependence, which has been analyzed (21) in terms of rate constants for protonated (k_a) and depronated (k_b) forms of the protein. For each of the blue proteins, $k_a >$ k_b, and a pK slightly above 6 is found. One possible explanation, then, is that a protonated residue near the blue copper center slightly assists electron transfer from $Fe(EDTA)^{2-}$ in these cases. We hasten to add, however, that we do not place much confidence in conclusions based on relatively small pH effects.

The equation of relative Marcus theory (22), $k_{12} = (k_{11}k_{22}K)^{\frac{1}{2}}$, may be applied to estimate the self-exchange rate constants (k_{11}) for the three blue proteins based on their reactions with $Fe(EDTA)^{2-}$. This procedure better evaluates the electron transfer reactivity of each blue copper center, as it corrects for differences in driving force. As expected, then, the reactivity gaps mentioned above are greatly widened, according to the following calculated (23) k_{11} (M^{-1} sec^{-1}) values: stellacyanin (6.2 $\times 10^5$) > plastocyanin (4.6 $\times 10^1$) > azurin (2.2 $\times 10^{-2}$). Correction for electrostatic charge effects does not alter the reactivity order. By estimating the charges on the proteins at pH 7 from sequence data, electrostatics-corrected self-exchange rate constants (k_{11}^{corr}) have been calculated (23) as follows (M^{-1} sec^{-1}): stellacyanin (2.7 $\times 10^5$) > plastocyanin (1.8 $\times 10^1$) > azurin (1.1 $\times 10^{-2}$).

Unfortunately, experimental k_{11} values are not available for any one of the three blue proteins. It has been established, however, that the self-exchange rates for azurin (24) and plastocyanin (25) are both slow on the NMR time scale at 25°. It is not likely that the predicted k_{11} values from the $Fe(EDTA)^{2-}$ reactions will correspond to the measured self-exchange rate constants. Indeed, agreement between calculated and measured k_{11} values is only to be expected if the activation requirements of both partners in a cross reaction are exactly the same as those used in their respective self-exchanges. Access to a buried, or at best partially exposed, outer-sphere redox center should vary substantially for different reactants, and it is likely that calculated k_{11} values will span a wide range in such cases.

Based on the $Fe(EDTA)^{2-}$ results, the blue copper center in stellacyanin appears to be much more accessible than that situated in either azurin or plastocyanin. Thus it should be profitable to compare the electron transfer reactivities of these three proteins with a variety of redox agents. Kinetic studies of the oxidation of the three blue proteins by $Co(phen)_3^{3+}$ have been made (26), and the results together with those for other redox agents are set out in Table IV. The electrostatic corrections to the predicted k_{11} values are modest both for the large charge on plastocyanin and the small one on azurin, as the protein self-exchange and the cross reaction work terms compensate. The reactivity of the stellacyanin redox center as calculated from the cross reaction with $Co(phen)_3^{3+}$ is much greater than the corresponding k_{11}^{corr} for azurin or plastocyanin, in accord with our earlier finding based on k_{12} [$Fe(EDTA)^{2-}$] values. The rather good agreement between k_{11}^{corr} values based on $Fe(EDTA)^{2-}$ and $Co(phen)_3^{3+}$ provides strong evidence that the redox center in stellacyanin is accessible to reactants in solution. Such good agreement between k_{11}^{corr} values contrasts markedly with the situation in azurin and plastocyanin, where $k_{11}^{corr}[Fe(EDTA)^{2-}] < k_{11}^{corr}[Co(phen)_3^{3+}]$.

Table IV. Electron Transfer Cross-Reaction and Self-Exchange Rate Constants for Blue Copper Proteins (25°, μ0.1M, pH 7) [a]

Blue Protein	Redox Agent	k_{12} $(M^{-1} sec^{-1})$	Calcd. k_{11} $(M^{-1} sec^{-1})$	Calcd. k_{11}^{corr} $(M^{-1} sec^{-1})$
Stellacyanin $(Z_{ox} = 0)$	$Fe(EDTA)^{2-}$	4.3×10^5 [b]	6.2×10^5	2.7×10^5
	$Co(phen)_3^{3+}$	1.8×10^5	7.1×10^5	3.1×10^5
Plastocyanin $(Z_{ox} = -9)$	$Fe(EDTA)^{2-}$	8.2×10^4 [c]	4.6×10^1	1.8×10^1
	$Co(phen)_3^{3+}$	4.9×10^3 [c]	2.5×10^5	2.6×10^3
	cyt c(II)	1×10^6	3×10^7	5×10^5
Azurin $(Z_{ox} = -1)$	$Fe(EDTA)^{2-}$	1.3×10^3 [c]	2.2×10^{-2}	1.1×10^{-2}
	$Co(phen)_3^{3+}$	3.2×10^3 [c]	4.9×10^4	1.7×10^4
	cyt c(III)	1.1×10^3	3×10^4	1×10^4
	cyt c 551(III)	6.1×10^6 [d]	1×10^9	1×10^9

[a] From Refs. *20, 21, 23, 26*.
[b] $\mu = 0.5M$.
[c] $\mu = 0.2M$.
[d] $\mu = 0.05M$.

The calculated self-exchange rate of $1.1 \times 10^{-2} M^{-1} sec^{-1}$ for azurin based on $Fe(EDTA)^{2-}$ is also notably lower than the value of 1×10^4 $M^{-1} sec^{-1}$ obtained (*21*) from the cross reaction with horse heart cytochrome c. Further, the azurin self-exchange rate based on the cross reaction with a possible physiological partner, cytochrome c 551, is calculated (Table IV) to be extremely high ($1 \times 10^9 M^{-1} sec^{-1}$). Thus, our analysis of azurin reactivity indicates that a large variation in protein–reagent interaction occurs in the transition state, with kinetic access to the blue copper center decreasing according to cytochrome c 551 > cytochrome c ~ $Co(phen)_3^{3+}$ > $Fe(EDTA)^{2-}$. As various physical studies have shown that the blue copper is sequestered in a hydrophobic region away from solvent (*27, 28, 29*), such a wide variation in electron transfer reactivity is not surprising. (A similarly large variation in calculated k_{11} values for *Chromatium vinosum* HiPIP has been taken as evidence that different pathways of electron transfer to the relatively buried redox center (the $Fe_4S_4S_4^*$ cluster) are used by different reagents (30).) It is probable that $Fe(EDTA)^{2-}$, which is least reactive, cannot penetrate the hydrophobic region and is forced to transfer an electron to copper over a distance that could be as great as three or four angstroms. With the cytochromes, especially cytochrome c 551, electron transfer is greatly facilitated, presumably by a pathway in which the heme edge and the copper center are brought into close proximity. The fact that cytochrome c 551 is the best electron transfer agent yet found for azurin supports the proposal (31) that these two proteins are physiological partners.

The electrostatics-corrected self-exchange rate for plastocyanin based on $Co(phen)_3^{3+}$ is $2.6 \times 10^3 \, M^{-1} \, sec^{-1}$. The k_{11}^{corr} value for plastocyanin based on the cytochrome c cross reaction, $5 \times 10^5 \, M^{-1} \, sec^{-1}$, is substantially smaller than the uncorrected value $(3 \times 10^7 \, M^{-1} \, sec^{-1})$. Taking either value, however, it is apparent that both cytochrome c and Co-$(phen)_3^{3+}$ are better electron transfer agents for plastocyanin than is $Fe(EDTA)^{2-}$.

Acknowledgments

We thank F. C. Anson and F. J. Grunthaner for helpful discussions. Research on blue copper proteins at the California Institute of Technology has been supported by the National Science Foundation. This paper is contribution no. 5366 from the Arthur Amos Noyes Laboratory of Chemical Physics.

Literature Cited

1. Holwerda, R. A., Wherland, S., Gray, H. B., *Annu. Rev. Biophys. Bioeng.* (1976) **5**, 363.
2. Milne, P. R., Wells, J. R. E., Ambler, R. P., *Biochem. J.* (1974) **143**, 691.
3. Sailasutá, N., Anson, F. C., Gray, H. B., unpublished data.
4. Ambler, R. R., Brown, L. H., *Biochem. J.* (1967) **104**, 784.
5. Yamanaka, T., in "The Biochemistry of Copper," J. Peisach and W. Blumberg, Eds., p. 275, Academic, New York, 1966.
6. Peisach, J., Levine, W. G., Blumberg, W. E., *J. Biol. Chem.* (1967) **242**, 2847.
7. Reinhammar, B., *Biochim. Biophys. Acta* (1972) **275**, 245.
8. Malkin, R., Malmström, B. G., *Adv. Enzymol.* (1970) **33**, 177.
9. Fee, J. A., *Struc. Bonding* (1975) **23**, 1.
10. McMillin, D. R., Holwerda, R. A., Gray, H. B., *Proc. Natl. Acad. Sci. USA* (1974) **71**, 1339.
11. McMillin, D. R., Rosenberg, R. C., Gray, H. B., *Proc. Natl. Acad. Sci. USA* (1974) **71**, 4760.
12. Solomon, E. I., Rawlings, J., McMillin, D. R., Stephens, P. J., Gray, H. B., *J. Am. Chem. Soc.* (1976) **98**, 8046.
13. Solomon, E. I., Hare, J. W., Gray, H. B., *Proc. Natl. Acad. Sci. USA* (1976) **73**, 1389.
14. Hare, J. W., Ph.D. Thesis, California Institute of Technology, 1976.

15. Solomon, E. I., Clendening, P. J., Gray, H. B., Grunthaner, F. J., *J. Am. Chem. Soc.* (1975) **97**, 3878.
16. Markley J. K., Ulrich, E. L., Berg, S. P., Krogmann, D. W., *Biochemistry* (1975) **14**, 4428.
17. Hare, J. W., Solomon, E. I., Gray, H. B., *J. Am. Chem. Soc.* (1976) **98**, 3205.
18. Wurzbach, J., Grunthaner, P. J., Dooley, D. M., Gray, H. B., Grunthaner, F. J., Gay, R. R., Solomon, E. I., *J. Am. Chem. Soc.* (1976) **99**, 1257.
19. Moscowitz, A., in "Optical Rotatory Dispersion," C. Djerassi, Ed., pp. 150–177, McGraw-Hill, New York, 1960.
20. Wherland, S., Holwerda, R. A., Rosenberg, R. C., Gray, H. B., *J. Am. Chem. Soc.* (1975) **97**, 5260.
21. Rosenberg, R. C., Wherland, S., Holwerda, R. A., Gray, H. B., *J. Am. Chem. Soc.* (1976) **98**, 6364.
22. Marcus, R. A., *J. Phys. Chem.* (1967) **67**, 853.
23. Wherland, S., Gray, H. B., unpublished results.
24. Hill, H. A. O., private communication.
25. Beattie, J. K., Fensom, D. J., Freeman, H. C., Woodcock, E., Hill, H. A. O., Stokes, A. M., *Biochim. Biophys. Acta* (1975) **405**, 109.
26. McArdle, J. V., Coyle, C. L., Gray, H. B., Yoneda, G. S., Holwerda, R. A., *J. Am. Chem. Soc.*, in press.
27. Boden, N., Holmes, M. C., Knowles, P. F., *Biochem. Biophys. Res. Commun.* (1974) **57**, 845.
28. Avigliano, L., Finazzi-Agro, A., Mondovi, B., *FEBS Lett.* (1974) **38**, 205.
29. Finazzi-Agro, A., Rotilio, G., Avigliano, L., Guerrieri, P., Boffi, V., Mondovi, B., *Biochemistry* (1970) **9**, 2009.
30. Rawlings, J., Wherland, S., Gray, H. B., *J. Am. Chem. Soc.* (1976) **98**, 2177.
31. Antonini, E., Finazzi-Agro, A., Avigliano, L., Guerrieri, P., Rotilio G., Mondovi, B., *J. Biol. Chem.* (1970) **245**, 4847.

RECEIVED July 26, 1976.

9

Electron Transfer Reactions of Cytochrome *c*

N. SUTIN

Brookhaven National Laboratory, Upton, NY 11973

Crystallographic studies have shown that the heme group of cytochrome c *lies in a crevice of the protein with an edge of the porphyrin ring located at the molecule surface. The cytochrome* c *self-exchange rate, as well as the rates of cytochrome* c *oxidation–reduction reactions with metal complexes and metalloproteins, are reviewed and interpreted in terms of a model in which electron transfer occurs through the exposed edge of the heme group. Comparison of* $\Delta H\ddagger$ *and* $\Delta S\ddagger$ *for metal complex–metal complex, metal complex–metalloprotein, and metalloprotein–metalloprotein reactions indicates that the metalloprotein reactions feature significantly different activation parameters. The use of the Marcus equations and corrections to these due to work terms and non-adiabaticity are discussed.*

Cytochrome *c*, a small heme protein (mol wt $\sim$ 12,400) is an important member of the mitochondrial respiratory chain. In this chain it assists in the transport of electrons from organic substrates to oxygen. In the course of this electron transport the iron atom of the cytochrome is alternately oxidized and reduced. Oxidation–reduction reactions are thus intimately related to the function of cytochrome *c*, and its electron transfer reactions have therefore been extensively studied. The reagents used to probe its redox activity range from hydrated electrons (*1, 2, 3*) and hydrogen atoms (*4*) to the complicated oxidase (*5, 6, 7, 8*) and reductase (*9, 10, 11*) systems. This chapter is concerned with the reactions of cytochrome *c* with transition metal complexes and metalloproteins and with the electron transfer mechanisms implicated by these studies.

Electron Exchange Reactions

The simplest electron transfer reaction that cytochrome *c* can undergo, at least in principle, is the self-exchange reaction. The rate of this reaction is compared with other electron exchange reactions in Table I. The electron exchange between ferrocytochrome *c* and ferricytochrome *c* is not particularly fast. For example, the cytochrome *c* electron exchange is much less rapid than electron exchange of the bipyridine or phenanthroline complexes of iron or ruthenium. Indeed the electron exchange rate of cytochrome *c* is very similar to that of the ruthenium hexaammine couple. The $\Delta H\ddagger$ for the cytochrome couple at ionic strength $0.1M$ (the "physiological" ionic strength) is somewhat more positive and the $\Delta S\ddagger$ less negative than for the hexaammine ruthenium exchange and other comparable systems. A question arises as to why the cytochrome *c* molecule is so complicated when one can obtain relatively rapid electron transfer with very simple metal complexes.

The structure of cytochrome *c* determined by Dickerson and his colleagues (*23, 24, 25, 26*) is depicted in Figure 1. The heme group, which lies in a crevice of the essentially globular protein, is covalently bonded to the protein by thioether bridges between the porphyrin ring and two cysteine residues in the peptide chain. The iron atom is situated

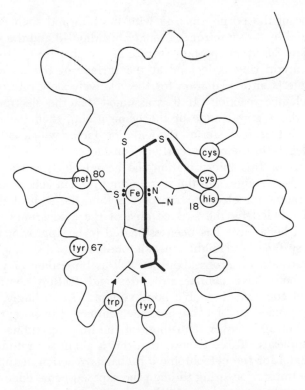

Figure 1.   Skeleton of horse heart ferricytochrome c.
Adapted from Ref. 26, Figure 8.

Table I.   Kinetic Parameters for

Reaction[a]	k $(M^{-1} sec^{-1})$
$V(H_2O)_6^{2+} + V(H_2O)_6^{3+}$	1×10^{-2}
$V(H_2O)_6^{2+} + V(H_2O)_6^{3+}$	3×10^{-3}
$Fe(H_2O)_6^{2+} + Fe(H_2O)_6^{3+}$	4.2
$Ru(NH_3)_6^{2+} + Ru(NH_3)_6^{3+}$	8.2×10^2
$Ru(NH_3)_6^{2+} + Ru(NH_3)_6^{3+}$	3.6×10^3
$Ru(NH_3)_5BzIm^{2+} + Ru(NH_3)_5BzIm^{3+}$	5×10^4
$Co(phen)_3^{2+} + Co(phen)_3^{3+}$	4×10^1
$Fe(EDTA)^{2-} + Fe(CyDTA)^{-}$	3.0×10^4
$Fe(CN)_6^{4-} + Fe(CN)_6^{3-}$	25
$Fe(CN)_6^{4-} + Fe(CN)_6^{3-}$	9.6×10^3
$Hh(II) + Hh(III)$	1.2×10^3
$Hh(II) + Hh(III)$	1×10^4
$Fe(phen)_3^{2+} + Fe(phen)_3^{3+}$	$\sim 10^7$
$Fe(phen)_3^{2+} + Fe(phen)_3^{3+}$	3×10^8
$Ru(phen)_3^{2+} + Ru(phen)_3^{3+}$	$\sim 10^7$

[a] Hh, cytochrome c from horse-heart.

in the plane of the porphyrin ring with the fifth and sixth coordination sites occupied by a ring nitrogen atom of histidine-18 and the sulfur atom of methionine-80. An important feature of the structure is that an edge of the porphyrin ring is located at the surface of the molecule. This feature suggests an explanation for the relatively slow rate of the cytochrome exchange reaction. If it is assumed that the electron exchange occurs via the exposed edge of the heme group, then the cytochrome reactions will feature a steric or orientation factor which is not present to such a degree in the other reactions considered.

In terms of this interpretation the actual electron transfer occurs through the heme edge, and the surrounding protein acts as an insulator. The steric factor for electron transport is then the area of the exposed heme edge divided by the surface area of the cytochrome c molecule. The exposed heme area has been estimated to be approximately 3% of the protein surface. We might expect a totally exposed symmetrical heme group to transport electrons about as well as bipyridine or phenanthroline, since both have similar structures and feature coordination to nitrogen and conjugated double bond systems. Accordingly, we expect the exchange reactions of horse-heart cytochrome c to be approximately $(3 \times 10^{-2})^2$ or 10^{-3} slower than those of the iron bipyridine or phenanthroline complexes. Table I shows that this is about the rate ratio found.

The model for the cytochrome c exchange reaction at this point has the electron transfer occuring through the exposed heme edge. The function of the protein is to provide the correct reduction potential of the metal center and also to provide the necessary specificity in terms of

Electron Exchange Reactions (25°C)

Medium	$\Delta H^{\ddagger}$ (kcal mol^{-1})	$\Delta S^{\ddagger}$ (e.u.)	References
μ 2.0	12.6	−25	12
μ 0.10	—	—	12, 13[b]
μ 0.55	9.3	−25	14
μ 0.013	10.3	−11	15
μ 0.10	—	—	13,[b] 15
—	—	—	33
μ 0.1	5.1	−34	16
—	4.0	−25	17
$\mu \rightarrow 0$	8.5	−24	18
μ 0.1M K$^+$	—	—	18
μ 0.1, pH 7	12.4	−3	19, 20
μ 1.0, pH 7	6.4	−19	20
$\mu \sim 0.2$	~ 2	~ -20	21
1.84M Na$_2$SO$_4$	—	—	22
$\mu \sim 0.2$	~ 2	~ -20	21

[b] Corrected for ionic strength.

structure and charge distribution so that the cytochrome *c* can recognize its natural reductase and oxidase.

The next question is whether the charge distribution in the vicinity of the exposed heme edge is consistent with the proposed electron transfer mechanism. While horse-heart cytochrome *c* carries no net charge at pH 10, it carries a net positive charge at neutral pH (*27*). The x-ray structure of horse-heart cytochrome *c* shows that most of the positively charged residues are located on the surface of the molecule (*26*). In particular, lysine residues 13, 27, and 79 are positioned so that they impart a net positive charge to the region of the protein in the vicinity of the exposed heme edge. Indeed, as Dickerson and Timkovich point out (*26*), all the cytochromes *c* whose three-dimensional structures are known have a ring of positive charge around the heme crevice. Studies of the effect of ionic strength on the electron transfer reactions of cytochrome *c* are consistent with this charge distribution. The results of these studies are plotted in Figure 2. The rate of reaction of cytochrome *c* with Co-(phen)$_3^{3+}$ (curve A) and its self-exchange reaction (curve D) both increase with increasing ionic strength. By contrast, the reaction rates of cytochrome *c* with the negatively charged reactants Fe(EDTA)$^{2-}$ (curve B) and Fe(CN)$_6^{3-}$ (curve C) decrease with increasing ionic strength. Interestingly enough, ionic strength seems to have only a very small effect on the reaction rate of horse-heart ferrocytochrome *c* with ferricytochrome *c* 551, a bacterial cytochrome from *Pseudomonas aeruginosa* (curve E). The latter cytochrome is believed to carry a net negative charge at neutral pH (*34*).

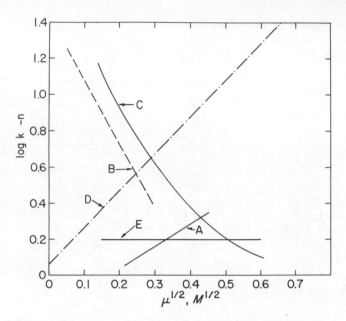

Figure 2. Ionic strength dependence of the electron transfer rate constants. (A) Co(phen)$_3^{3+}$ + Hh(II) (28); (B) Fe(EDTA)$^{2-}$ + Hh(III) (29); (C) Fe(CN)$_6^{3-}$ + Hh(II) (31) (see also (32)); (D) Hh(II) + Hh(III) (19); (E) Hh(II) + Ps(III) (31).

If it is assumed that the Brønsted-Bjerrum equation (Equation 1)

$$\log_{10}k = \log_{10}k_0 + z_A z_B \sqrt{\mu} \tag{1}$$

obtains in these systems (and it very probably does not) then formal charges of approximately 1+ and 2+ can be assigned to the active sites of reduced and oxidized cytochrome c, respectively (Table II). The magnitudes of these charges are not unreasonable and appear consistent with the structural data.

Table II. Slope of Brønsted–Bjerrum Plot for Some Horse-Heart Cytochrome c Reactions (25°C)

Reaction[a]	Slope	Charge Hh(II)	Hh(III)	Reference
Hh(II) + Co(phen)$_3^{3+}$	+1.2	+0.4		28
Fe(EDTA)$^{2-}$ + Hh(III)	−3.4		+1.7	29
Hh(II) + Hh(III)	+2.2	+1.0	+2.2	19
Hh(II) + Ps(III)	~0			31

[a] Hh, horse-heart cytochrome c; Ps, $Pseudomonas$ cytochrome c 551.

Cross-reactions

Next we turn to the interpretation of the rate constants for electron-transfer reactions of cytochrome c that are accompanied by a net chemical change (Tables III and IV). The rate constants for the reaction of cytochrome c with both negatively charged ($Fe(CN)_5L^{3-}$ and $*Ru[-(OSO_3\emptyset)_2phen]_3^{4-}$) and positively charged ($Fe(bipy)_2(CN)_2^+$ and $*Ru(bipy)_3^{2+}$) complexes can be very great.

Marcus has shown (44, 45) that a relatively simple relation exists between the rate constants for reactions accompanied by a net chemical change ($\Delta G° \neq 0$) and those for the component self-exchange reactions. Provided $\Delta G°$ is not too negative, this relation is:

$$k_{12} = (k_{11}k_{22}K_{12})^{1/2} \tag{2}$$

where k_{12} and K_{12} are the rate and equilibrium constants, respectively, for an electron transfer reaction accompanied by a net chemical change, and k_{11} and k_{22} are the corresponding exchange rate constants. An analogous relation is expected to hold for the activation enthalpies and entropies (46).

$$\Delta H_{12}\ddagger = 0.5(\Delta H_{11}\ddagger + \Delta H_{22}\ddagger + \Delta H_{12}°) \tag{3}$$

$$\Delta S_{12}\ddagger = 0.5(\Delta S_{11}\ddagger + \Delta S_{22}\ddagger + \Delta S_{12}°) \tag{4}$$

In deriving Equation 2 it is assumed that the work terms required to bring together the various pairs of reactants are cancelled. This may be a reasonable assumption when the cross-reaction involves similarly charged reactants but can result in a serious underestimate of the cross-reaction rate when oppositely charged reactants are involved. This is because the electrostatic interaction between the reactants in the cross-reaction is attractive whereas the interaction between the reactants is repulsive in the self-exchange reactions. The cross-reaction will, as a consequence, proceed faster than predicted by the simple square-root relation in which differences between the work terms are neglected. It is not always easy to correct for the differences in the electrostatic contributions to the work terms, although some progress is being made in this regard (47, 48). There are also nonelectrostatic contributions to the work terms, and neglect of these may result in an overestimate of the cross-reaction rate in certain systems (46). The latter situation arises when one exchange involves a hydrophilic pair of reactants and the other exchange reaction a hydrophobic pair of reactants. Under these conditions the cross-reaction features an unfavorable hydrophilic–hydrophobic interaction and will proceed more slowly than predicted by Equation 2.

Table III.   Observed and Calculated Rate
Involving Cytochrome c and

Reaction	$E°$ (mv)[b]
$Hh(II) + Fe(CN)_6^{3-}$	260,[c] 420[d]
$Hh(II) + Fe(CN)_5CNS^{3-}$ $Hh(II) + Fe(CN)_5N_3^{3-}$	260,[c] 240[c]
$Hh(II) + Fe(CN)_5PØ_3^{2-}$	260,[c] 540[c]
$Hh(II) + Fe(CN)_5NH_3^{2-}$	260,[c] 330[c]
$Hh(II) + Fe(bipy)(CN)_4^-$	260,[c] 550[c]
$Fe(EDTA)^{2-} + Hh(III)$	120,[g] 260[c]
$*Ru[(OSO_3\phi)_2phen]_3^{4-} + Hh(III)$	~ -900,[h] 260[c]

[a] At $0.1M$ ionic strength, pH 6.8–7.2, and 25°C.
[b] The $E°$ values are the reduction potentials for the two reactant couples.
[c] Ref. 37.
[d] Calculated from $K = 450$, Ref. 35.

Table IV.   Observed and Calculated Rate
Involving Cytochrome c and

Reaction	E_o (mv)
$*Ru(bipy)_3^{2+} + Hh(III)$ $Hh(II) + Fe(bipy)_2(CN)_2^+$	-830,[b] 260[c]
$Ru(NH_3)_6^{2+} + Hh(III)$	51,[d] 260[c]
$Ru(NH_3)_5BzIm^{2+} + Hh(III)$	150,[e] 260[c]
$Hh(II) + Co(phen)_3^{3+}$	260,[c] 370[c]

[a] At $0.1M$ ionic strength, pH 6.8–7.2, and 25°C.
[b] Ref. 40.
[c] Ref. 37. $\Delta H°$ and $\Delta S°$ for $Hh(III) + ½H_2 \rightleftharpoons Hh(II) + H^+$ are -14.4 kcal mol^{-1} and -28 e.u., respectively, at $0.1M$ ionic strength, pH 7.0, 25°C.
[d] Ref. 42.
[e] Ref. 41.
[f] The reductant is the charge-transfer excited state of the Ru(II) complex.

Constants for Electron Transfer Reactions
Negatively Charged Reactants[a]

	$k(25°)$ $(M^{-1} sec^{-1})$	$\Delta H\ddagger$ $(kcal\ mol^{-1})$	$\Delta S\ddagger$ $(e.u.)$	References
obsd.	1.2×10^7	~ 0	-26	31
obsd.	8.0×10^6	1.1	-24	32
calcd.	8×10^4			
obsd.	1.0×10^7	1.2	-23	32
obsd.	9.0×10^5	2.9	-22	32
calcd.	2×10^3			f
obsd.	3.0×10^7	1.2	-21	32
calcd.	2×10^6			f
obsd.	2.5×10^6	2.4	-22	32
calcd.	4×10^4			f
obsd.	1.6×10^8			32
calcd.	3×10^7			f
obsd.	2.6×10^4	6.0	-18	29
calcd.	7×10^4			
obsd.	1.2×10^9	4.4	$+2$	33[h]

[e] Ref. 36.
[f] Exchange rate data from Ref. 36.
[g] Ref. 38.
[h] The reductant is the charge-transfer excited state of the Ru(II) complex.

Constants for Electron Transfer Reactions
Positively Charged Reactants[a]

	$k(25°)$ $(M^{-1} sec^{-1})$	$\Delta H\ddagger$ $(kcal\ mol^{-1})$	$\Delta S\ddagger$ $(e.u.)$	References
obsd.	$< 2 \times 10^8$			33[i]
obsd.	1.9×10^8			32
obsd.	3.8×10^4	2.9	-28	39
calcd.	1.2×10^5	7	-14	g
obsd.	5.8×10^4			58
calcd.	6.3×10^4			58
obsd.	1.5×10^3	11.3	-6	28
calcd.	2.0×10^3	11.1	-6	h

[g] Using $\Delta H°$ and $\Delta S°$ from Refs. 37 and 43. $\Delta H°$ and $\Delta S°$ for $Ru(NH_3)_6^{3+} + \frac{1}{2}H_2 \rightleftharpoons Ru(NH_3)_6^{2+} + H^+$ are -5.8 kcal mol^{-1} and -14.2 e.u., respectively, at 0.5–1.0M ionic strength, 25°C.

[h] Using $\Delta H°$ and $\Delta S°$ of 4.66 kcal mol^{-1} and $+24.4$ e.u., respectively, for the $Hh(II) + Co(phen)_3^{3+} \rightleftharpoons Hh(III) + Co(phen)_3^{2+}$ reaction in 0.05M phosphate, pH 6.8, 25°C, Ref. 41.

[i] Ref. 58.

A second assumption made in the derivation of Equation 2 is that all of the reactions involved in the rate comparisons are adiabatic, that is, that the electron transfer occurs with unit probability once the reactants have attained the appropriate nuclear configurations. As has been discussed previously (44, 45, 46, 49), the actual electron transfer occurs in the intersection region of the reactants' potential energy curve with the products' potential energy curve. Essentially what is assumed is that the splitting in the intersection region is large enough so that the system remains on the lower potential energy surface on passing through the intersection region.

The probability that the system will remain on this lower potential energy surface can be calculated from the Landau–Zener formula (Equation 5). In this expression, $2E_{I,II}$ is the splitting at the intersection,

$$p = 1 - \exp\left[\frac{-4\pi^2 E_{I,II}{}^2}{hv|s_I - s_{II}|}\right] \tag{5}$$

s_I and s_{II} are the slopes of the zero-order potential energy surfaces at the intersection ($s_I = -s_{II}$ for an exchange reaction), and v is the velocity with which the point representing the system moves through the intersection region. For typical conditions it is found that $p \sim 1$ for interactions $E_{I,II}$ of more than 0.5 kcal mol^{-1} (50). Under these conditions the reactions will be adiabatic, and the square root relation is expected to hold provided $E_{I,II}$ is not too large. However, for small $E_{I,II}$:

$$p = \frac{4\pi^2 E_{I,II}{}^2}{hv|s_I - s_{II}|} \tag{6}$$

and Equation 2 will as a consequence also be satisfied by those nonadiabatic reactions (small $E_{I,II}$) for which the interaction energy in the cross-reaction is approximately equal to the geometric mean of the interaction energies in the corresponding exchange reactions (Equations 7, 8).

$$(E_{I,II})_{12} = [(E_{I,II})_{11}(E_{I,II})_{22}]^{1/2} \tag{7}$$

$$p_{12} = (p_{11}p_{22})^{1/2} \tag{8}$$

In other words, the square-root relation may also hold for certain classes of nondiabatic reactions (51). Conversely, the fact that a series of reactions obeys the square-root relation does not require that all of the reactions involved be adiabatic.

Conformity with Equation 2 also does not rule out more complex mechanisms (28). This may be illustrated by considering the oxidation of ferrocytochrome c by Co(phen)$_3{}^{3+}$. Let us assume that the reactive form of ferrocytochrome c is not the native protein but is instead a conformer that exists in rapid equilibrium with the native form.

$$Hh(II) \rightleftharpoons Hh(II)^* \qquad\qquad K_{II}$$

$$Hh(III) + Hh(II)^* \rightleftharpoons Hh(II)^* + Hh(III) \qquad k_{11}^*$$

$$Co(phen)_3^{3+} + Hh(II)^* \rightleftharpoons Co(phen)_3^{2+} + Hh(III) \qquad k_{12}^*, K_{12}^*$$

In terms of the above scheme, the observed rate constant for the oxidation of ferrocytochrome by $Co(phen)_3^{3+}$ is given by:

$$
\begin{aligned}
k_{12} &= K_{II} k_{12}^* \\
&= K_{II}(k_{11}^* k_{22} K_{12}^*)^{1/2} \\
&= K_{II}\left(\frac{k_{11}}{K_{II}} k_{22} \frac{K_{12}}{K_{II}}\right)^{1/2} \\
&= (k_{11} k_{22} K_{12})^{1/2}
\end{aligned}
$$

where k_{11} and K_{12} are the observed exchange rate constant and equilibrium constant, respectively. Evidently the pre-equilibrium constant cancels, and the normal square-root relation obtains. Agreement with Equation 2 thus does not preclude a rapid conformational change on ferrocytochrome c (and/or ferricytochrome c) prior to the electron transfer step.

With these reservations in mind, we can compare observed and calculated rate constants. Where appropriate, the full Marcus expression, which includes the log f term (Equation 9), has been used in the calculation of the electron transfer rates.

$$\log f_{12} = \frac{(\log K_{12})^2}{4 \log (k_{11} k_{22}/Z^2)} \tag{9}$$

Table III shows that the calculated rates are 10^2–10^3 times faster than the observed rates for the complexes which carry a relatively high negative charge ($Fe(CN)_6^{3-}$ and $Fe(CN)_5N_3^{3-}$) and that the agreement between the observed and calculated rates improves as the charge on the cyanoiron(III) complex decreases. This trend suggests that the enhanced rates for these complexes are caused by electrostatic effects, and indeed there is good evidence that ferricyanide is strongly bound to cytochrome c (52, 53, 54, 55), presumably in the vicinity of the heme group (53, 54, 55). Also, the calculated rate constant for the $Fe(EDTA)^{2-}$–Hh(III) reaction is larger than the observed value. This may be explained by postulating that, for this reaction, the nonelectrostatic contributions outweigh the electrostatic contributions to the work terms. The effect is, however, small.

Noncancellation of the electrostatic work terms is presumably less important for the reactions of cytochrome c with the positively charged complexes shown in Table IV. Indeed it is apparent from Table IV that there is excellent agreement between the observed and calculated rate constants and activation parameters for the $Ru(NH_3)_5BzIm^{2+}$–Hh(III) and for the Hh(II)–Co(phen)$_3^{3+}$ reactions. This result is not unexpected in view of the fact that, of the reactants considered, $Ru(NH_3)_5BzIm^{2+}$ and Co(phen)$_3^{3+}$ are both positively charged and, at least to some extent, hydrophobic. These properties should favor cancellation of the electrostatic as well as of the nonelectrostatic contributions to the work terms for the cross-reaction since the heme group of cytochrome c is hydrophobic, and the vicinity of its exposed edge is positively charged. Conversely, the excellent agreement of the observed and calculated rates is strong evidence that the Hh(II)–Hh(III) exchange, like these two cross-reactions, features electron transfer through the exposed heme edge. Moreover, good overlap of the porphyrin with the benzimidazole and phenanthroline π-systems is likely, and this should ensure that the cross-reactions will be adiabatic (or at least that $p_{12} = (p_{11}p_{22})^{1/2}$).

Kinetic data for electron transfer between two metalloproteins are presented in Table V. The rate constants and activation parameters for the Ps(II)–Ps(III) and Az(I)–Az(II) exchange reactions were calculated from the kinetic data for the first three reactions (for which $K \simeq 1$, $\Delta H° \simeq 0$, $\Delta S° \simeq 0$; in addition, the rate constant for the Hh(II)–Ps(III) reaction is independent of ionic strength (31)). The calculated exchange data were then used to predict the kinetic parameters for the Ps(II)–Az(II) reaction. As is evident from Table V, the agreement of the observed and predicted parameters is satisfactory, particularly since the Ps(II)–Az(II) reaction has a relatively complex mechanism (57) involving conformational changes on both Ps(III) and Az(I).

Table V. Observed and Calculated Rate Constants for Electron Transfer Reactions betwen Two Metalloproteins [a]

Reaction		$k(25°)$ $(M^{-1} sec^{-1})$	$\Delta H^{\ddagger}$ $(kcal\ mol^{-1})$	$\Delta S^{\ddagger}$ $(e.u.)$	References
Hh(II) + Hh(III)	obsd.	1.2×10^3	12.4	−3	19, 20
Hh(II) + Ps(III)	obsd.	7.9×10^4	12	0	31
Az(I) + Hh(III)	obsd.	1.7×10^3	13.4	1	56
Ps(II) + Ps(III)	calcd.	5.2×10^6	11.6	3	
Az(I) + Az(II)	calcd.	2.4×10^3	14.4	5	
Ps(II) + Az(II)	calcd.	1.1×10^5	13.0	4	
Ps(II) + Az(II)	obsd.	6.1×10^6	7.8	−1	57

[a] At pH 6.8–7.2 and 25°C. The symbols Hh, Az, and Ps denote horse-heart cytochrome c, azurin, and *Pseudomonas* cytochrome c 551, respectively.

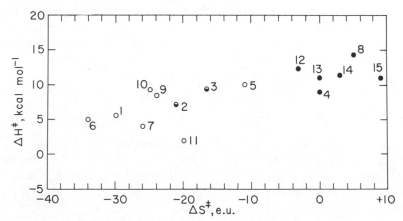

Figure 3. Plot of $\Delta H\ddagger_{corr.}$ *vs.* $\Delta S\ddagger_{corr.}$ *for electron transfer reactions.*
(○), metal complex–metal complex reactions; (◑), metal complex–metalloprotein reactions; (●), metalloprotein–metalloprotein reactions.
(1) $Ru(NH_3)_6^{2+} + Co(phen)_3^{3+}$ *(58); (2)* $Ru(NH_3)_6^{2+} + Hh(III)$ *(39);*
(3) $Hh(II) + Co(phen)_3^{3+}$ *(29, 30); (4) Ferrocyt* $c_1 + Hh(III)$ *(9, 10);*
(5) $Ru(NH_3)_6^{2+} + Ru(NH_3)_6^{3+}$ *(15); (6)* $Co(phen)_3^{2+} + Co(phen)_3^{3+}$ *(16);*
(7) $Fe(EDTA)^{2-} + Fe(EDTA)^-$ *(17); (8) Az(I) + Az(II), this work; (9)*
$Fe(CN)_6^{4-} + Fe(CN)_6^{3-}$ *(18); (10)* $Fe(H_2O)_6^{2+} + Fe(H_2O)_6^{3+}$ *(14); (11)*
$Fe(phen)_3^{2+} + Fe(phen)_3^{3+}$ *(21); (12) Hh(II) + Hh(III) (19); (13) Hh(II)*
+ Ps(III) (31); (14) Ps(II) + Ps(III) this work; (15) Ps(II) + Az(II) (57).

The comparatively high activation enthalpies and especially the relatively positive activation entropies for the metalloprotein electron transfer reactions (Table V) are noteworthy. This feature is further illustrated in Figure 3. The $\Delta H\ddagger$ and $\Delta S\ddagger$ values for the cross-reactions included in the plot have been corrected by $\Delta H°/2$ and $\Delta S°/2$; very exothermic reactions have not been included. In addition, in an attempt to exclude nonadiabatic processes, only relatively rapid reactions were included. Rather arbitrarily, rapid was taken to mean $k > 1\ M^{-1}\ sec^{-1}$. After these corrections, the plot shown in Figure 3 gives the relation between the intrinsic electron transfer barriers for three different types of reaction. From the plot it is evident that the metal complex–metal complex reactions feature very negative values of $\Delta S\ddagger$ and moderately low values of $\Delta H\ddagger$. By contrast, the metalloprotein–metalloprotein reactions are associated with much more positive $\Delta S\ddagger$ values and relatively high $\Delta H\ddagger$ values while the metal complex–metalloprotein reactions have activation parameters which are intermediate in character. A similar trend has also been noted for electron transfer reactions involving some copper proteins (57). This pattern may result from the more extensive hydration of the protein molecules; it is likely that in the protein–protein reactions, dehydration accompanies formation of the precursor complex. In addition, the degrees of hydration of the oxidized and reduced forms of a metalloprotein may differ significantly so that water rearrangement might also play an important role in the activation process. These factors could result in making the overall $\Delta S\ddagger$ for the metalloprotein–metalloprotein reactions more positive than the $\Delta S\ddagger$ values for the reactions involving the less extensively hydrated metal complexes (57).

Cytochrome c, Dead or Alive?

The above discussion has focused attention on electron transfer through the exposed edge of the heme group. This model is suggested from experience with small metal complexes and is substantiated by the rate constants and ionic strength dependence of the cytochrome c cross-reactions.

The activation parameters for the cytochrome c reactions require, however, that this view be enlarged by considering the differing extents of solvent interaction for metal complexes and metalloproteins. In addition, cytochrome c, like other metalloproteins, differs from the small metal complexes in the number of tertiary structural arrangements— conformations—which are thermally accessible to it. In this respect cytochrome c is much more alive (dynamic) than even so large a small metal complex as $Fe(phen)_3^{2+}$ or a heme molecule. This vitality is conferred on the metal center by the protein sheath, and the two regions of the metalloprotein are coupled. Conformational changes may result in or result from changes in the ligands bound to the iron.

Evidence for the dynamic nature of ferricytochrome c is provided by anation (53, 54, 55) and Cr(II) reduction (60, 61) studies. The anation studies provide good evidence that cytochrome c can undergo a crevice-opening conformational change in which the iron–methionine sulfur bond is broken. Crevice opening (adjacent attack) has also been implicated in the Cr(II)–Hh(III) reaction in chloride media (60). In the adjacent attack mechanism, electron transfer from the Cr(II) to the Fe(III) is postulated to occur via a chloride bridge. This interpretation is consistent with the location of the chromium in the Cr(III)–Hh(II) product formed at pH 4. Grimes et al. (62) have shown that the chromium cross-links residues 40 to 53 and residues 61 to 73, probably binding at asparagine 52 and tyrosine 67, both of which form part of the crevice region. Further evidence that the crevice-opened conformer is more reactive than native ferricytochrome c is provided by the kinetics of the reduction of carboxymethylated cytochrome c by Cr(II). Although the kinetics of this reaction are complex, the modified cytochrome c (in which the iron–sulfur bond has been broken) is about an order of magnitude more reactive than is the native form (63).

Cytochrome c undergoes conformational changes with changing pH (*64, 65, 66*). The reactivity changes associated with these conformational changes are quite complex. The species formed by the addition of one heme-linked proton to native ferricytochrome c is more reactive than the native form toward reduction by chromium(II) ($k \geqslant 4 \times 10^5 \, M^{-1}$ sec^{-1} compared with $3 \times 10^3 \, M^{-1}$ sec^{-1} for reduction of native cytochrome c (*60*)). Interestingly, the ferri- and ferrocytochrome c species formed at pH $\sim$ 4 are also more reactive than native cytochrome c toward outer-sphere (Ru(NH$_3$)$_6^{2+}$ (*67*) and Co(phen)$_3^{3+}$ (*30*)) electron transfer agents. By contrast, the ferricytochrome c species formed at low (< 2.5) and high (> 9) pH are relatively unreactive (*28, 30, 61*). The rate constants for the reaction of the native and high pH forms of Hh(III) with Fe(EDTA)$^{2-}$ are 2.6×10^4 and $2.7 \times 10^1 \, M^{-1}$ sec^{-1}, respectively (*29*). Although it is generally believed that in the high pH form of ferricytochrome c the heme iron is bound to a nitrogen atom of lysine-79 rather than to the methionine-80 sulfur (*65*), replacement of the coordinated sulfur by hydroxide ion is another possibility (*68*).

The pH dependence of cytochrome c oxidation–reduction reactions and the studies of modified cytochrome c thus demonstrate that the coordination environment of the iron and the conformation of the protein are relatively labile and strongly influence the reactivity of the metalloprotein toward oxidation and reduction. The effects seen may originate chiefly from alterations in the thermodynamic barriers to electron transfer, but the conformation changes are expected to affect the intrinsic barriers also. One such conformation change is the opening of the heme crevice referred to above. The anation and Cr(II) reduction studies provide an estimate of $\sim$ 60 sec^{-1} for this process in Hh(III) at 25°C (*59*). To date, no evidence has been found for a rapid heme-crevice opening step in ferrocytochrome c.

Even in the absence of adjacent attack, the heme-crevice opening step could be critical to electron transfer in another manner. Crevice opening and iron–sulfur bond rupture may trigger conformation changes (possibly coupled to the movement of the iron out of the plane of the porphyrin ring system) that render the Hh(III) more susceptible to reduction (*69*). Although the spontaneous crevice-opening reaction may be too slow to be involved in electron transport in respiring mitochondria, a process of this kind could still be involved in vivo since binding of Hh(III) to its reductase could accelerate the crevice-opening conformation change. Other, more subtle, conformation changes are, of course, also possible. As discussed earlier, agreement with the Marcus theory does not necessarily preclude such conformation changes, which may occur before, after, or in conjunction with the electron transfer process.

Electron Hopping and Porphyrin Cation Radical Mechanisms

Two other mechanisms that have been proposed for cytochrome c reactions are noteworthy. Dickerson et al. (*70*) have suggested a cytochrome reduction mechanism in which the electron hops from one aromatic residue to the next. Specifically, they have proposed that transfer of the reducing electron occurs from tyrosine-74 to tryptophan-59, to tyrosine-67 after a conformation change, to tetrapyrrole ring 4 of the heme, and finally to the iron. Other pathways have also been proposed (*62, 71*). The electron-hopping mechanism has now been abandoned by its original proponents (*26*). It first ran into difficulties over the relatively large amount of energy required to add an electron to or to remove one from a tyrosine or phenylalanine side chain. The fatal blow was the observation that the nonaromatic residue leucine occupies the position analogous to tyrosine-74 in cytochrome c 550 (*72*).

Another mechanism that has been proposed involves the formation of a porphyrin π-cation radical (Equation 10) (*73, 74*). Objections to

$$\mathrm{Fe^{III}\ cyt} \rightleftharpoons \mathrm{Fe^{II}\ cyt^+} \overset{+\,e}{\underset{-\,e}{\rightleftharpoons}} \mathrm{Fe^{II}\ cyt} \qquad (10)$$

this mechanism include the fact that, in the absence of π-acceptor axial ligands, the porphyrin is more difficult to oxidize or to reduce than is the metal center (*75*). Moreover, the data in Table I show that the bipyridine and phenanthroline complexes of iron, as well as those of ruthenium and osmium, can undergo rapid electron exchange without the intermediate oxidation or reduction of the ligands. The rapid electron transfer in these metal complexes is in part a consequence of the mixing of the metal t_{2g} orbitals with the π^* orbitals of the aromatic ligands (*75*). That such delocalization occurs in ferricytochrome c also is established by NMR studies which show the presence of appreciable unpaired electron density on the heme periphery (*76, 77*).

Electron and nuclear tunneling remain as other possibilities. However, the above discussion has shown that the exposed edge of the heme group provides an accessible and relatively low energy pathway for electron transfer to and from the iron atom of cytochrome c. Although the accessibility and/or the reactivity of the heme may be improved in certain situations by spontaneous or induced conformation changes of the protein, the heme group still has the edge over the other electron transfer mechanisms that have been proposed.

Acknowledgments

The author is indebted to Carol Creutz for many helpful discussions on cytochrome c and related reactions. This work was performed under the auspices of the U.S. Energy Research and Development Administration.

Literature Cited

1. Land, E. J., Swallow, A. J., *Biochim. Biophys. Acta* (1974) **368**, 86.
2. Lichtin, N. N., Shafferman, A., Stein, G., *Biochim. Biophys. Acta* (1973) **314**, 117.
3. Wilting, J., Van Buuren, K. J. H., Braams, R., Van Gelder, B. F., *Biochim. Biophys. Acta* (1975) **376**, 285.
4. Lichtin, N. N., Shafferman, A., Stein, G., *Biochim. Biophys. Acta* (1974) **357**, 386.
5. Nicholls, P., *Biochim. Biophys. Acta* (1974) **346**, 261.
6. *Ibid.* (1976) **430**, 13.
7. *Ibid.* (1976) **430**, 30.
8. Yu, C. A., Yu, L., King, T. E., *J. Biol. Chem.* (1975) **250**, 1383.
9. *Ibid.* (1973) **248**, 528.
10. *Ibid.* (1973) **248**, 3366.
11. Smith, L., Davies, H. C., Nava, M., *J. Biol. Chem.* (1974) **249**, 2904.
12. Krishnamurty, K. V., Wahl, A. C., *J. Am. Chem. Soc.* (1958) **80**, 5921.
13. Jacks, C. A., Bennett, L. E., *Inorg. Chem.* (1974) **13**, 2035.
14. Silverman, J., Dodson, R. W., *J. Phys. Chem.* (1952) **56**, 846.
15. Meyer, T. J., Taube, H., *Inorg. Chem.* (1968) **7**, 2369.
16. Neumann, H. M., quoted in Farina, R., Wilkins, R. G., *Inorg. Chem.* (1968) **7**, 514.
17. Wilkins, R. E., Yelin, R. E., *Inorg. Chem.* (1968) **7**, 2667.
18. Campion, R. J., Deck, C. F., King, P., Jr., Wahl, A. C., *Inorg. Chem.* (1967) **6**, 672.
19. Gupta, R. K., Koenig, S. H., Redfield, A. G., *J. Mag. Res.* (1972) **7**, 66.
20. Gupta, R. K., *Biochim. Biophys. Acta* (1973) **292**, 291.
21. Chan, M., Wahl, A. C., "Book of Abstracts" 167th Meeting, ACS, Los Angeles (1974) Inorg. 97.
22. Ruff, I., Zimonyi, M., *Electrochim. Acta* (1973) **18**, 515.
23. Dickerson, R. E., Takano, T., Eisenberg, D., Kallai, O. B., Samson, L., Cooper, A., Margoliash, E., *J. Biol. Chem.* (1971) **246**, 1511.
24. Swanson, R., Trus, B. L., Mandel, N., Mandel, G., Kallai, O. B., Dickerson, R. E., *J. Biol. Chem.* (1977) **252**, 759.
25. Takano, T., Trus, B. L., Mandel, N., Mandel, G., Kallai, I. B., Swanson, R., Dickerson, R. E., *J. Biol. Chem.* (1977) **252**, 776.
26. Dickerson, R. E., Timkovich, R., in "The Enzymes," P. B. Boyer, Ed., 3rd ed., Vol. XIA, p. 397, Academic, New York, 1975.
27. Barlow, G. H., Margoliash, E., *J. Biol. Chem.* (1966) **241**, 1473.
28. McArdle, J. V., Gray, H. B., Creutz, C., Sutin, N., *J. Am. Chem. Soc.* (1974) **96**, 5737.
29. Hodges, H. L., Holwerda, R. A., Gray, H. B., *J. Am. Chem. Soc.* (1974) **96**, 3132.
30. Brunschwig, B. S., Sutin, N., *Inorg. Chem.* (1976) **15**, 631.
31. Morton, R. A., Overnell, J., Harbury, H. A., *J. Biol. Chem.* (1970) **245**, 4653.
32. Cassatt, J. C., Marini, C. P., *Biochemistry* (1974) **13**, 5323.
33. Creutz, C., unpublished data.
34. Ambler, R. P., *Biochem. J.* (1963) **89**, 341.
35. Brandt, K. G., Parks, P. C., Czerlinski, G. H., Hess, G. P., *J. Biol. Chem.* (1966) **241**, 4180.
36. Stasiw, R., Wilkins, R. G., *Inorg. Chem.* (1969) **8**, 156.
37. Margalit, R., Schejter, A., *Eur. J. Biochem.* (1973) **32**, 492.

38. Schwarzenbach, G., Heller, J., *Helv. Chim. Acta* (1951) **34**, 576.
39. Ewall, R. X., Bennett, L. E., *J. Am. Chem. Soc.* (1974) **96**, 940.
40. Navon, G., Sutin, N., *Inorg. Chem.* (1974) **13**, 2159.
41. Ciana, A., Crescenzi, V., unpublished data.
42. Lim, H. S., Barclay, D. J., Anson, F. C., *Inorg. Chem.* (1972) **11**, 1460.
43. Lavallee, D. K., Lavallee, C., Sullivan, J. C., Deutsch, E., *Inorg. Chem.* (1973) **12**, 570.
44. Marcus, R. A., *J. Chem. Phys.* (1965) **43**, 679.
45. Marcus, R. A., *Annu. Rev. Phys. Chem.* (1964) **15**, 155.
46. Marcus, R. A., Sutin, N., *Inorg. Chem.* (1975) **14**, 213.
47. Haim, A., Sutin, N., *Inorg. Chem.* (1976) **15**, 476.
48. Wherland, S., Gray, H. B., *Proc. Natl. Acad. Sci. U.S.A.* (1976) **73**, 2950.
49. Sutin, N., *Annu. Rev. Nucl. Sci.* (1962) **12**, 285.
50. Sutin, N., in "Inorganic Biochemistry," G. L. Eichhorn, Ed., Vol. 2, p. 611, American Elsevier, New York, 1973.
51. Sutin, N., *Acc. Chem. Res.* (1968) **1**, 225.
52. Stellwagen, E., Shulman, R. G., *J. Mol. Biol.* (1973) **80**, 559.
53. Stellwagen, E., Cass, R. D., *J. Biol. Chem.* (1975) **250**, 2095.
54. Power, S. D., Choucair, A., Palmer, G., *Biochem. Biophys. Res. Commun.* (1975) **66**, 103.
55. Miller, W. G., Cusanovich, M. A., *Biophys. Struct. Mech.* (1975) **1**, 97.
56. Greenwood, C., Finazzi Agro, A., Guerrieri, P., Avigliano, L., Mondovi, B., Antonini, E., *Eur. J. Biochem.* (1971) **23**, 321.
57. Rosen, P., Pecht, I., *Biochemistry* (1976) **15**, 775.
58. Creutz, C., Sutin, N., *J. Biol. Chem.* (1974) **249**, 6788.
59. Sutin, N., Yandell, J. K., *J. Biol. Chem.* (1972) **247**, 6932.
60. Yandell, J. K., Fay, D. P., Sutin, N., *J. Am. Chem. Soc.* (1973) **95**, 1131.
61. Przystas, T. J., Sutin, N., *Inorg. Chem.* (1975) **14**, 2103.
62. Grimes, C. J., Piezkiewicz, D., Fleischer, E. B., *Proc. Natl. Acad. Sci. U.S.A.* (1974) **71**, 1408.
63. Brittain, T., Wilson, M. T., Greenwood, C., *Biochem. J.* (1974) **141**, 455.
64. Fung, D., Vinogradov, S. N., *Biochem. Biophys. Res. Commun.* (1968) **31**, 596.
65. Gupta, R. K., Koenig, S. H., *Biochem. Biophys. Res. Commun.* (1971) **45**, 1134.
66. Lanir, A., Aviram, I., *Arch. Biochem. Biophys.* (1975) **166**, 439.
67. Adegite, A., Sutin, N., unpublished data.
68. Pettigrew, G. W., Aviram, I., Schejter, A., *Biochem. Biophys. Res. Commun.* (1976) **68**, 807.
69. Creutz, C., Sutin, N., *Proc. Natl. Acad. Sci. U.S.A.* (1973) **70**, 1701.
70. Dickerson, R. E., Takano, T., Kallai, O. B., Samson, L., in "Structure and Function of Oxidation Reduction Enzymes," A. Akerson, A. Ehrenberg, Eds., p. 69, Pergamon, Oxford, 1972.
71. Myer, Y. P., *Biochemistry* (1972) **11**, 4195.
72. Timkovich, R., Dickerson, R. E., Margoliash, E., quoted in "The Enzymes," P. B. Boyer, Ed., 3rd ed., Vol. XIA, p. 397, Academic, New York, 1975.
73. Dolphin, D., Felton, R. H., *Acc. Chem. Res.* (1974) **7**, 26.
74. Dolphin, D., Niem, T., Felton, R. H., Fujita, I., *J. Am. Chem. Soc.* (1975) **97**, 5288.
75. Brown, G. M., Hopf, F. R., Meyer, T. J., Whitten, D. G., *J. Am. Chem. Soc.* (1975) **97**, 5385.
76. Redfield, A. G., Gupta, R. K., *Cold Spring Harbor Symp. Quant. Biol.* (1971) **36**, 405.
77. Wuthrich, K., *Proc. Natl. Acad. Sci. U.S.A.* (1969) **63**, 1071.

RECEIVED July 26, 1976.

The Mechanism of Oxygen Reduction in Oxidases—Studies with Laccase and Cytochrome Oxidase

BO G. MALMSTRÖM

Department of Biochemistry, University of Göteborg and Chalmers Institute of Technology, Fack, S-402 20 Göteborg 5, Sweden

The role of copper and iron in the enzymic reduction of dioxygen has been studied with laccase and cytochrome c oxidase. Both contain four metal ions in their functional units. Anaerobically each oxidase molecule can accept 1 e^-/ metal ion from reducing substrates, so the metals are in the Cu(II) and Fe(III) states. Only two metal ions per molecule are detectable by EPR, however. The EPR-nondetectable ions form an antiferromagnetically coupled binuclear complex, suggested to reduce oxygen directly to H_2O_2, thereby by-passing the energetically unfavorable formation of O_2^-. The presumed H_2O_2 intermediate appears to decompose by 1-e^- steps, as a paramagnetic intermediate is formed with laccase. With ^{17}O this has been shown to represent an oxygen radical, possibly O^-, having unusual relaxation properties.

Energy in most living cells is provided by combustion of organic molecules through complicated oxidative pathways in which dioxygen is the terminal acceptor of electrons. At temperatures where life exists, dioxygen is a rather inert substance, however, and cell respiration consequently requires specific catalysts. The key enzyme is undoubtedly cytochrome *c* oxidase (*see* Refs. 1 and 2 for recent reviews), which has been estimated to be responsible for more than 90% of the oxygen consumption in the biosphere. This oxidase utilizes the full oxidizing capacity of oxygen, which is consequently reduced to two molecules of water in a four-equivalent process: $O_2 + 4e^- + 4H^+ \rightarrow 2H_2O$. The mechanism of the enzymic catalysis of this reaction provides one of the major problems of bioinorganic chemistry.

Unfortunately cytochrome oxidase is an insoluble enzyme associated with many other respiratory catalysts in subcellular organelles called mitochondria. This makes it difficult to obtain in a form amenable to rigorous physico-chemical studies by spectroscopic and kinetic techniques. There are, however, a few water-soluble enzymes which catalyze the same reaction but which have the advantage that they are easily obtained in pure form (*see* Ref. 3 for a review). The most studied one is undoubtedly laccase, prepared either from the tree *Rhus vernicifera* or from the fungus *Polyporus versicolor*. This chapter will review recent work with laccase providing some detailed understanding of the catalytic mechanism of oxygen reduction. With cytochrome oxidase it is more difficult to provide data allowing a unique interpretation, but some results suggesting mechanistic analogies between this oxidase and laccase will be presented.

The Idea of Multi-Electron Transfer

With both cytochrome oxidase and laccase the oxidation of the reducing substrate involves a 1-e$^-$ transfer to a high valence form of a metal in the enzyme (*1, 2*). It would seem a natural hypothesis that the reduced metal ion is then reoxidized by oxygen in 1-e$^-$ reaction, thus forming the hyperoxide radical O_2^-, generally called superoxide by biochemists. As pointed out by George (*4*) more than 10 years ago, such a mechanism would, however, be expected to be slow with oxidases characterized by a high reduction potential. For example, the primary electron acceptor in fungal laccase (the so-called type 1 Cu^{2+}) has a standard reduction potential of about 780 mV (*5*). It can then be estimated that a reoxidation mechanism involving hyperoxide formation would have a minimum activation energy of 106 kJ mol^{-1} (*3*), and the reaction would thus be extremely slow.

At the same time that these difficulties were first pointed out, several investigators noted that both cytochrome oxidase and laccase contain four metal ions in their functional units, and they suggested that these metal ions in their reduced form donate 4 e$^-$ to oxygen in a cooperative process (*see* Ref. 6 for a review). Work in my own laboratory soon excluded this, however, in the case of laccase, as the metals are present in distinct chemical environments in an asymmetric structure (*6*). Cytochrome oxidase has generally also been considered asymmetric, and despite recent doubts (*7*) this concept now appears well established (*2*). I suggested consequently that oxygen reduction instead involves consecutive 2-e$^-$ transfers. This idea received some support from simple kinetic experiments (*8, 9*), but at the time of these studies the interpretation was challenged in the case of cytochrome oxidase by other investigators favoring four separate 1-e⁻ steps (*10*). In this paper I will summarize evidence indicating that the energetically unfavorable formation of hyperoxide is by-passed by a two-equivalent reduction but that the intermediate so formed is further reduced in 1-e⁻ processes.

The EPR-Nondetectable Metal Ions

It has been shown that both laccase (*11*) and cytochrome oxidase (*12*) can, under anaerobic conditions, accept 1 e⁻/metal ion from reducing substrates, so the metals are expected to be in the $Cu(II)$ and $Fe(III)$ states, respectively. In both cases all four metal ions should thus be paramagnetic and detectable by EPR. Despite this, it is well established with laccase that only two Cu^{2+} ions are seen by EPR (*3*). Susceptibility measurements have shown that the EPR-nondetectable ions are indeed nonparamagnetic (*13*). This finding together with recent chemical evidence (*14*) favors our suggestion that they form a binuclear pair of $Cu(II)$ ions coupled antiferromagnetically.

It has long been considered that only two out of the four metal ions in the cytochrome oxidase unit are detectable by EPR, but this concept has, in fact, lacked a firm experimental basis until recently. The reason for this is that the total intensities of anisotropic EPR spectra, such as those given by Fe^{3+}, are difficult to estimate. All published work on cytochrome oxidase prior to 1976 is based on an equation containing a significant error. The correct expression has been given by Aasa and Vänngård (*15*), who point out that the earlier error is caused by the fact that the integrated intensity takes a different form in a field-swept spectrometer compared with a frequency-swept one. With the correct formulation, the low spin heme signal in the oxidized enzyme corresponds closely to one heme (*16*). The signal around $g = 2$, generally ascribed to Cu^{2+}, also corresponds to one unpaired spin, even if its origin remains uncertain (*2*). Thus, it now appears well established that cytochrome oxidase, like laccase, contains two EPR-nondetectable ions. Recent susceptibility results (*17*) indicate antiferromagnetic coupling between one $Fe(III)$ and one $Cu(II)$ ion.

Oxidation–Reduction Properties

The EPR-nondetectable ions in laccase function as a cooperative 2-e⁻ unit (*5*). With cytochrome oxidase redox titrations based on the heme absorption bands (*2*) indicate the presence of a high and a low potential site (380 and 220 mV, respectively). On the other hand, the quasi equilibrium established in the rapid initial transfer of electrons from reduced cytochrome *c* to the primary electron acceptor in the oxidase, cytochrome *a*, indicates a potential of 285 mV for this site (*18*).

This finding can be explained in terms of heme–heme interaction, a phenomenon first suggested by Nicholls (19) and now well established by other observations (2). Wilson et al. (20) have suggested that certain ligands induce a cooperativity between two metal centers, similar to that in laccase so that these function as a 2-e⁻ unit.

Oxygen Intermediates

A few years ago optical intermediates in the reactions between the reduced enzymes and dioxygen were detected with laccase (21) and with the related protein ceruloplasmin (22). It was tentatively suggested that they represent H_2O_2 or one of its ions bound to a metal ion in the enzymes. With cytochrome oxidase, the search for a functional intermediate also has presented greater problems, and it is only recently that Chance and associates (23) have been able to report substantial progress

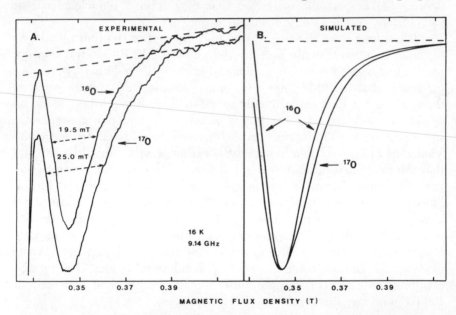

Figure 1.   Experimental (left) and simulated (right) EPR spectra of the laccase–oxygen intermediate obtained with $^{16}O_2$ and $^{17}O_2$, respectively.

(A). High-field part of the EPR spectrum of fungal laccase A, recorded at 9.15 GHz. The enzyme was anaerobically reduced with four electron equivalents of ascorbic acid and mixed with oxygen-saturated buffer, 100% ^{16}O and 91.8% ^{17}O, respectively. The reaction was quenched after 30 min. The protein concentration was 270 μM in 50 mM sodium acetate buffer at pH 5.5. The spectra were recorded at 16 K and a microwave power of 170 mW, which are non-saturating conditions for the intermediate signal. (B). Simulated spectra of the signals in (A) assuming rhombic g-tensor ($g_x = 1.906$, $g_y = 1.939$, $g_z = 2.16$) and anisotropic ^{17}O hyperfine splitting. The ^{17}O spectrum (91.8% ^{17}O, $I = 5/2$; 8.2% ^{16}O, $I = 0$) was simulated assuming interaction with one ^{17}O nucleus with $A_x = 5.5$ mT and $A_y = A_z = 0$. Lorentzian line-shape and a linewidth of 15 mT were used.

by the ultilization of a low temperature tracing method. It is unfortunately difficult to assign electronic structures to the various optical species observed, and they may, in fact, not involve oxygen directly but rather different states of the metal ions in the enzymes. With laccase we have recently obtained the first definitive demonstration of an oxygen intermediate in an oxidase reducing oxygen to water. A brief description of these results follows.

When fully reduced laccase is mixed with oxygen, three of the metal sites (type 1 Cu^+ and the reduced 2-e^- acceptor) have become reoxidized while the fourth site (type 2 Cu^+) remains reduced (*24*). The intermediate so formed has the optical properties earlier assigned to an H_2O_2 compound, but this interpretation is inconsistent with the transfer of 3 e^- from reduced enzyme sites to oxygen. Indeed, we recently showed (*25*) that the intermediate is paramagnetic but that it has unusual relaxation properties, so that it cannot be observed above 25 K. By carrying out the reaction with $^{17}O_2$, it can directly and unambiguously be demonstrated that the intermediate represents an oxygen radical, as shown in Figure 1. The O^- radical can have relaxation properties (*26*) similar to those of the intermediate (*25*). In addition, the ^{17}O splitting for this radical is similar to the hyperfine splitting estimated from the data in Figure 1 (about 5 mT).

Acknowledgments

I would like to thank all collaborators who have provided original results for this survey. Our investigations have been supported by grants from Statens naturvetenskapliga forskningsråd and Knut och Alice Wallenbergs stiftelse.

Literature Cited

1. Caughey, W. S., Wallace, W. J., Volpe, J. A., Yoshikawa, S., in "The Enzymes," P. D. Boyer, Ed., Vol. 13, Academic, New York, 1976.
2. Malmström, B. G., *Q. Rev. Biophys.* (1973) **6**, 389–431.
3. Malmström, B. G., Andréasson, L.-E., Reinhammar, B., in "The Enzymes," P. D. Boyer, Ed., Vol. 12, pp. 507–579, Academic, New York, 1975.
4. George, P., *Oxidases Relat. Redox Syst.* (1965) **1**, 3–33.
5. Reinhammar, B., *Biochim. Biophys. Acta* (1972) **275**, 245–259.
6. Malmström, B. G., *Symmetry Funct. Biol. Syst. Macromol. Level, Proc. Nobel Symp. 11th* (1969) 153–163.
7. Tiesjema, R. H., Muijsers, A. O., van Gelder, B. F., *Biochim. Biophys. Acta* (1973) **305**, 19–28.
8. Antonini, E., Brunori, M., Greenwood, C., Malmström, B. G., *Nature* (1970) **228**, 936–937.
9. Malmström, B. G., Finazzi Agrò, A., Antonini, E., *Eur. J. Biochem.* (1969) **9**, 383–391.
10. Chance, B., Erecinska, M., Change, II, E. M., *Oxidases Relat. Redox Syst., Proc. Symp.* (1973) **2**, 851–867.

11. Fee, J. A., Malkin, R., Malmström, B. G., Vänngård, T., *J. Biol. Chem.* (1969) **244**, 4200–4207.
12. Heineman, W. R., Kuwana, T., Hartzell, C. R., *Biochem. Biophys. Res. Commun.* (1972) **49**, 1–8.
13. Moss, T. H., Vänngård, T., *Biochim. Biophys. Acta* (1974) **371**, 39–43.
14. Briving, C., Deinum, J., *FEBS Lett.* (1975) **51**, 43–46.
15. Aasa, R., Vänngård, T., *J. Magn. Reson.* (1975) **19**, 308–315.
16. Aasa, R., Albracht, S. P. J., Falk, K.-E., Lanne, B., Vänngård, T., *Biochim. Biophys. Acta* (1976) **422**, 260–272.
17. Falk, K.-E., Vänngård, T., Angström, J., *FEBS Lett.*, in press.
18. Andréasson, L.-E., *Eur. J. Biochem.* (1975) **53**, 591–597.
19. Nicholls, P., *Struct. Funct. Cytochromes Proc. Symp.* (1968) 76–86.
20. Wilson, D. F., Erecińska, M., Lindsay, J. G., Leigh, Jr., J. S., Owen, C. S., in "Enzymes; Electron Transport Systems," P. Desnuelle and A. M. Michelson, Eds., pp. 195–210, North-Holland, Amsterdam, 1975.
21. Andréasson, L.-E., Brändén, R., Malmström, B. G., Vänngård, T., *FEBS Lett.* (1973) **32**, 187–189.
22. Manabe, T., Manabe, N., Hiromi, K., Hatano, H., *FEBS Lett.* (1972) **23**, 268–270.
23. Chance, B., Saronio, C., Leigh, Jr., J. S., *Proc. Natl. Acad. Sci. USA* (1975) **72**, 1635–1640.
24. Andréasson, L.-E., Brändén, R., Reinhammar, B., *Biochim. Biophys. Acta,* (1976) **438**, 370–379.
25. Aasa, R., Brändén, R., Deinum, J., Malmström, B. G., Reinhammar, B., Vänngård, T., *FEBS Lett.* (1976) **61**, 115–119.
26. Sander, W., *Naturwissenschaften* (1964) **51**, 404.

RECEIVED July 26, 1976.

Electron Transfer Pathways in Blue Copper Proteins

ISRAEL PECHT, OLE FARVER, and MICHEL GOLDBERG

Department of Chemical Immunology, The Weizmann Institute of Science, Rehovot, Israel

Several structural and functional aspects of blue copper proteins have been investigated. The systematic study of the ultraviolet spectroscopic properties of blue single copper proteins (azurins and stellacyanin) brings evidence for sulfur as a ligand of the copper ion. The energy profile of the electron transfer equilibrium between Ps. Acruginosa azurin and $Fe(CN)_6^{3-/4-}$ was obtained by combined analysis of chemical relaxation times and amplitudes along with microcalorimetry and spectrophotometric titrations. The superoxide (O_2^-) reduces the type 1 $Cu(II)$ in Rhus laccase. A relatively fast, full reoxidation of the partially reduced enzyme takes place in the presence of oxygen. Rhus laccase reacts specifically and with high affinity with H_2O_2 to form a stable product which may be a functional intermediate.

Two central themes are encountered in studying the mechanism of action of redox proteins:

(1) The detailed pathways of redox equivalents to and from the active centers

(2) The intramolecular events in the multicenter oxidases, which enable the functional cooperation of the different active sites in carrying out reduction or oxidation of specific substrates, notably those where multiple electron transfer steps are favored pathways

The function of the electron-mediating proteins which contain a single redox active site (e.g., rubredoxin, azurins, flavodoxins, plastocyanins) is mainly related to the first aspect. Still, the pronounced specificity encountered in their function in biological energy conversion processes indicates that their redox center, often a transition metal ion, is embedded in an evolutionarily optimized polypeptide envelope. The detailed three-dimensional structures of an increasing number of these electron mediators, along with kinetic studies of their reactions, have led to new insights into the redox pathways of the c-type cytochromes and some of the non-heme iron proteins (1). Unfortunately no direct structural information is yet available for those electron-mediating proteins where copper serves as the active center. It is remarkable that all members of this group have their single copper bound in a site which confers upon them the unique blue state. They are widely found in nature, ranging from the plastocyanins in the photosynthetic apparatus to stellacyanin and umecyanin, which are found in nonphotosynthetic plant tissues, to the large family of azurins from different bacterial sources (2).

This group of blue single copper proteins is a convenient system for probing structural features of the metal-binding site. Early suggestions that a sulfur ligand is involved in the coordination sphere (2) received further support recently from the spectral investigations of derivatives in which the copper was replaced by $Co(II)$ ions (3) and from resonance Raman and ESCA studies on the native proteins (4, 5, 6). We have adopted a different approach to this problem, monitoring the intrinsic probes of these proteins, by examining the spectral properties in the uv range. Characteristic alterations were found which are probably caused by the $Cu(I)$–S chromophore along with changes from conformational differences between the reduced and oxidized states of the protein.

The electron transfer mechanism of azurin, a well known example for this type of proteins, has been systematically studied using the chemical relaxation method and a well defined inorganic outer sphere redox couple. In parallel, the investigations of the reaction with its presumed physiological partner, cytochrome c, were pursued (7). The specificity of the interaction between azurin and cytochrome c P-551 is expressed in higher specific rates and in the control of the electron transfer equilibrium by conformational transitions of both proteins.

In the blue oxidases, the type 1 copper is only one of at least three active centers. The striking feature of this class of proteins, apart from the complexity of the different copper binding sites per se is the intricate relationship among these sites. Equilbrium measurements produced evidence for the effect of external ligand binding to type 2 $Cu(II)$ on the redox potentials of both type 1 and type 3 sites (8). From the kinetic data it became clear that rates of reduction and oxidation of one site are rather sensitive to the redox states of the other sites (9). This is functionally relevant, since all three redox sites of laccase are supposed to undergo reversible valence changes during the catalytic cycle of the enzyme. We have examined the reduction and oxidation of *Rhus* laccase under conditions where a rather small fraction of its sites was reduced using O_2^- radicals as reductant in oxygen-saturated solutions. The type 1

Cu(II) ions were reduced and fully reoxidized within 1–2 sec, although under these conditions no more than a single reduction equivalent may be in those molecules. This is not the case when reduction of these sites take place anaerobically by, for example, the CO_2^- radical ion.

Interaction between oxidized fungal laccase and hydrogen peroxide added in excess, occurs via its type 2 Cu(II) site. This binding is manifested by both the ESR and absorption spectra, yet the affinity of this complex is rather low, and it decomposes rapidly (10). In contrast we found that tree laccase forms a relatively stable and high affinity complex with one mole of H_2O_2 which may be reduced together with the enzyme sites and could be of mechanistic significance in the reduction of dioxygen to water.

Experimental

Materials. Laccase and stellacyanin were prepared from the acetone extract of *Rhus vernicifera* lacquer according to the procedure of Reinhammar (11) and were kept at −20° in salt-free solutions. Azurin from *Pseudomonas aeruginosa* and *Alcaligenes faecalis* was isolated and purified by the method of Ambler (12) and was stored at 4° in 0.05M sodium acetate buffer (pH 3.9). The protein solutions were extensively dialyzed against potassium phosphate buffer before being used in the experiments and, when necessary, were concentrated by vacuum dialysis. The purity of the proteins was determined by checking the respective absorption ratios between the blue band and the 280-nm band; they were always in good agreement with the literature values. In addition, the quality of the laccase preparation was assayed by measuring its enzymatic activity with N,N-dimethyl-p-phenylenediamine (11).

Stellacyanin and the two azurins were reduced by the following methods—by ascorbate, followed by anaerobic dialysis; by hydrogen, using platinum black as catalyst (7); by sodium borohydride; or by dithionite. All of these methods led to the same product as concluded from the absorption spectra of the native, reduced, and reoxidized proteins. Apoproteins were prepared by dialysis against 0.05M NaCN in 0.1M phosphate buffer pH 7.4, followed by repeated dialysis against the buffer. The concentrations of the native proteins were determined using $\epsilon_{614} = 5700 \ M^{-1} \ cm^{-1}$ for laccase (13), $\epsilon_{605} = 4080 \ M^{-1} \ cm^{-1}$ and $\epsilon_{280} = 23,200 \ M^{-1} \ cm^{-1}$ for stellacyanin (13), $\epsilon_{625} = 5700 \ M^{-1} \ cm^{-1}$ (14), and $\epsilon_{280} = 10,700 \ M^{-1} \ cm^{-1}$ (15) for *Ps. aeruginosa* azurin and $\epsilon_{625} = 4620 \ M^{-1} \ cm^{-1}$ for *A. faecalis* azurin (16).

All materials used were of analytical grade. All solutions were prepared in doubly distilled water. The concentration of hydrogen peroxide stock solution was determined iodometrically (17) before and after each peroxide titration.

Methods. Measurements of absorption spectra and spectrophotometric titrations were carried out on a Cary 15 spectrophotometer equipped with a thermostated cell compartment. Circular dichroic spectra were recorded on a Cary 60 spectropolarimeter equipped with a 6001 CD attachment. Fluorescence spectra were measured on a Hitachi–

Perkin Elmer MPF-3 spectrofluorometer. X- and Q-band measurements of EPR spectra were carried out at liquid nitrogen and liquid helium temperatures. Microcalorimetric measurements were performed on a LKB 10700 batch microcalorimeter. Temperature-jump relaxation kinetics were measured using a double beam instrument (18) with a cell adapted for anaerobic work. The relaxation signals were fed into an H.P. 2100 computer and analyzed as described in Ref. 7. The pulse radiolysis exepriments were carried out on the 5-MeV linear accelerator at the Hebrew University. Details of the system have been published previously (19).

SPECTROSCOPIC STUDIES. All measurements were made at 25° in 0.05M or 0.1M potassium phosphate buffer, pH 7.0. Each spectrum was recorded with several samples, which generally differed in concentration, and was scanned two or three times. Circular dichroism is expressed as molar ellipticity $[\theta]$ in units of degrees cm^2 · dmole^{-1}.

RELAXATION KINETICS. The details of the experimental procedure have been described earlier (14). 0.1M phosphate buffer, pH 7.0, containing 2×10^{-5} M EDTA was used in all relaxation experiments. These were performed with solutions of different initial reagent composition— either ferrocyanide was added to oxidized azurin or ferricyanide to reduced azurin. Temperature jumps of 2.9° or 4.7° were applied to the reaction solution. The subsequent transmission changes were monitored at 625 nm (absorption of oxidized azurin) or 420 nm (absorption of ferricyanide). Each plotted value of the relaxation time or amplitude represents the average of at least four measurements.

PULSE RADIOLYSIS. A detailed account of the experimental procedure has been given elsewhere (19). Solutions of tree laccase were prepared in triply distilled water and contained tert-butanol as scavenger for OH radicals. The solution was saturated with argon or oxygen by prolonged bubbling (> 20 min) in large glass syringes equipped with standard capillary taper joints. Bubbling was done prior to the addition of the required volume of concentrated protein solution, thus minimizing denaturation by foaming.

ANAEROBIC OXIDATION–REDUCTION TITRATIONS. Oxidation–reduction titrations of Rhus laccase were carried out in a specially constructed optical cell described in Ref. 20. The solutions were freed from oxygen by alternative evacuation and flushing with water-saturated argon from which traces of oxygen were removed by passing the gas through four columns of methylviologen. During the titrations a slight excess argon pressure was maintained to avoid diffusion of oxygen into the cell. Titrant was added through a serological cap with a Hamilton micro syringe. The titrated solutions were stirred by a small magnetic bar.

Experimental values were corrected for dilution of protein on adding titrant, for residual absorbance of the fully reduced chromophore, and when necessary, for absorbance of the oxidized and reduced forms of the titrants used.

Spectroscopy of Blue Single Copper Proteins in the Ultraviolet Region

The nature of the blue copper site might be conceived as the result of a compromise between the free energy requirements for the preferred coordination structure of the metal ion in its two redox states and the optimal conformation of the polypeptide forming it. It is therefore expected that changes in the state of the copper would also be expressed in some of the properties of amino acid residues, notably those which are related to the binding site. Thus an earlier study from this laboratory has shown a correlation between the redox state of type 1 copper and the intrinsic fluorescence of *Rhus* laccase (*20*). Here we have compared the uv spectral properties of three blue single-copper proteins (*Pseudomonas aeruginosa* azurin, *Alcaligenes faecalis* azurin, and *Rhus* stellacyanin) in their oxidized, reduced, and copper-free states, using absorption, difference absorption, circular dichroism, and fluorescence measurements. The comparison between the two azurin species is of special significance as they are homologous proteins, yet certain residues which are of particular interest (e.g., the single tryptophan) occupy different positions in the amino acid sequence (*21*).

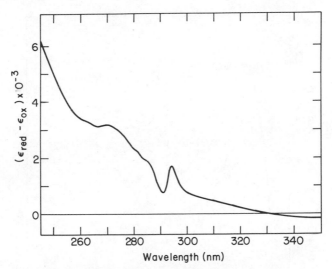

Figure 1. *Reduced-minus-oxidized difference absorption spectrum of Ps.* aeruginosa *azurin. Sample and reference cell contained 6.8 × 10⁻⁵ M solutions of reduced and oxidized protein, respectively. Hydrogen, with platinum black as catalyst, was used as reductant (7). Medium: 0.1M potassium phospate buffer, pH 7.0*

Figure 1 shows the difference absorption spectrum between reduced and oxidized *Ps. aeruginosa* azurin. As is well known, in the visible region the typical blue band disappears upon reduction, and no other changes are observed. However, pronounced and complex changes are seen in the uv region. Upon reduction a decrease is found above 330 nm, whereas below this wavelength the reduced azurin has the higher extinc-

Table I.   Ultraviolet Extinction of Reduced Copper Proteins

| | | Wavelength (nm) | | | |
Protein	Value Measured $(M^{-1} cm^{-1})$	250	260	270	280
Ps. aeruginosa azurin	$\epsilon_{red} - \epsilon_{ox}$	5050	3450	3200	2350
A. faecalis azurin	$\epsilon_{red} - \epsilon_{ox}$	4250	3050	2950	2250
Stellacyanin	$\epsilon_{red} - \epsilon_{ox}$	5150	4300	4250	3500
Cu^+ — thionein[a]	ϵ_{red}	3650	3050	2650	2200

[a] Values calculated from spectral data reported by Rupp and Weser (24).

tion. This is consistent with the result of a preliminary experiment reported by Yamanaka et al. (22). The peak in the difference spectrum at 294 nm is caused by a small red shift of the well resolved fine structure peak of tryptophan at 292 nm (22). A monotonous increase in the extinction of the reduced protein relative to the oxidized takes place towards

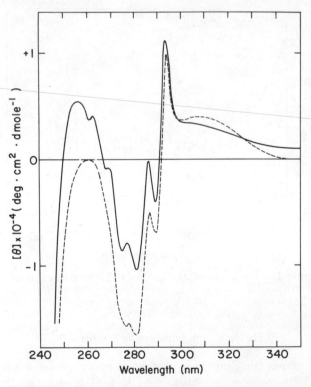

Figure 2.   CD spectra of oxidized (———) and reduced
(– – –) Ps. aeruginosa azurin. The samples contained
1.2×10^{-4} M protein in 0.05M potassium phosphate,
pH 7.0. Reduction was achieved with ascorbate, followed by extensive dialysis. Optical pathlength, 10
mm.

lower wavelengths, producing a difference of $\sim 5000\ M^{-1}\ cm^{-1}$ at 250 nm. This increase is modulated by slight, but distinct, variations in the fine structure of the aromatic absorption bands (275–290 nm) and by a shoulder at 270 nm. Essentially similar difference spectra between reduced and oxidized protein have also been measured for *A. faecalis* azurin and stellacyanin. The isosbestic points are somewhat shifted to the blue (324 nm for the former, 304 nm for the latter), and the peak at 294 nm is only a flat shoulder, which is simply a consequence of the tryptophan transition at 292 nm not being resolved in these proteins. But a marked increase of the extinction difference with decreasing wavelength and a shoulder at 270 nm is in accord with the previous findings. Moreover, the ($\epsilon_{red} - \epsilon_{ox}$) values found are altogether quite similar (Table I). The somewhat higher values for stellacyanin at 260 and 270 nm are caused by a much more pronounced shoulder at 270 nm.

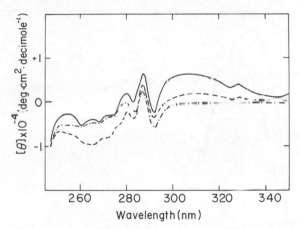

Figure 3. CD spectra of oxidized (——), reduced (– – –), and copper-free (– · –) stellacyanin. Protein solutions: 5 × 10⁻⁵ M in 0.1M potassium phosphate, pH 7.0. The reduced sample was prepared by catalytic reduction with hydrogen; the oxidized sample by aerobic reoxidation of the same reduced sample. The apoprotein was prepared as described under "Experimental." Optical pathlength, 10 mm.

CD spectra of the oxidized and reduced form have been measured for all three proteins; the spectra of two of them are presented in Figures 2 and 3. The complex set of dichroic bands in the region of the aromatic transitions is in all cases fully conserved upon reduction, with only the magnitude of the ellipticities being affected. The small but significant differences at 310 and at 293 nm as well as the large increase of the ellipticity at 280 nm found for *Ps. aeruginosa* azurin (Figure 2) are exactly analogous to changes found for reduced *A. faecalis* azurin compared with the oxidized protein. The CD spectra of oxidized *Ps. aeruginosa* azurin and *A. faecalis* azurin (Figures 2 and 4) are essentially similar. They exhibit exactly the same band pattern, except for some small shifts and differ only in the ellipticities of the various bands. The only major difference in the CD properties of these two proteins is the response of the positive band at 260 nm to reduction. Whereas in *A. faecalis* azurin this band changes little upon reduction, a pronounced change towards negative ellipticity occurs in the case of *Ps. aeruginosa* azurin (Figure 2). The effect of reduction on the CD spectrum of stellacyanin (Figure 3) is also limited to the amplitudes of the different bands, their position remaining practically unaltered. The increase of more than two-fold in the ellipticity of the broad negative band around 265 is probably related to the same transition which produces a pronounced shoulder at 270 nm in the reduced-minus-oxidized difference absorption spectrum.

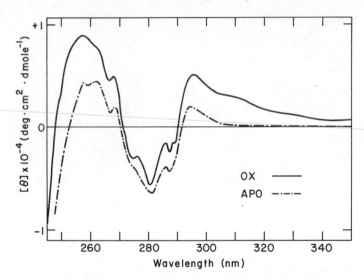

Figure 4.   CD spectra of native (———) and copper-free (– · –) A. faecalis *azurin. The apoprotein was prepared as described under "Experimental." Solutions: 9×10^{-5} M in 0.1M potassium phosphate, pH 7.0. Optical pathlength, 10 mm.*

The spectral properties of the three proteins in the uv are strongly influenced by the redox state of the copper. Although this is probably true for all blue copper proteins, very little attention has been paid to these changes. The overall similarity of the redox-related changes in absorption and CD spectra strongly suggests that most of the changes are common to all three proteins and are therefore related to some common structural property, most probably involving the copper site. The results of various spectroscopic and chemical studies have been interpreted to indicate that blue copper is coordinated to sulfur (*3, 4, 5, 6, 23*). The findings of this study are consistent with such a concept. The general course of the reduced-minus-oxidized difference absorption spectra is similar to the absorption spectrum of the Cu(I)–S chromophore, as seen from extinction values (Table I) derived from Cu(I)–thionein spectra (*24*). However, the chemical form of the sulfur ligand in the blue proteins is not fully established, but it has generally been proposed to be the sulfhydryl group of a cysteine residue. Yet one should also consider the possibility that methionine may serve as a ligand. Some support for this hypothesis comes from a model study where the chelate complex of a thiaether with Cu(II) has been found to exhibit an intense blue absorption band, similar to that found in blue copper proteins (*26*). In nine different azurin species (from *Pseudomonas, Bordetella,* and *Alcaligenes* (*21*)) and in eight different plastocyanin species (bean, potato, and green algae (*27*)), methionine 121 (in the plastocyanin numbering Met 97) is invariant, and the sequence around it contains some further residues which have conserved their aromatic or hydrophobic character [Tyr 108 (85), Phe/Tyr 110 (87), Phe/Tyr 111 (88), Leu/Val/Ile 125 (101), Leu/Val 127 (103)] apart from the invariant Gly 123 (99) and the single cysteine residue 112 (89). This may indicate the involvement of both cysteine and methionine in the coordination sphere of the blue copper, as far as they are available.

The additional effects in the aromatic region of the difference spectrum (250–300 nm) are probably caused by aromatic transitions which are influenced by the redox state of the copper. The shoulder at 270 nm, which occurs in all three proteins, could result from an increase in tyrosine absorption. In this context, it is interesting to recall that Tyr 108 (azurin numbering), which is relatively close to the proposed copper ligands Cys 112 and Met 121, is completely invariant both in azurin and plastocyanin and may therefore be an obligatory constituent of the copper site.

The redox-induced changes in the CD spectrum of the aromatic region do not seem to be related to a copper chromophore because they are not uniform in the three proteins. Instead, they probably arise from differences in the direct or indirect (conformational) effects of the redox state of the copper on the dichroism of various aromatic transitions. The comparatively small changes in the 285–295-nm range suggest that the tryptophan transitions are less affected than those of tyrosine and phenylalanine. The changes around 250 nm may involve disulfide groups which produce Cotton effects at this wavelengths.

For certain blue single-copper proteins it was proposed long ago that a Cu(II) or Cu(II)-related band is present between 300–350 nm (*28*), but it has attracted only very limited attention in spite of the potential implications for the study of blue copper-containing oxidases. A summary of the spectral measurements in the 300–350-nm region is presented in Figure 5. Removing the copper almost abolishes the absorption in this region. This is confirmed by the CD spectra where the problem of achieving spectrally very clean solutions or matching the concentrations of holo- and apoproteins is less crucial than for absorption or difference absorption spectra. The extinction of the native proteins in this region is in the range 200–700 M^{-1} cm^{-1} compared with extinctions of about 1×10^4 (azurin) or 2.3×10^4 M^{-1} cm^{-1} (stellacyanin) at 280 nm. Such experimental problems made it difficult to obtain reliable data on the shape of the Cu(II) band below 305 nm, where the aromatic absorption begins.

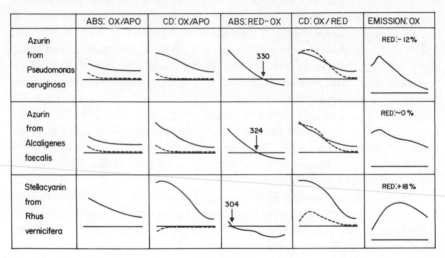

Figure 5. Optical properties of three blue copper proteins in the range 300–350 nm.

The various protein samples were prepared as indicated under "Experimental." Medium: 0.05 or 0.1M potassium phosphate, pH 7.0, 25°. (——) oxidized protein, (– – –) reduced or copper-free protein. Abscissa: The wavelength range extends from 300 to 350 nm. Ordinate: All absorption spectra were drawn to the same scale, as was done for the CD spectra. Fluorescence emission spectra: the scale is different for each protein. All spectra were recorded in the ratio mode with $\lambda_{ex} = 280$ nm. The effect of reduction is indicated as percentage change of the maximum emission intensity of the oxidized protein.

From measurements where reconstituted azurin was compared with apo-azurin, we estimate that the copper-dependent extinction does not exceed 450 M^{-1} cm^{-1} at 280 nm, but it remains to be determined whether it passes through a maximum between this wavelength and 320 nm. A likely candidate for the ligand, which in conjunction with Cu(II) gives rise to the absorption in this region, is peptide nitrogen. Various complexes involving the coordination of a deprotonated peptide nitrogen to Cu(II) were found to absorb in this region (15, 29). The moderately intense band has been ascribed to a ligand–metal charge transfer transition. An interesting model system exhibiting a similar band is the Cu(II)–poly-L-histidine complex formed in neutral-to-slightly-acid solutions (*30*). This gains special significance in view of NMR data of single blue copper proteins (*31*) suggesting both peptide and imidazole nitrogens as ligands to the copper.

Considering now the reduced-minus-oxidized different absorption spectra, it becomes clear that the isosbestic point shifts to a shorter wavelength when going from the azurins to stellacyanin, because the absorbance of oxidized stellacyanin in this region is higher than that of the oxidized azurins. Underlying, of course, is the assumption that the Cu(I) spectrum has a similar intensity in all three proteins, at least in this region. The response of the fluorescence emission intensity to the reduction of the copper can be explained on the same basis. *Ps. aeruginosa* azurin has its emission centered around 308 nm. When the protein is reduced, the absorbance in this wavelength region increases. Therefore any quenching effect originating from internal nonradiative energy transfer between the excited tryptophan residue and the copper chromophore is expected to become larger, thereby decreasing the quantum yield. Exactly the opposite happens with stellacyanin. Here the reduced protein absorbs less in the region of the emission maximum, enhancing its intensity. The emission spectrum ($\lambda_{ex} = 280$ nm) of *A. faecalis* azurin hardly changes on reduction, probably because quenching and enhancing effects approximately compensate. Recall that the single tryptophan residue of *A. faecalis* azurin is not homologous to the single tryptophan of *Ps. aeruginosa* azurin (*21*), a fact which may contribute both to the different fluorescence spectra of the native proteins and to the different response to reduction.

Removal of the copper ion only moderately effects the aromatic part of the CD spectrum (Figures 3, 4), a spectral region which is in general less affected by removing the copper than by reducing it. This indicates that the structural integrity of blue single-copper proteins depends little on the presence of the copper ion as far as these aromatic residues are concerned. A similar result has been obtained from the comparison of the tryptophan fluorescence in native and copper-free *Ps. fluorescence* azurin (*23*). The conclusion is that the interaction between the copper ion and the protein, as discussed at the outset, is largely dominated by the conformation of the latter.

Electron Transfer Profile of *Ps. aeruginosa Azurin*

An effective approach to resolving the electron pathway to and from the redox center of azurin is the systematic investigation of its equilibria and kinetics of interactions with inorganic redox couples. Hexacyanoferrate (II/III) is a well defined redox couple, known to react via an outer sphere path. Furthermore, it is related in its electronic structure to Fe(II/III) heme, the redox center of all cytochromes, including cytochrome c P551, the natural partner of azurin (7). Thus we have studied the azurin–hexacyanoferrate (II/III) system by temperature-jump relaxation spectroscopy (32), spectrophotometric titrations, and microcalorimetry.

The relaxation spectrum consists of a single relaxation mode. The relaxation time, τ, was measured up to high concentrations of protein $(1 \times 10^{-3} M)$ and hexacyanoiron $(8 \times 10^{-2} M)$. When azurin (I) was titrated with ferricyanide, τ^{-1} leveled off with increasing concentration (Figure 6b). This behavior reveals that the electron transfer between azurin and $Fe(CN)_6^{4-/3-}$ is not a simple one-step process as the single relaxation observed might have indicated. The limiting dependence on ferricyanide concentration suggested a kinetic scheme involving the fast reversible formation of an azurin (I)–ferricyanide complex followed by a slower electron transfer step. The formation of a corresponding azurin (II)–ferrocyanide complex is to be expected from the principle of microscopic reversibility, yet the concentration dependence of the relaxation time provided no evidence for it; when azurin(II) was titrated with increasing amounts of ferrocyanide, τ^{-1} showed a monotonous increase up to very high concentrations (Figure 6a). Direct evidence for the involvement of an azurin (II)–ferrocyanide complex was obtained by analyzing the relaxation amplitudes. These constitute a further source of information about the mechanism of reaction, apart from the kinetic data derived from the analysis of the relaxation times (33).

Assuming the following Scheme A, the "normal" enthalpy of the relaxation mode observed, $\tilde{\Delta H}$, which can be calculated from the relaxa-

$$
Az(II) + Fe(II) \underset{}{\overset{K_1}{\rightleftharpoons}} Az(II) \cdot Fe(II) \underset{k_{-3}}{\overset{k_{+3}}{\rightleftharpoons}} Az(I) \cdot Fe(III) \overset{K_2}{\rightleftharpoons} Az(I) + Fe(III) \tag{A}
$$

$$
\text{(fast)} \qquad\qquad \text{(slow)} \qquad\qquad \text{(fast)}
$$

tion amplitudes, is linearly related to the individual enthalpy changes, as described by Equation 1, where $\Delta H_i°$ is the individual reaction en-

$$
\tilde{\Delta H} = \frac{1}{K_1 \cdot ([Az(II)] + [Fe(II)]) + 1} \Delta H_1° +
$$

$$
\frac{K_2}{K_2 + [Az(I)] + [Fe(III)]} \Delta H_2° + \Delta H_3° \tag{1}
$$

thalpy of the i-th step, and K_1 and K_2^{-1} are the association constants of steps 1 and 2, respectively (14). In the experiments involving addition of ferrocyanide to oxidized azurin(II), the concentrations of azurin(I) and

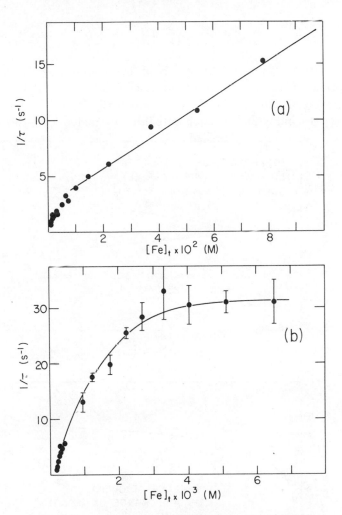

Figure 6. Ps. aeruginosa *azurin–hexacyanoiron (II/III) equilibrium–concentration dependence of the reciprocal relaxation time. (a) Azurin (II) reacted with ferrocyanide (24–860-fold excess), 6.5° C. (b) Azurin (I) (2–5 × 10⁻⁴ M) reacted with ferricyanide, 16.8° C. τ^{-1} has a minimal value at $[Fe]_t \cong [Az]_t$ and increases again at lower $[Fe]_t$. Points in the region $[Fe]_t < [Az]_t$ have been omitted for clarity.*

ferricyanide formed remained small, even at the large excess of ferrocyanide used, because of the position of the overall equilibrium ($K_{overall} = 1.1 \times 10^{-2}$ at 25°, pH 7.0, 0.1M potassium phosphate). Since they did not reach a level sufficient for significant complex formation between them, Equation 1 can be simplified to Equation 2. Figure 7 shows that Equa-

$$\tilde{\Delta H} = \frac{1}{1 + K_1 \,[\mathrm{Fe(II)}]} \, \Delta H_1^\circ + \Delta H_2^\circ + \Delta H_3^\circ \qquad (2)$$

tion 2 is an appropriate description of the experimentally determined dependence of $\tilde{\Delta H}$ on the ferrocyanide concentration. At low $[\mathrm{Fe(II)}]$ the variation of $\tilde{\Delta H}$ with $[\mathrm{Fe(II)}]$ is practically linear (Figure 7). The values for the individual reaction enthalpies extracted from this and simi-

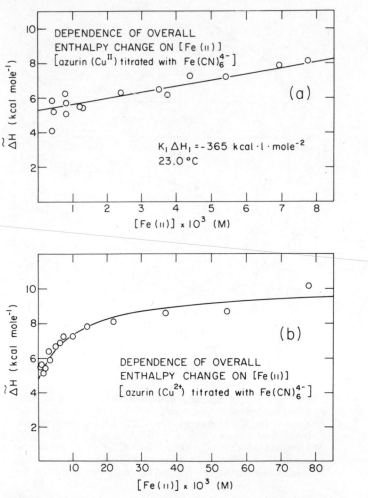

Figure 7. *Relaxation amplitude analysis: variation of $\tilde{\Delta H}$ with $[Fe(CN)_6^{4-}]$. $\tilde{\Delta H} =$ "normal" enthalpy of the relaxation mode observed (14). (a) At 23.5°C, plot according to linearized form of Equation 2. Total azurin 2–7×10^{-5} M. (b) At 6.5°C, plot according to Equation 2. Total azurin 2–11×10^{-5} M. The line drawn is the best fit to the experimental points with $K_1 = 1.0 \times 10^2$ M^{-1} and $\Delta H_1^\circ = -5.5$ kcal/mole.*

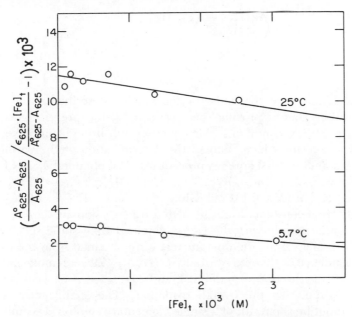

Figure 8. The variation of the apparent equilibrium constant with $[Fe(CN)_6^{4-}]$. Total azurin: $4-6 \times 10^{-5}$ M. Abscissa: $[Fe(CN)_6^{4}] \cong [Fe]_t$ (total concentration of hexacyanoiron). Ordinate: apparent equilibrium constant K_{app}.

lar plots are: $\Delta H_1° = -5.5$ kcal/mole, $\Delta H_2° = 7.7$ kcal/mole, and $\Delta H_3° = 6.7$ kcal/mole. Further support for the postulated formation of an azurin(II)–ferrocyanide complex came from the static titrations of azurin(II) with ferrocyanide. The apparent overall equilibrium constant for Scheme A is given in Equation 3 where $K_3 = k_{+3}/k_{-3}$. Under the concentration conditions we used, Equation 3 simplifies to Equation 4. In

$$K_{app} = K_1 K_2 K_3 \frac{1 + ([Az(I)] + [Fe(III)])/K_2 + [Az(I)][Fe(III)]/K_2^2}{1 + ([Az(II)] + [Fe(II)])K_1 + [Az(II)][Fe(II)]K_1^2}$$

(3)

$$K_{app} = K_1 K_2 K_3 \frac{1}{1 + K_1[Fe(II)]} \cong K_1 K_2 K_3 (1 - K_1[Fe(II)]) \quad (4)$$

Figure 8, K_{app} is plotted against $[Fe(II)]$, and it shows a behavior corresponding to Equation 4. The association constant K_1 ($57 M^{-1}$ at $25°$), obtained from the ratio of negative slope to intercept, is in good agreement with the value derived from the relaxation amplitude)$54 M^{-1}$ at the same temperature). Equation 5 relates relaxation time, equilibrium con-

$$\frac{1}{\tau} = k_{+3}K_1 \frac{[Az(II)] + [Fe(II)]}{K_1([Az(II)] + [Fe(II)]) + 1} +$$

$$k_{-3} \frac{[Az(I)] + [Fe(III)]}{K_2 + [Az(I)] + [Fe(III)]} \quad (5)$$

centrations, and rate constants for the scheme described above. Using simplifications and approximations justified by the prevailing equilbria and concentration conditions, this equation was brought into the appropriate linear forms which enabled the determination of the various constants by standard least squares procedures. We obtained $k_{red} = K_1k_{+3} = 3.4 \times 10^2 \ M^{-1} \ s^{-1}$, $k_{ox} = k_{-3}K_2^{-1} = 2.7 \times 10^4 \ M^{-1} \ s^{-1}$, $k_{-3} = 45 \ s^{-1}$, and $K_2 = 1.65 \times 10^{-3} \ M$ (25°, 0.1M potassium phosphate, pH 7, I = 0.22). Overall equilibrium constants were calculated from the kinetic data and found to agree satisfactorily with the values obtained from static spectrophotometric titrations. Activation parameters were obtained from the temperature dependence of the rate constants: $\Delta H^{\ddagger}_{red} = 5.9$ kcal/mole and $\Delta S^{\ddagger}_{red} = -27$ eu for the overall reduction, and $\Delta H^{\ddagger}_{ox} = -4.1$ kcal/mole and $\Delta S^{\ddagger}_{ox} = -52$ eu for the overall oxidation. The significantly negative $\Delta H^{\ddagger}_{ox}$ cannot be the result of a single elementary step but is composed of the negative binding enthalpy and the positive activation enthalpy of the actual electron transfer step.

The proposed mechanism gains even further support from the fact that the whole set of thermodynamic and activation parameters, obtained by four different approaches (temperature dependence of equilibrium and rate constants, amplitudes of chemical relaxation and microcalorimetry) and related to the same or to different parts of the reaction leads to a self-consistent energy profile for this system (Figure 9). In terms of free energy, the reaction is nearly symmetrical about the transition state of the electron transfer within the complex, which clearly constitutes the rate-limiting step. The activation barriers for the overall reaction in both directions is largely entropic; in the overall oxidation the entropy barrier even offsets the favorable enthalpy term. This negative enthalpy contribution to the overall activation energy is caused by the unusual relative value of the enthalpy of the electron transfer transition state, which is about midway between reactants and products.

The low association constants show that the binding of hexacyanoferrate to azurin prior to electron transfer is relatively weak. Still, the formation of presumably electrostatic complexes between the negatively charged hexacyanoferrate ion and azurin is somewhat surprising, as the overall net charge of azurin at pH 7.0 is also negative (pI = 4.9). This probably implies the existence of a patch of positively charged residues on the protein surface where hexacyanoferrate ions may bind and the electron transfer takes place. A similar situation is proposed to prevail in the reduction of azurin (II) by $Fe(EDTA)^{2-}$, which has been studied by Wherland et al. (*34*). The very low overall $\Delta H^{\ddagger}_{red}$ (-2 kcal/mole) and the largely negative overall $\Delta S^{\ddagger}_{red}$ (-37 eu) for this reaction, quite similar to the activation parameters for the reduction by $Fe(CN)_6^{4-}$, may indicate that the $Fe(EDTA)^{2-}$ reaction also includes a distinct intermediate complex formation step with a negative association enthalpy, followed by the actual electron transfer.

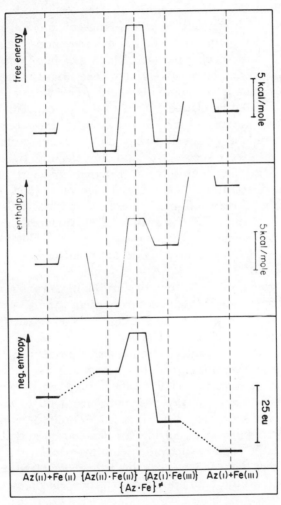

Figure 9.   Free energy, enthalpy, and entropy profiles of the azurin (I/II)–Fe(CN)$_6^{4-/3-}$ system. Reference state: 1M reactants, 0.1M potassium phosphate, pH 7.0, I = 0.22. Free energy given for 298°K. For this temperature the entropy scale is equivalent to the energy scales.

It is conceivable that this reaction pattern may be more common, particularly among electron mediating proteins, like the azurins and the plastocyanins, which are involved in metabolic systems. One of the main elements of the specificity required for carrying out their physiological function is the ability to bind specifically at the sites of receipt or delivery of the electrons. In many cases, however, this functional binding capability may be difficult to detect experimentally. Inorganic redox couples are limited in the repertoire of interacting modes as compared with proteins which can offer a whole range of interactions of varying steric and chemical nature combined in different modes so as to produce higher affinity for a specific partner. Binding of inorganic agents tends therefore to be weak, and relatively high concentrations are necessary to detect any deviation from simple bimolecular behavior. Moreover, if the electron transfer step is relatively fast (a situation encountered when the net driving force in terms of free energy is large), the rate at which such a deviation could be observed is beyond the time resolution of most techniques used. This proves true also for electron transfer reactions between proteins. The high intrinsic reactivity of most redox metalloproteins, as deduced from their self-exchange rates (35), leads to high rates of electron transfer and makes it difficult to observe the leveling off. Hexacyanoiron enabled the observation of intermediate complex formation with azurin because of its particular properties (high charge, moderate intrinsic reactivity, and a redox potential leading to a small net driving force).

The electron transfer within the azurin–hexacyanoiron complex is characterized by unusually high entropy barriers ($\Delta S^{\ddagger}_{+3} = -17$ eu, $\Delta S^{\ddagger}_{-3} = -39$ eu). This may be the result of high steric or orientational requirements for the electron transfer. The copper site of azurin is known to be deeply buried, and resonant electron transfer probably requires an extended system of overlapping orbitals connecting the redox site to the protein surface. If, as one expects in view of the Franck–Condon requirements, the conducting pathway does not correspond to an equilibrium configuration, the necessary distortions from equilibrium positions may have a low probability of occurring simultaneously, and this would be reflected in a large negative activation entropy. However, the possibility remains open that the reaction is nonadiabatic, its low transmission probability giving rise to the high apparent activation entropies.

The Reaction of Partially Reduced Rhus laccase with Oxygen

An obvious mechanistic assumption for the reduction of the dioxygen molecule by laccase is its binding to the enzyme. This raises the question at what reduction state of the laccase molecule will the interaction with oxygen take place. We have investigated this problem in a system where controlled partial reduction of laccase was carried out in the presence of oxygen. This can be achieved by pulse radiolysis of oxygen-containing solutions, where in a diffusion-controlled reaction, all initially produced reducing equivalents ($e_{aq}^- H$) are coverted into the superoxide radical ion, O_2^-. This radical reduces laccase, as shown by the decrease of the absorption at the blue band (614 nm) (Figure 10). The striking feature of this system is that the presence of oxygen in the solution leads to the immediate reoxidation of this site.

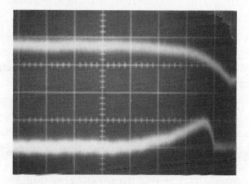

Figure 10. Rhus laccase—reduction by O_2^- and reoxidation by oxygen observed at 614 nm. Solution contained 1.5 × 10^{-5}M enzyme, 1.2 × 10^{-3} M oxygen, 1% tert-butanol, pH 6.9, 25°. Upper trace: Reduction of type 1 Cu(II) by O_2^-. Sensitivity, 2.4 mOD/division; sweep rate, 5 msec/division. Lower trace: Reduction and reoxidation of type 1 Cu(II). Sensitivity, 3.2 mOD/division; sweep rate, 100 msec/division.

The reduction has been followed over a range of concentrations including the presence of excess laccase (2×10^{-6} to $2.5 \times 10^{-5} M$) and at comparable or smaller concentrations relative to oxygen. The decrease in absorption at 614 nm to a lower level was always a first-order process, independent of either initial protein or oxygen concentration with the specific rate of $1.3 \pm 0.2 \times 10^2$ sec^{-1} ($\sim 25°$, $I \sim 0$). The reduction yield was low. Even at three-fold excess of reductant over enzyme, only 7% of the type 1 copper was reduced. This may be caused by effective competition by carbohydrate residues bound to the protein, which constitute about 45% of its weight, and to the dismutation of the O_2^- radicals. Preincubation of laccase with 2.5-mM fluoride ions, which bind to type 2 Cu(II), lowered the specific reduction rate by more than 50%. These findings suggest that the rate-limiting step in the reduction of type 1

Cu(II) is intramolecular. The effect of fluoride on the reduction rate is consistent with both a direct involvement of type 2 Cu(II) in the reduction or an indirect effect mediated via a change in conformation or in redox potential of the type 1 Cu(II). The type 2 copper ion could be the primary electron-accepting site of the laccase molecule, as has been proposed for the reduction of the enzyme by hydroquinone (36), the first-order process observed being therefore the electron transfer from type 2 Cu to type 1 Cu(II). The particaption of type 3 Cu(II) instead of type 2 Cu(II) is not excluded, but no associated change of its absorption band at 330 nm could be observed during the redox cycle described for the 614-nm band.

The reoxidation, taking place under conditions of a very large oxygen excess, followed a first-order pattern and was independent of the dioxygen concentration. Furthermore, no dependence of the specific rate on the amount of total or reduced enzyme could be observed. The rate constant was $2.5 \, sec^{-1}$ ($\sim 25°$, $I \sim 0$). The relative extent of reoxidation varied from 60 to 100% and did not depend either on oxygen concentration or on the amount of reduced enzyme. The variation in the degree of reoxidation seems to be caused by qualitative differences in the protein samples used. Work is in progress to clarify this point. The presence of fluoride ions decreased the reoxidation rate constant by more than one order of magnitude and the extent of reoxidation by at least 50%. When *Rhus* laccase was reduced pulse radiolytically with e_{aq}^- or CO_2^- radicals under anaerobic conditions, no reoxidation of the type 1 Cu was observed (37). When the reaction was carried out at relatively low oxygen concentration ($1 \times 10^{-4} \, M$), the major part of the type I Cu(II) reduction was caused by e_{aq}. Yet full reoxidation is also observed as under higher oxygen concentrations. These results show that we are indeed observing two distinct processes of reduction and reoxidation of type 1 copper at least, excluding their interpretation as a reversible formation of laccase $-O_2^-$ complex which subsequently decomposes because of the decay of O_2^- in the solution.

In the presence of molecular oxygen, and only then, the type 1 copper is reoxidized in a first-order process. The type 1 copper is generally thought to be deeply embedded inside the protein, therefore a direct interaction between oxygen and this copper site is improbable. This, together with the fact that no laccase molecule contained more than one reduction equivalent suggests the following:

(1) Molecular oxygen is able to interact with some site of the enzyme (type 2 or type 3 copper) in a fast reaction, after this site has received one electron from type 1 copper.

(2) The rate-determining step is probably the intramolecular electron transfer or a conformational change coupled to it.

At the present stage we have no information about the electron transfer from the type 1 copper. The marked effect of fluoride on the reoxidation suggests the participation of the type 2 copper in the reaction, but the involvement of type 3 copper cannot be excluded. The effect of fluoride may be direct, by simply inhibiting the interaction between type 2 (or type 3) copper and molecular oxygen, or indirect, by modifying the coordination of the copper site and thereby affecting its reactivity towards oxygen.

The Interaction between Rhus laccase and Hydrogen Peroxide

The role of the peroxide stage in the reduction of dioxygen by laccases and related oxidases is of long standing interest (2, 38, 39). In the formal sense, at least, this intermediate is obvious in the conversion of oxygen to water. Still, no direct evidence for its existence has been obtained yet. Among the different possible intermediates to be considered, the peroxide stage is the one expected to be thermodynamically most favorable. Recent values for the redox potentials of dioxygen reduction intermediates obtained mainly from pulse radiolysis studies ($E^\circ_{O_2-H_2O_2} = +0.69$ V (0.27 at pII 7); $E^\circ_{O_2-O_2^-} = -0.33$ V) reaffirm the view that a two-electron pathway is more advantageous than a sequence of single electron steps (40, 41). The stabilization energy provided by the binding of dioxygen and its reduction intermediate to the enzyme may also be an important factor in promoting the two-electron pathway (42). We have studied the reaction of *Rhus* laccase, in its oxidized and reduced states, with hydrogen peroxide and have found evidence for a specific high affinity interaction between the reagent and the enzyme which leads to a stable, spectroscopically distinct product (Figure 11, insert).

The oxidative titration of fully reduced *Rhus* laccase by H_2O_2 is shown in Figures 11 and 12. The 614-nm absorbance caused by the type 1 Cu(II) achieves its final value after adding about 4.0 redox equivalents (2 moles H_2O_2), as expected on the basis of oxidative titrations with other oxidants. In contrast, the near-uv band from the type 3 copper first reaches about the value expected for the fully oxidized enzyme, but then increases further until it levels off when the third mole of H_2O_2 has been added. The new band shown in Figure 11 is similar in general shape to that attributed to the type 3 site, but the maximum is slightly blue shifted.

Figure 12 shows the titration data in the form of a Nernst plot of the 330-nm absorbance against the 614-nm absorbance, including values up to full reoxidation of the type 1 chromophore. The slope n of the straight line is 1. This is in contrast to the value obtained from titrations of oxidized laccase with hydroquinone as reductant, which gave $n = 2$

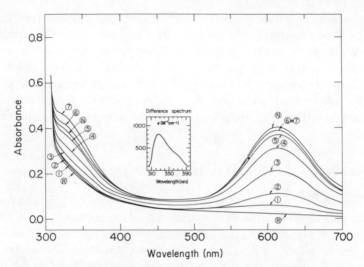

Figure 11. *Spectral changes upon anaerobic titration of reduced* Rhus *laccase with hydrogen peroxide.*

(N) Native laccase, 7.0×10^{-5} M, pH 7.0. (R) Fully reduced laccase after adding 1.54×10^{-4} M ascorbic acid. (1)–(7) Spectra obtained 20 min after each successive addition of 2.8×10^{-5} M H_2O_2. Addition of another 2.8×10^{-5} M H_2O_2 increased the absorbance at 330 nm by 0.02 units (not shown). Further addition of H_2O_2 did not change the spectrum. Insert: the difference spectrum of peroxide-treated laccase–native laccase in the near-uv.

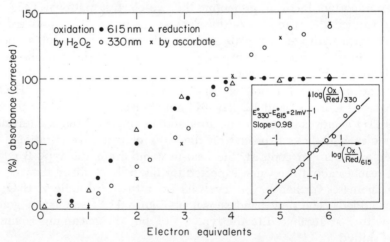

Figure 12. *Anaerobic titration of* Rhus *laccase.*

Oxidation by hydrogen peroxide (● and ○) followed by reduction by ascorbate (△ and ×). The abscissa represents electron equivalents present per molecule of laccase. Insert: Nernst plot of the type 3 copper against type 1 copper. Ox/Red represents the calculated ratio between concentration of the oxidized and reduced chromophores. For calculating Ox_{330} the extinction valve for native oxidized laccase at 330 nm was used.

(Figure 13), and with previously reported reductive titrations using one-electron donors (*8, 43*). The finding that in the reductive titrations n is always 2 led to the concept that the type 3 site is a two-electron acceptor. The fact that $n = 1$ for the titration of reduced laccase with H_2O_2 may indicate that under these conditions the type 3 and type 1 are in a single electron equilibrium. This would mean that while the oxidized type 3 Cu(II) pair behaves as a strongly cooperative two-electron acceptor, it undergoes uncoupling in the reduced protein and functions then as a pair of independent single-electron exchanging sites. These may be equivalent or nonequivalent, as, under certain specific conditions, both cases are consistent with the stoichiometry and the course of the oxidative titration.

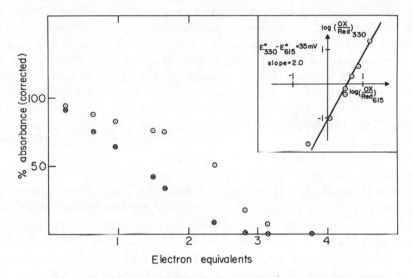

Figure 13.   Anaerobic titration of Rhus *laccase by hydroquinone.*
Results from two titrations of 1.2 × 10⁻⁴ M and 1.6 × 10⁻⁴ M protein, pH 7.0. The hydroquinone solution (0.01M) contained 10⁻⁴ M EDTA. The abscissa represents electron equivalents added per molecule of laccase. Insert: Nernst plot of the type 3 copper vs. the type 1 copper. A° is the corrected optical absorption of the fully oxidized enzyme. A is the absorption, corrected for the background absorption, measured during the titration. The straight line is drawn with a slope = 2.

For consistency with the experimental data, an equivalent site model requires that the absorption characteristics of each of the two type 3 copper ions be independent of their state of coupling. For a nonequivalent site model one has to assume that the absorption at 330 nm is entirely caused by the Cu(II) ion with the higher redox potential, so that the observed equilibrium can be assigned to the oxidation–reduction of this site. In both cases we take into account that the addition of four oxidizing equivalents restore the original absorption of the enzyme, and we assume that the redox potential of the type 2 copper is not affected by the titration pattern.

Since there can be only one true equilibrium state, at least one of the two functional states of coupling should be of nonequilibrium nature (44), leading to the observed chemical hysteresis. This suggests a coupling–uncoupling of the type 3 pair linked to metastable conformational states in the protein.

When oxidized laccase was reacted with one mole of H_2O_2, both under anaerobic and aerobic conditions, the same spectral changes as in the final phase of the titration of reduced laccase with peroxide were observed in the near-uv. For this spectral change to be complete, an equimolar amount of H_2O_2 is sufficient even at enzyme concentrations as low as $2 \times 10^{-6} M$. This change developed slowly and at these concentrations about 20 min were required for the maximal value to be attained. It remained stable even at room temperature for more than 24 hr. In order to check whether this spectral change indeed results from a specific interaction between the enzyme and peroxide, we investigated the effect of several strong oxidants on the spectral properties of the oxidized enzyme. This was examined with peroxydisulfate (with and without catalytic amounts of copper), $(PtCl_6)^{2-}$ and $(IrCl_6)^{2-}$ ions, and also with varying concentrations of oxygen. No other reagent except for H_2O_2 led to this spectral behavior, although when reacted with the reduced enzyme, full reoxidation of both type 1 and type 3 sites was achieved as judged by their absorptions. The stability of the new species was examined by two efficient catalysts for H_2O_2 decomposition, namely platinum black and bovine catalase. Neither of these catalysts caused any significant change of the extra extinction at 330 nm. Two kinds of experiments were carried out in order to establish the reversible nature of the enzyme–peroxide interaction. Peroxide-treated enzyme was reduced with excess ascorbate and opened to air to provide turnover conditions. The spectrum of the solution after exhaustion of all the substrate corresponded to that of native laccase without the extra absorption at 330 nm. In further experiments, reductive titrations of the peroxide-treated enzyme with ascorbate ions were performed. Laccase which had been treated with one mole of H_2O_2 was incubated with 0.01 μM catalase overnight, in order to decompose any residual H_2O_2, and then titrated with ascorbate. The results of this titration are also shown in Figure 12. As can be seen the full reduction of the 330-nm band required six redox equivalents.

We checked whether the reaction between oxidized laccase and H_2O_2 is affected by the presence of fluoride ions, which are known to bind strongly to type 2 $Cu(II)$. No effect could be found even at fluoride concentration of $10 \times 10^{-3} M$. Moreover, the EPR spectra of peroxide-treated and native enzyme were indistinguishable even when measured at Q-band and 10°K.

A plausible rationalization of our observations is the formation of an enzyme–peroxide complex at the type 3 site. The fact that it is fully formed even at rather low concentrations of reagents ($2 \times 10^{-6} M$ in each) and the stability observed over many hours in the presence of highly efficient and specific catalysts for peroxide decomposition require an exceptional affinity of interaction.

In contrast to the tree laccase–H_2O_2 complex described here, the type 2–H_2O_2 complex described by Brandén et al. (*10*) for the fungal enzyme is of significantly lower stability, as it is not formed in an observable amount at 0.03mM enzyme and 0.1mM H_2O_2. It is interesting, that although the spectrum reported for the fungal laccase–H_2O_2 complex (maximum absorbance at 400 nm) is in general different from that measured for the tree laccase, a difference spectrum in the near-uv region can be calculated which is very similar to the difference spectrum for the tree enzyme. This, together with the observation that in fungal laccase fluoride inhibits the appearance of the 400-nm band, but not the intensity increase around 330 nm, strongly suggests that Brandén et al. (*10*) were also observing H_2O_2 interactions with type 3 site. The differences in behavior between the two laccases may be, amongst others, the consequence of the markedly higher potentials of the fungal enzyme copper sites.

Makino and Ogura (*44*) reported changes after adding hydrogen peroxide to oxidized laccase which were similar to those found in the present study. However, the increase in absorbance around 330 nm was interpreted as being caused by some reduced type 3 copper, present in the native preparation, being oxidized by hydrogen peroxide.

The most intriguing and significant aspect of the peroxide complex of *Rhus* laccase naturally lies in its possible role in the functional reduction of dioxygen. Under conditions of excess reductant, the steady-state concentration of this complex is expected to be rather low. However, when the amount of reductant is limiting, it is conceivable that half-reduced species, upon reacting with oxygen, would produce this complex as an intermediate at a more significant concentration. The high binding energy for peroxide (> 10 kcal/mole) should be of advantage in overcoming the barriers in the reduction step. Preliminary experiments where half-reduced laccase was reacted with dioxygen seem to support this hypothesis. We have found that a rapidly formed intense transient absorption decays with a half-life of ~ 27 min (at 25°) to a stable spectrum strikingly similar to the laccase–peroxide spectrum, with the same specific extinction of the extra band at ~ 330 nm. The EPR spectrum at 10°K is the same as for native or peroxide-treated laccase, and at least five reducing equivalents are needed for the full reduction of this new species.

It has recently been reported (45) that reoxidation of fully reduced *Rhus* laccase with oxygen is characterized by a second-order rate constant of about $5 \cdot 10^6 \, M^{-1} \, s^{-1}$ for both type 1 and type 3 copper. We have investigated the kinetics of the reaction between reduced *Rhus* laccase and hydrogen peroxide spectrophotometrically under anaerobic conditions at 10° and 25°C by following the changes in absorbance at 330 and 615 nm (39). The rate of reoxidation of the type 1 Cu(I) was found to be first order, independent of either concentration of H_2O_2 or the state of reduction of other sites in the enzyme (10°C, $k = 4.6 \cdot 10^{-3} \, s^{-1}$; 25°C, $k \sim 0.015 \, s^{-1}$). The oxidation of the type 3 site by hydrogen peroxide is significantly faster ($\tau_{1/2} < 10 \, s$ at 25°C). At 10°C we found the rate to be first order both in reduced type 3 site and in hydrogen peroxide, with an overall second-order rate constant $k = 1.8 \times 10^3 \, M^{-1} \, s^{-1}$. These findings indicate that the primary step in this reaction is the reoxidation of the type 3 site by H_2O_2, in parallel with a slower intramolecular oxidation of the type 1. The reoxidation of type 2 copper cannot be monitored spectrophotometrically, but from the overall stoichiometry as well as from the above equilibrium titrations, it is obvious that this site is also involved in the reaction. A more extensive kinetic investigation of the reduction of H_2O_2 is presently being carried out.

Concluding Remarks

We have used a range of different physical and chemical approaches in the effort to better understand how the different blue copper proteins function. With the relatively simpler, electron-mediating proteins like azurin, the ultraviolet chromophores were shown to be informative in terms of copper–protein interactions. These proteins are also a useful system for detailed examination of the electron transfer pathways to and from their single copper site.

The elaborate mechanism by which blue oxidases react with dioxygen to produce water was tackled by studying the possible role of H_2O_2. We have observed the formation of a stable and high affinity complex between tree laccase and H_2O_2. Moreover, the finding that the oxidation of the reduced enzyme with H_2O_2 follows a pattern which is different from that operative in the reduction of the oxidized enzyme may have important implications for the mechanism of action of laccase.

Literature Cited

1. Dickerson, R. E., Timkovich, R., in "The Enzymes," Boyer, P. B., Ed., 3rd ed., Vol. XI, Academic, New York, 1975.

2. Fee, J. A., *Struct. Bonding* (1975) **23**, 1.
3. McMillin, D. R., Rosenberg, R. C., Gray, H. B., *Proc. Nat. Acad. Sci. U.S.A.* (1974) **71**, 4760.
4. Siiman, O., Young, N. M., Carey, P. R., *J. Am. Chem. Soc.* (1976) **98**, 744.
5. Solomon, E. I., Clendening, P. J., Gray, H. B., *J. Am. Chem. Soc.* (1975) **97**, 3878.
6. Miskowski, V., Tang, S.-P. W., Spiro, T. G., Shapiro, E., Moss, T. H., *Biochemistry* (1975) **14**, 1244.
7. Rosen, P., Pecht, I., *Biochemistry* (1976) **15**, 775.
8. Reinhammar, B., Vänngard, T., *Eur. J. Biochem.* (1971) **18**, 463.
9. Andréasson, L.-E., Malmström, B. G., Strömberg, C., Vänngard, T., *Eur. J. Biochem.* (1973) **34**, 434.
10. Brandén, R., Malmström, B. G., Vänngard, T., *Eur. J. Biochem.* (1971) **18**, 234.
11. Reinhammar, B., *Biochim. Biophys. Acta* (1970) **205**, 35.
12. Ambler, R. P., Brown, L. H., *Biochem. J.* (1967) **104**, 784.
13. Malmström, B. G., Reinhammar, B., Vänngard, T., *Biochim. Biophys. Acta* (1970) **205**, 48.
14. Goldberg, M., Pecht, I., *Biochemistry* (1976) **15**, 4197.
15. Tang, S.-P. W., Coleman, J. E., Myer, Y. P., *J. Biol. Chem.* (1968) **243**, 4286.
16. Rosen, P., Pecht, I, *Israel J. Med. Sci.* (1977) in press.
17. Kolthoff, I. M., Sandell, E. B., "Textbook of Quantitative Inorganic Analysis," 3rd edition, p. 600, Macmillan, New York, 1961.
18. Rigler, R., Rabl, C. R., Jovin, T. M., *Rev. Sci. Instrum.* (1974) **45**, 580.
19. Faraggi, M., Pecht, I., *J. Biol. Chem.* (1973) **248**, 3146.
20. Goldberg, M., Pecht, I., *Proc. Natl. Acad. Sci. U.S.A.* (1974) **71**, 4684.
21. Ambler, R. P., in "Recent Developments in the Chemical Study of Protein Structures," A. Previero, J.-F. Pechere, and M.-A. Coletti-Previero, Eds., p. 289, Inserm, Paris, 1971.
22. Yamanaka, T., Kijimoto, S., Okunuki, K., *J. Biochem.* (Tokyo) (1963) **53**, 256.
23. Finazzi-Agrò, A., Rotilio, G., Avigliano, L., Guerrieri, P., Boffi, V., Mondovi, B., *Biochemistry* (1970) **9**, 2009.
24. Rupp, H., Weser, U., *FEBS Lett.* (1974) **44**, 293.
25. Briving, C., Deinum, J., *FEBS Lett.* (1975) **51**, 43.
26. Jones, T. E., Rorabacher, D. B., Ochrymowycz, L. A., *J. Am. Chem. Soc.* (1975) **97**, 7485.
27. Ryden, L., Lundgren, J.-O., *Nature* (1976) **261**, 344.
28. Peisach, J., in "The Biochemistry of Copper," J. Peisach, P. Aisen, and W. E. Blumberg, Eds., p. 404, Academic, New York, 1966.
29. Zuberbuhler, A., Kaden, Th., *Helv. Chim. Acta* (1968) **51**, 1805.
30. Levitzki, A., Pecht, I., Berger, A., *J. Am. Chem. Soc.* (1972) **94**, 6844.
31. Markley, J. L., Ulrich, E. L., Berg, S. P., Krogmann, D. W., *Biochemistry* (1975) **14**, 4428.
32. Eigen, M., De Maeyer, L., in "Techniques of Chemistry," A. Weissberger, Ed., 3rd ed., vol. II, part II, p. 63, John Wiley, New York, 1974.
33. Thusius, D., Foucault, G., Guillain, F., in "Dynamic Aspects of Conformational Changes in Biological Macromolecules," C. Sadron, Ed., p. 271, D. Reidel, Dordrecht/Boston, 1973.
34. Wherland, S., Holwerda, R. A., Rosenberg, R. C., Gray, H. B., *J. Am. Chem. Soc.* (1975) **97**, 5260.
35. Beattie, J. K., Fensom, D. J., Freemann, H. C., Woodcock, E., Hill, H. A. O., Stokes, A. M., *Biochim. Biophys. Acta* (1975) **405**, 109.
36. Holwerda, R. A., Gray, H. B., *J. Am. Chem. Soc.* (1974) **96**, 6008.
37. Pecht, I., Goldberg, M., in "Fast Processes in Radiation Chemistry and Biology," G. E. Adams, M. Fielden, and B. D. Michael, Eds., p. 277, John Wiley, London, 1975.

38. Pecht, I., Abstracts from Sixth FEBS Meeting, Madrid, 1969, 150.
39. Farver, O., Goldberg, M., Lancet, D., Pecht, I.. *Biochem. Biophys. Res. Commun.* (1976) **73**, 494.
40. Ilan, Y. A., Meisel, D., Czapski, G., *Israel J. Chem.* (1974) **12**, 891.
41. Wood, P. M., *FEBS Lett.* (1974) **44**, 22.
42. Jencks, W. P., *Adv. Enzymol.* (1975) **43**, 219.
43. Reinhammar, B., *Biochim. Biophys. Acta* (1972) **275**, 245.
44. Revzin, A., Neumann, E., Katchalsky, A., *J. Mol. Biol.* (1973) **79**, 95.
45. Makino, N., Ogura, Y., *J. Biochem.* (Tokyo) (1971) **69**, 91.
46. Andréasson, L.-E., Brandén, R., Reinhammar, B., *Biochim. Biophys. Acta* (1976) **438**, 370.

RECEIVED July 26, 1976.

Mechanism of Autoreduction of Ferric Porphyrins and the Activation of Coordinated Ligands

GERD N. LA MAR and JOHN DEL GAUDIO

Department of Chemistry, University of California, Davis, CA 95616

The autoreduction of ferric porphyrins in the presence of certain ligands is accompanied by the formation of ligand radicals, suggesting that these reactions may serve as models for the activation of substrates by peroxidases. NMR and ESR spectroscopy demonstrate that the autoreduction of the dicyanotetraphenylporphinatoferrate(III) yields the dicyano ferrous porphyrin and the cyanide radical by a mechanism thought to involve homolytic bond cleavage. Similar reactions are observed with n-hexane thiol and piperidine. The reoxidation by molecular oxygen of the ferrous complex of CN^- or tributyl phosphine leads directly to the low spin ferric complex, as opposed to the expected oxo-dimer. A mechanism involving the formation of the superoxide anion is proposed, suggesting that this reaction may model the hemoprotein activation of molecular oxygen.

Iron porphyrins constitute the active site of an important class of redox metalloenzymes. This class includes the cytochromes which are involved in electron transfer processes (*1*); peroxidases, whose main function is to oxidize substrates at the expense of hydrogen peroxide (*2*); and oxygenases, which catalyze the incorporation of oxygen into substrates via the activation of molecular oxygen (*3*). Although the functions of these enzymes are quite varied, they all cause the iron atom to undergo valency changes during the operation of the enzyme. The mechanism by which an electron is transferred to and from the iron atom is poorly understood.

In the case of a simple redox enzyme such as cytochrome *c* where both sites of the heme iron are ligated by peptide side chains, two pathways have been suggested by which an electron can travel to or from the iron atom—a direct electron transfer from the enzyme reductase to an exposed edge of the porphyrin π cloud followed by a rapid transfer to the iron or, alternatively (4), electron transfer via an axial ligand of the iron porphyrin (5). Both reduction mechanisms involve rapid ligand-to-metal charge transfer as a critical step in the process. A related class of redox hemoproteins, including oxygenases, peroxidases, and cytochrome P450, has only one polypeptide side-chain ligand bound to the heme iron, with the sixth site available for coordinating a substrate which is to be activated. Thus these enzymes are involved not only in one-electron valency changes of the iron, but these changes are also coupled to the oxidation or reduction of the substrates. In the case of oxygenases and cytochrome P450, this activation of molecular oxygen is thought to involve formation of the superoxide ion (6) by a reaction such as Reaction 1. The superoxide ion as the activated form of molecular oxygen has been substantiated (6).

$$E–Fe:O_2 \rightarrow E–Fe^{III} + O_2^- \tag{1}$$

Peroxidases peroxidize a variety of substrates at the expense of hydrogen peroxide. These peroxidases oxidize amines, AH_2, for which product analysis has suggested a free radical intermediate (2). The currently accepted valency changes of the enzyme can be represented by the scheme in Reactions 2–5. Reaction 2 represents the conversion

$$\text{perox } (Fe^{III}) + H_2O_2 \rightarrow \text{compound I } (Fe^{IV}, \text{por}^+) \tag{2}$$

$$\text{compound I} + AH_2 \rightarrow \text{compound II } (Fe^{IV}) + AH^· \tag{3}$$

$$\text{compound II} + AH_2 \rightarrow \text{perox } (Fe^{III}) + AH^· \tag{4}$$

$$AH^· \rightarrow \text{products} \tag{5}$$

of the enzyme to a highly oxidized state which is capable of one two-electron or two one-electron oxidations (7) (por$^+$ represents the porphyrin radical). In Reactions 3 and 4, activation of the amine is proposed to consist of one-electron oxidations of the amine. However, the one-electron oxidation of coordinated ligands by iron porphyrins had not been demonstrated at the time we initiated this research. ESR investigations of the oxidation of some substrates by peroxidases have provided direct evidence for transient free radical intermediates (8, 9). However, it has not been determined whether these substrates coordinate to the iron or whether they are secondary products of activation by the enzyme. The identification and investigation of model systems which can activate substrates by one-electron redox reactions can be expected to provide valuable insight into the mechanism of the heme enzymes.

Scattered reports have appeared in the literature indicating that ferric porphyrins can autoreduce in solutions containing certain potential ligands (*10, 11, 12*). Probably the best known example is the formation of bis(piperidine)tetraphenylporphorinatoiron(II), TPPFe^{II}(pip)$_2$, by the addition of piperidine to chlorotetraphenylporphyrinatoiron(III), Reaction 6. Although this reaction was first reported in 1967 (*13*), the

$$\text{TPPFe}^{III}\text{Cl} \overset{\text{pip}}{\to} \text{TPPFe}^{II}(\text{pip})_2 \tag{6}$$

identity of the reducing agent and the mechanism of the reaction remained obscure until very recently. Our recent preliminary communication (*14*) on the autoreduction of ferric porphyrins demonstrated that this reaction involves reduction of the complex by a presumably coordinated ligand, with the substrate being oxidized to a free radical. The radicals were readily detected by the appearance of an ESR signal during the reaction involving cyanide ion, piperidine, or *n*-hexanethiol as substrates.

Inasmuch as this redox reaction provides a method for activating substrates which may coordinate to iron porphyrins and hence provides some kind of a model for peroxidase activity, we have continued our work on the redox system most amenable to spectroscopic investigation, namely the cyanide ion oxidation. We have further discovered that oxidation of the reduced bis-cyano ferrous complex by molecular oxygen does not yield the expected oxo-bridged dimer (*15*) but instead appears to proceed by a mechanism that suggests formation of the superoxide ion, O_2^-. Since this latter process would represent the one-electron reduction of molecular oxygen by a ferrous porphyrin complex and could conceivably shed light on the mechanism of oxygen activation (*6*), the reoxidation by molecular oxygen of the autoreduced ferric complexes was also investigated spectroscopically.

Thus the two reactions which we propose to clarify are the autoreduction (Reaction 7) (where *P* is a general porphyrin) and the subsequent reoxidation (Reaction 8). Although a number of substrates (i.e.,

$$\text{PFe}^{III}X \underset{?}{\overset{L}{\to}} \text{PFe}^{II}L_2 + L^{\cdot} \tag{7}$$

$$\text{PFe}^{II}L_2{}^n + O_2 \to \text{PFe}^{III}L_2{}^{n+1} \tag{8}$$

amines, thiols, phosphines, cyanide), reacted according to Reaction 7, as evidenced by the change from ferric to ferrous optical and NMR spectra, our work emphasizes the reactions for $L = CN^-$ because the nature of the complexes in both oxidation states can be determined. The preliminary data on the more complex piperidine reaction are presented only to indicate that the autoreduction mechanism involving a one-electron redox reaction appears common for the various substrates.

Experimental

Materials. The TPPFeCl was prepared and purified by literature methods (*16, 17*). Octaethyl porphyrin was a gift from H. H. Inhoffen, and the hemin dimethyl ester [PP(IX)DMEFe^{III}Cl] was purcased from Sigma Chemical Co.

The solvents used for NMR and ESR measurements were commercial sources of deuterated solvents while spectro-grade solvents were used for visible spectra. Solvents were dried by storing over molecular sieves.

The amines, tributyl phosphine, *n*-hexanethiol, NaCN, and KCN were obtained from standard commercial sources. The CN$^-$ source in CDCl$_3$ and CD$_2$Cl$_2$ was NBu$_4$CN which was prepared by literature methods (*18*).

Spectroscopic Measurements. NMR SPECTRA. The NMR spectra were recorded at 298°C with a JEOL-PS100 FT NMR spectrometer operating at 99.5 MHz. The autoreductions were done directly in the NMR tube, generally with 0.4-ml samples of 10-m*M* iron porphyrin solutions containing a 20–100 molar excess of the substrate. Samples were routinely prepared in a nitrogen atmosphere.

Anaerobic samples were prepared by degassing the solvent by three freeze-thaw cycles while solids and NMR tubes were degassed by placing under vacuum, then storing in a nitrogen atmosphere. If oxygen is not carefully excluded, the ferrous porphyrin obtained by the autoreduction may be rapidly oxidized to the oxo-bridged dimer.

A photochemical contribution to the autoreduction reaction was investigated by preparing two identical NMR samples of TPPFe^{III}(CN)$_2{}^{1-}$ in dry DMSO containing excess CN^{-1}; one sample was shielded from and the other exposed to the normal fluorescent light in the laboratory. The NMR were recorded, and it was noted that the exposed sample autoreduced about 50% faster. Under the conditions of the experiment, thermal effects on the exposed sample are expected to be negligible.

The ferrous porphyrins were reoxidized by introducing oxygen into the NMR tube of the autoreduced sample. To detect water as a product of the reoxidation of TPPFe^{II}(CN)$_2{}^{2-}$, it was necessary to completely exclude the possibility of atmospheric contact (dry DMSO rapidly absorbs water from the atmosphere). These experiments were done in an NMR tube fitted with a ground glass stopcock. This allowed the addition of oxygen into the NMR tube by vacuum line techniques, completely eliminating atmospheric contact. Experiments done on blanks of dry DMSO showed no water peak.

ESR SPECTRA. The ESR spectra were recorded with a Varian E-4 EPR spectrometer. The autoreductions and reoxidations were followed in the ESR cavity generally with 0.1-ml samples of the same concentration as the above NMR samples. In the case of TPPFe^{III}(CN)$_2{}^{1-}$ the ESR and NMR of the autoreduction were recorded on the same sample by supporting the ESR tube in an NMR tube, thus verifying that the ESR signal arose as the autoreduction proceeded.

VISIBLE SPECTRA. The visible spectra were obtained with a Cary 14 recording spectrophotometer by standard techniques. The spectrum of TPPFe^{III}(CN)$_2^{1-}$ in DMSO containing excess CN$^-$ was obtained under oxygen to insure that autoreduction had not occurred. The spectrum of TPPFe^{II}(CN)$_2^{2-}$ was obtained on a sample whose NMR spectrum had indicated complete reduction.

Results and Discussion

Autoreduction of Fe(III) Porphyrins. OXIDATION OF PIPERIDINE. The autoreduction of TPPFeCl with neat piperidine is rapid, however the reaction rate can be decreased by dilution with CDCl$_3$ or DMSO. The NMR spectrum of TPPFeCl in CDCl$_3$, on addition of piperidine, shows resonances consistent with the presence of high spin Fe(III) and low spin TPPFe^{II}(Pip)$_2$. A low spin TPPFe^{III}(Pip)$_2$ species was not observed. It is likely that the latter complex rapidly reduces at room temperature.

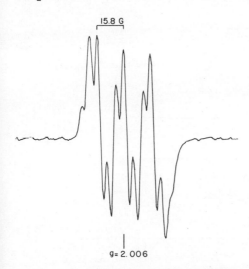

15.8 G

g = 2.006

Figure 1.   ESR signal obtained by adding piperidine to a CDCl$_3$ solution of TPPFeCl

The reduction was carried out in an ESR cavity and found to be accompanied by the appearance of a strong ESR signal exhibiting a triplet structure indicative of an amine radical (Figure 1). This suggests that the reduction of the iron porphyrin proceeds by a one-electron oxidation of a coordinated piperidine. Horseradish peroxidase activates aniline (2), and the activation is thought to occur via the generation of an aniline radical, (C$_6$H$_5$NH·). There is a similarity between Reaction 4 of the peroxidase scheme, which in the case of aniline would involve the one-electron oxidation of aniline, perhaps by coordination to an Fe(IV) porphyrin, compared with Reaction 7 which most likely involves the one-electron oxidation of piperidine coordinated to an Fe(III) porphyrin.

The piperidine reaction is not completely characterized. The conditions for the generation of the radical in Figure 1 have not been clearly defined, and intermediates observed during the reduction have not been identified. These difficulties have been overcome in the cyanide system, which proved to be more amenable to spectroscopic investigation.

OXIDATION OF CYANIDE ION. In the case of CN⁻ as substrate, identification of reactants and products proved feasible. Addition of excess KCN to a 10-mM TPPFeCl solution in DMSO gives the low spin ferric bis-cyanide complex, Reaction 9. The NMR of TPPFe(CN)$_2^{1-}$ has been

$$TPPFeCl + 2CN^- \rightleftharpoons TPPFe(CN)_2^{1-} + Cl^- \qquad (9)$$

previously characterized (*19*) and is illustrated in A of Figure 2. The spectrum is time dependent, converting from that typical of low spin Fe(III) to one characteristic of diamagnetic Fe(II), E in Figure 2. The electronic spectrum of the initial and final solution is shown in Figure 3.

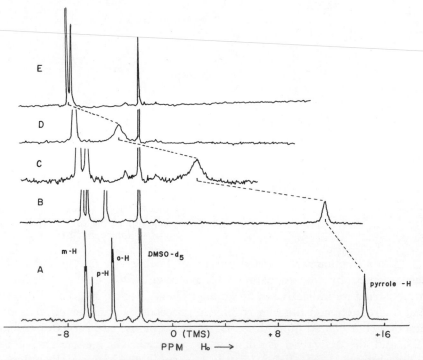

Figure 2. Proton NMR traces showing the autoreduction of TPPFe^{III}-(CN)$_2^{1-}$ in DMSO at 25°C. (A) TPPFe^{III}(CN)$_2^{1-}$; (B–D) increasing reduction; (E) final product TPPFe^{II}(CN)$_2^{2-}$.

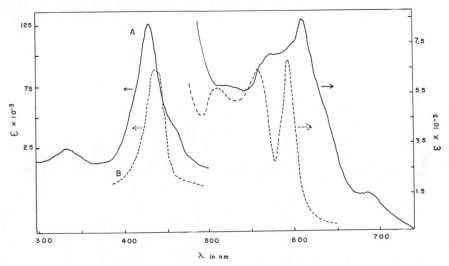

Figure 3. Electronic spectrum of (A) TPPFe^{III}(CN)$_2$$^{1-}$ in DMSO; (B) after complete reduction

Spectrum B, the final solution, is also characteristic of an Fe(II) porphyrin (20) and verifies that reduction of the iron has occurred.

In the case of the NMR spectra, the peaks move with an average resonance observed for each position (B, C, and D of Figure 2). The observation of an averaged chemical shift indicates that the same entity exists in both oxidation states and establishes the product of the reduction to be the ferrous bis-cyano complex (i.e., Reaction 10). Rapid electron exchange averages the two spectra according to Reaction 11. The chemical shift difference gives a lower limit to the rate of electron exchange, with $k \gg 1.5 \times 10^4$ sec.

$$\text{TPPFe(CN)}_2{}^{1-} + e^- \rightarrow \text{TPPFe(CN)}_2{}^{2-} \qquad (10)$$

$$\text{TPPFe}^{II}(\text{CN})_2{}^{2-} + \text{TPPFe}^{*III}(\text{CN})_2{}^{1-} \underset{k_1}{\overset{k_{1'}}{\rightleftarrows}} \qquad (11)$$
$$\text{TPPFe}^{III}(\text{CN})_2{}^{1-} + \text{TPPFe}^{*II}(\text{CN})_2{}^{2-}$$

The preparation of the iron porphyrin TPPFeCl-d_{20} (21), where the phenyl groups are completely deuterated, has allowed the further characterization of the bis-cyano Fe(II) and Fe(III) porphyrins. In the spectrum of TPPFe^{III}(CN)$_2$$^{1-}$ in DMSO, the resonance at +15.45 ppm upfield from TMS is verified as the pyrrole H. Comparing the spectrum of TPPFe^{II}(CN)$_2$$^{2-}$ and TPPFe^{II}(CN)$_2$$^{2-}$-$d_{20}$ shows the peak at −7.82 ppm from TMS to be the pyrrole H, and integration indicates that the

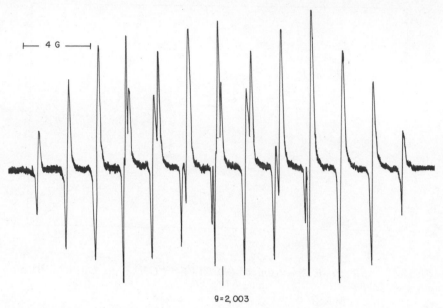

Figure 4. The ESR spectrum observed during the autoreduction of TPPFe^III $(CN)_2{}^{1-}$ in DMSO

low field shoulder at -7.85 ppm contains eight protons, consistent with its assignment as the phenyl ortho protons.

The nature of the reducing agent was provided by ESR spectroscopy. By carrying out the reduction of $TPPFe(CN)_2{}^{1-}$ in DMSO within the cavity of an ESR spectrometer, a strong signal appeared as the reduction proceeded. This complex signal, Figure 4, is identical to that previously reported during the anodic oxidation of tetraphenylarsonium cyanide in DMSO (22) and has been interpreted to represent the cyanide tetramer, A. The proposed mechanism of formation of this tetramer is reproduced in Reaction 12. Thus the cyanide radical $\cdot CN$ is being produced during the reduction of the ferric porphyrin, and the overall reaction consistent with both the NMR and ESR data can be written as Reaction 13.

$$CN^- \xrightarrow{-1e} CN\cdot \rightarrow (CN)_2 \xrightarrow{CN^-} (NC)_2C{=}N^- \xrightarrow[{-1e}]{CN^-} \left[\begin{array}{c} NC \\ \diagdown \\ NC \end{array} C{=}N \diagup\diagdown \begin{array}{c} CN \\ \end{array} \right]^{\cdot -}$$

$$\Big\lfloor_{CN^-} \uparrow_{-1e} $$

$$\xrightarrow{} (CN)_2{}^{\cdot -}$$

$$\hspace{8cm} A \qquad (12)$$

$$TPPFe^{III}(CN)_2{}^{1-} + CN^{-1} \rightarrow TPPFe^{II}(CN)_2{}^{2-} + \cdot CN \qquad (13)$$

FACTORS INFLUENCING THE AUTOREDUCTION OF TPPFe^{III}(CN)$_2$$^{1-}$.

Anaerobic Nature of the Reduction. Rigorous exclusion of molecular oxygen shows that the reduction proceeds anaerobically. This result excludes the possibility that the radicals observed during the autoreduction are being produced by some form of activated oxygen.

Autocatalytic Nature of the Reduction. The data in Figure 5 show the reduction rate to increase with time, indicating that the reduction is autocatalytic. This is most likely caused by the reduction of the ferric porphyrin by the intermediate radicals generated by Reaction 12, which are expected to be more potent reducing agents than the cyanide ion. Hence any detailed interpretation of the rates will be severely limited.

Solvent Effects. The autoreduction is solvent dependent, having been observed in CD$_3$CN and DMSO, but not in CDCl$_3$ and CD$_2$Cl$_2$. DMSO is the more suitable solvent, for solubility reasons, and the fol-

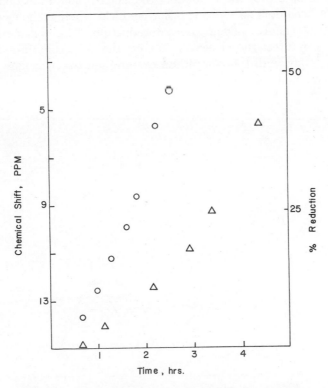

Figure 5. The effect of CN$^-$ concentration on the auto-reduction rate of a 10-mM solution of TPPFe^{III}(CN)$_2$$^{1-}$ in DMSO. $\triangle = 0.19$ M CN$^-$; $\bigcirc = 0.38$ M CN^{1-}. The increasing rate with time indicates that the autoreduction is also autocatalytic.

lowing discussions apply to the autoreduction of TPPFe(CN)$_2$$^{1-}$ in DMSO unless stated otherwise.

Effect of Light. The autoreduction occurs in the dark, that is by a thermal pathway, and further discussions are based only on results obtained with the exclusion of light. However, light accelerates the rate of autoreduction.

Effect of Water. The reduction rate increased as the concentration of water was decreased. For example, removing water ($\sim 50\, mM$) to levels where it is not observed by NMR increased the reduction rate almost three fold. Cyanide coordinated to ferric porphyrins acts as a hydrogen-bond acceptor towards water (23). Such an interaction would make the coordinated cyanide more difficult to oxidize (24) and hence would decrease the reaction rate.

Cyanide Ion Concentration. Increasing the cyanide ion concentration speeds up the autoreduction, as shown in Figure 5. Because of the autocatalytic nature of the reduction, the exact cyanide ion dependence has not yet been defined. Also it is not yet clear whether the cyanide ion is involved mechanistically in the autoreduction. The cyanide ion concentration can affect the observed rate via the autocatalytic mechanism or by competing with the complexed cyanide ion for hydrogen bonding with trace amounts of water.

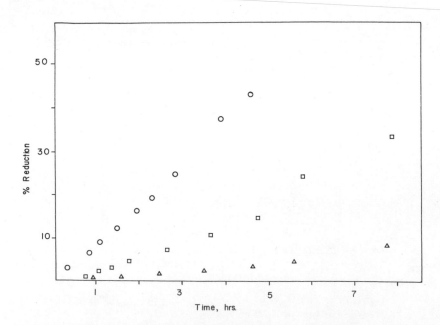

Figure 6. The reduction rate of various bis-cyano porphyrins. $\triangle = OEP$-$Fe^{III}(CN)_2{}^{1-}$, $\square = PP(IX)DMEFe^{III}(CN)_2{}^{1}$, $\bigcirc = TPPFe^{III}(CN)_2{}^{1-}$

Porphyrin Basicity. The reduction rate is decreased significantly as the porphyrin is made more basic (Figure 6). The bis-cyanide complex of ferric octaethyl porphyrin, the more basic porphyrin, is reduced the slowest while the bis-cyanide complex of ferric TPP, the least basic (*16*), is reduced the fastest. Increasing the porphyrin basicity places more electron density on the iron, making it more difficult to accept another electron.

Effect of Axial Ligand. In addition to cyanide ion and piperidine, autoreduction was observed for the following potential ligands—primary, secondary, and tertiary amines; pyridine; n-hexanethiol; and tributyl phosphine.

MECHANISM OF THE AUTOREDUCTION OF $TPPFe^{III}(CN)_2^{1-}$. The transfer of an electron from a cyanide ion to the Fe(III) can occur by at least three mechanisms.

(1) An outer sphere oxidation of free CN^{1-} by $TPPFe^{III}(CN)_2^{1-}$, which is mechanistically described by Reaction 13. The electrochemical oxidation of cyanide ion in acetonitrile occurs at potentials more positive than $+0.5$ volt vs. SCE (*22*). The electrochemical reduction of TPP-$Fe^{III}(CN)_2^{1-}$ in acetonitrile occurs at a potential of -0.5 volt vs. SCE (*25*). Thus an outer sphere mechanism is considered unlikely.

(2) A nucleophilic attack on the coordinated cyanide by a free cyanide, Reaction 14, which yields directly one of the precursors of the cyanide tetramer in Reaction 12. This mechanism is consistent with the

$$
\begin{array}{l}
\text{N} \\
\text{C} \leftarrow :\text{CN}^- \\
P\ \text{Fe}^{III} \rightarrow P\ \text{Fe}^{II} + [\text{CN}:\text{CN}]^{\cdot} \\
\text{C} \qquad\quad \text{C} \\
\text{N} \qquad\quad \text{N} \qquad\qquad \text{N} \\
\qquad\qquad \bigg|\ \dfrac{\text{CN}^-}{\text{rapid}} \rightarrow P\ \text{Fe}^{II}\ \begin{array}{l}\text{C}\\ \text{C}\\ \text{N}\end{array}
\end{array}
\tag{14}
$$

effect of the porphyrin basicity on the reaction rate, as well as with the cyanide ion dependence. However, as mentioned above, the cyanide ion dependence may result from other causes. The mechanism is inconsistent with the effect of water. Hydrogen bonding of water to a coordinated cyanide should enhance the rate by making the cyanide more susceptible to nucleophilic attack. Also the observed photochemical enhancement of the rate would not be expected with a mechanism involving nucleophilic attack. Thus, although this mechanism cannot be eliminated at this time, it is considered unlikely.

(3) Intramolecular one-electron transfer with subsequent dissociation (i.e., homolytic bond cleavage), as described by Reaction 15. This

$$
\begin{array}{c}
\mathrm{N} \\
\mathrm{C} \\
P\ \mathrm{Fe^{III}} \rightarrow P\ \mathrm{Fe^{II}} + \cdot\mathrm{CN} \\
\mathrm{C} \qquad \mathrm{C} \\
\mathrm{N} \qquad \mathrm{N}
\end{array}
\tag{15}
$$

$$
\begin{array}{c}
\ \mathrm{CN^-} \\
\Big\vert \quad\longrightarrow \\
\ \ \text{rapid}
\end{array}
\quad
\begin{array}{c}
\mathrm{N} \\
\mathrm{C} \\
P\ \mathrm{Fe^{II}} \\
\mathrm{C} \\
\mathrm{N}
\end{array}
$$

mechanism is consistent with the effect of the porphyrin basicity on the reaction and, more importantly, can be expected to be photochemically enhanced. Although the role of the cyanide ion concentration is unclear in this mechanism and could be involved in the autocatalytic mechanism, as well as in competition with trace amounts of water, as indicated above, homolytic bond cleavage can describe the autoreduction reaction for the variety of substrates investigated to date.

Our results indicate that the autoreduction cannot occur by a conventional outer sphere mechanism because of the gross mismatch of the electrochemical potentials. Experimental data available at this time are consistent with homolytic iron–carbon bond cleavage which may or may not involve a simultaneous nucleophilic attack on the coordinated cyanide. The homolytic metal–carbon bond cleavage may serve as a model for similar processes reported for vitamin B_{12} (26).

OTHER SYSTEMS. *Alkyl Thiols.* The reduction of hemin with ethanethiol has been suggested to occur by a free radical mechanism on the basis of product analysis (11). The reaction of n-hexanethiol with TPPFeCl in DMSO carried out in the ESR cavity gives rise to the signal illustrated in Figure 7. These are preliminary results, and the spectrum is of poor quality and probably reflects some saturation from the low steady-state concentration of the radical. Nevertheless the signal only appears during the autoreduction of the iron porphyrin and again indicates that the autoreduction occurs by a free radical pathway.

Pyridines, Amines, and Phosphines. We have not observed an ESR signal during the autoreduction with the above substrates. With the pyridines and phosphines so far investigated, the reduction probably has been too slow to generate appreciable concentrations of radical species. The autoreduction is much faster with primary and secondary amines, however the radicals produced are most likely too short-lived to detect with our present methods. We are investigating the applicability of rapid flow and spin trapping techniques.

Although the mechanistic details of the above systems have not yet been clarified, our results suggest that the autoreduction of ferric porphyrins by a free radical pathway that most likely involves the homolytic

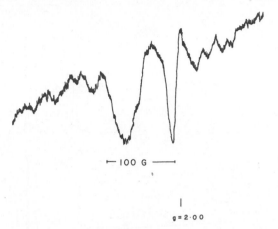

$$\vdash \text{100 G} \dashv$$

$$g = 2\cdot00$$

Figure 7. The ESR signal obtained upon the addition of n-*hexanethiol to a CDCl$_3$ solution of TPPFeCl*

cleavage of the iron substrate bond is a general feature of Fe(III) porphyrin chemistry.

Reoxidation of Fe(II) Porphyrins.   Model systems for the activation of molecular oxygen via coordination to an Fe(II) porphyrin have not been reported because of the rapid irreversible autooxidation of the Fe(II) to the Fe(III) oxo-bridged dimer (C) (Reaction 16). Since the

$$PFe^{II}(L)_2 \rightleftharpoons L + PFe^{II}L$$
$$\phantom{PFe^{II}(L)_2}\quad \vdash\!\!\!\longrightarrow PFe^{III}OFe^{III}P \qquad (16)$$
$$\phantom{PFe^{II}(L)_2 xxx} O_2$$
$$\qquad\quad (B) \qquad\qquad\qquad (C)$$

rate of this reaction is suppressed by excess ligand, L, an intermediate five-coordinated Fe(II) complex (B) is implicated (27). Sterically hindered Fe(II) porphyrins have been designed (27, 28) which prevent the formation of the oxo-bridged dimer, most likely by interfering with the bimolecular reaction between the Fe(II) oxygen adduct and a second Fe(II) porphyrin. The hindered porphyrins have been shown to reversibly bind molecular oxygen and behave as suitable model compounds for the active site of myoglobin and hemoglobin. However, to date these models have not been able to activate molecular oxygen.

CYANIDE COMPLEXES. We have investigated the oxidation of TPP-Fe^{II}(CN)$_2^{2-}$ in DMSO with molecular oxygen. The effect of introducing molecular oxygen into a 10-mM DMSO solution of TPPFe(II)(CN)$_2^{2-}$ containing excess KCN (prepared by allowing complete anaerobic auto-reduction of the ferric complex) is shown in Figure 8. The initial species

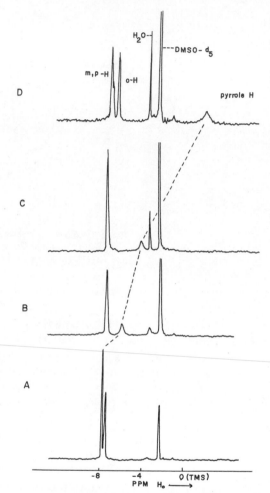

Figure 8. Proton NMR traces showing the reoxidation by oxygen of TPPFe^{II}(CN)₂²⁻ in DMSO at 25°C. (A) reduced complex; (B–D) traces with increasing time after addition of oxygen.

(A in Figure 8) is characteristic of the Fe(II) porphyrin, and the presence of an excess of KCN guarantees the bis-cyanide complex. On reaction with molecular oxygen all peak positions shift upfield to positions characteristic of the initial Fe(III) hemichrome (B, C, and D in Figure 8). Again the observation of averaged chemical shifts indicates that the same species exists in both oxidation states with rapid electron exchange averaging the spectra and establishes the oxidation product to be the bis-cyano Fe(III) complex, Reaction 17. This reaction has not been

$$TPPFe^{II}(CN)_2{}^{2-} + O_2 \rightarrow TPPFe^{III}(CN)_2{}^{1-} + O_2{}^{(?)-} \qquad (17)$$

followed to completion (about 80% completion) since the anaerobic autoreduction is competitive with the oxidation process. In the case of

$PP(IX)DMEFe^{II}(CN)_2{}^{2-}$, since the ferric form autoreduces slower, the oxidation with molecular oxygen in DMSO is more rapid and can be carried to completion. The NMR spectrum obtained for the oxidized product is identical to that of the well characterized $PP(IX)DMEFe^{III}$-$(CN)_2{}^{1-}$ and further establishes the product of the oxidation as the bis-cyano Fe^{III} species.

The NMR data in Figure 8 also indicate that water is an oxidation product. This is the only oxygen-containing species that has as yet been identified. The source of the proton in the water has not been established; however the oxidation of the phenyl-deuterated porphyrin TPP-$Fe^{II}(CN)_2{}^{2-}-d_{20}$ in DMSO-d_6 also gave a water peak, indicating that at least some of the protons were obtained from the pyrrole position. During these experiments precaution was taken to prevent any contact with the atmosphere. Since most of the intensity of the pyrrole peak is still present at the end of an autoreduction, reoxidation, and a second auto-reduction cycle, we suggest that proton obstraction from the pyrroles is only a minor pathway for deactivation of the activated oxygen.

Since the product of the oxidation is the bis cyanide Fe(III) species, which is again reduced by the excess cyanide, the reaction can be cycled, as represented by the following scheme:

$$H_2O \leftarrow O_2{}^{(?)-} \qquad O_2 \qquad \begin{pmatrix} PFe^{III}(CN)_2{}^{1-} \\ PFe^{II}(CN)_2{}^{2-} \end{pmatrix} \qquad \begin{pmatrix} CN^{1-} \\ CN \end{pmatrix}$$

The presence of the superoxide ion has not been confirmed. The ESR spectrum of the superoxide anion in frozen DMSO is known. Samples of reduced porphyrins have been frozen immediately after the introduction of molecular oxygen, however a superoxide anion signal has not been observed. It may be that the $O_2{}^-$ ion is too short lived under our experimental conditions to be observed. Since the oxidation is generating the $TPPFe(III)(CN)_2{}^{1-}$ species, which is then reduced by the excess CN^{1-} present, the cyanide tetramer ESR signal (Reaction 12) appears. Interestingly, superimposed on this is a three-line signal (Figure 9) which is not seen during the original reduction of $TPPFe(III)(CN)_2{}^{2-}$. The hyperfine splitting of the triplet is 1.5 gauss. Hyperfine splitting in the cyanide radical, obtained by uv irradiation of HCN in an argon matrix, was reported as 4.6 gauss and is thought to be surprisingly small (29). Hyperfine splitting in the NCO radical in the gas phase was reported as about 19 gauss (30). The origin of the three-line signal has not been determined. The addition of a DMSO solution of KO_2 solubilized with the dicyclohexyl-18-crown-6 cyclic ether to DMSO saturated with KCN (and more dilute solutions) did not reproduce the triplet nor

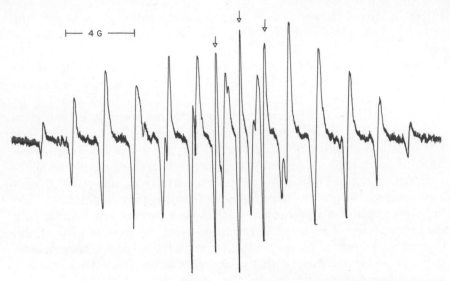

Figure 9.   The ESR signal observed during the oxidation of $TPPFe^{II}(CN)_2^{2-}$ in DMSO by oxygen

was the strong ESR signal (Reaction 12), indicative of CN formation, observed.

OTHER SYSTEMS. A second case where the oxidation of an Fe(II) porphyrin does not give the bridging oxo-dimer is the system $TPPFe^{II}$ in $CDCl_3$ with the strong field ligand $P(n\text{-}Bu)_3$ as the axial base (*31*). As mentioned above, the addition of excess $P(n\text{-}Bu)_3$ to a 10-mM solution of $TPPFe^{III}Cl$ in $CDCl_3$ causes the autoreduction of the porphyrin. On addition of molecular oxygen, the porphyrin peaks shift upfield with an averaged chemical shift, indicating the formation of a low spin Fe(III) complex. Although this system has not yet been fully chracterized, it is clear that the oxidation does not give the bridging oxo-dimer.

The autooxidation of the bis-piperidine complex $TPPFe^{II}(pip)_2$, obtained by the anaerobic autoreduction of $TPPFe^{III}Cl$ in $CDCl_3$ with piperidine (as with other amines), yields, as expected, the bridging oxo-dimer.

MECHANISM OF OXIDATION OF $TPPFE^{II}(CN)_2^{2-}$ WITH MOLECULAR OXYGEN. We have considered three mechanisms to account for the unusual lack of dimer formation in the autooxidation.

(1) An outer sphere oxidation of the Fe(II), Reaction 18.

$$O_2 + P \begin{matrix} N \\ C \\ Fe^{(II)} \\ C \\ N \end{matrix} \rightarrow P \begin{matrix} N \\ C \\ Fe^{III} \\ C \\ N \end{matrix} + O_2^{(?)-} \tag{18}$$

The oxidation of Fe(II) porphyrins is thought to involve a five-coordinate intermediate, Reaction 16. However, with cyanide ion as axial base it may be possible to transfer an electron from the Fe(II) to molecular oxygen by an electron transfer through the coordinated cyanide ion without the formation of an iron–oxygen bond. Although this mechanism appears attractive for the cyanide ion, the oxidation of the phosphine complex most likely occurs by the same mechanism, and electron transfer through the coordinated phosphine appears much less likely. Thus an outer-sphere mechanism is considered unlikely. A consideration of the standard potential, $E°$, for Reaction 18 appears to suggest that the Fe(II) porphyrin is an insufficiently strong reducing agent (by perhaps 0.2 volt) to reduce oxygen to O_2^- by an outer-sphere mechanism. However, $E°$ values are referenced to the standard-state concentrations of $1M$. Under our experimental conditions the concentration of the reactive species O_2^- is expected to be quite small, and consideration of the Nernst equation suggests that the potential for the reaction could be considerably larger than the standard $E°$ value. Thus an outer sphere mechanism cannot be eliminated on the basis of the standard potential $E°$. This argument is discussed more thoroughly in Ref. 32.

(2) Formation and cleavage of the oxo-bridged dimer, Reactions 19 and 20.

$$\begin{matrix} N \\ C \\ P\ \ Fe^{II} + O_2 \rightarrow P\ Fe^{III}OFe^{III}P \\ C \\ N \end{matrix} \qquad (19)$$

$$PFe^{III}OFe^{III}P + CN^- \xrightarrow{\ CN^-\ } \begin{matrix} N \\ C \\ P\ Fe^{III} \\ C \\ N \end{matrix} \qquad (20)$$

The oxo-bridged dimer is cleaved with excess KCN in DMSO, Reaction 20, i.e., under our experimental conditions, although the rate is very slow (*31*). The possibility of the oxo-bridged dimer as an intermediate can be discounted by considering the following experiment. A sample of PP(IX)DMEFe^{III}(CN)$_2^{2-}$ in DMSO saturated with KCN was allowed to autoreduce anaerobically until about 50% had been reduced, A in Figure 10. An equivalent amount of the dimer [PP(IX)DMEFe^{III}]$_2$O was added in the absence of oxygen (B in Figure 10). Then oxygen was introduced, causing the oxidation of the PP(IX)DMEFe^{II}(CN)$_2^{2-}$. As oxidation proceeded, as evidenced by the downfield movement of the methyl resonances (C and D in Figure 10), the spectrum of the oxo-

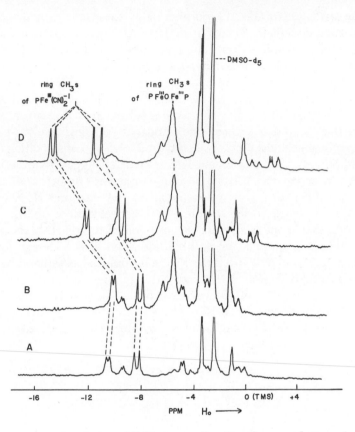

Figure 10. Proton NMR traces showing the reoxidation of PP(IX)DMEFe(CN)$_2{}^{2-}$ by oxygen in the presence of the oxo-bridged dimer, (PP(IX)DMEFe^{III})$_2$O. (A) trace for 50% autoreduced couple; (B) addition of oxo-bridged dimer in absence of O$_2$; (C) 10 min after adding oxygen to B; (D) 1 hr after adding oxygen to B.

bridged dimer remained unaltered. Hence the oxo-bridged dimer is not an intermediate. The oxo-bridged dimer (TPPFe)$_2$O is too insoluble in DMSO for a similar experiment.

(3) An inner sphere oxidation. It is likely that the first step in the oxidation of TPPFe(CN)$_2{}^{2-}$ involves replacement of coordinated CN^{1-} by oxygen as in the initial step of Reaction 16. Our results indicate that the strong field ligands, CN^{1-} and tributylphosphine prevent the formation of the oxo-bridged dimer; the one-electron oxidation of the iron suggests formation of the superoxide ion. The interaction of oxygen with a ferrous hemoprotein containing an appropriately strong field ligand could lead to the dissociation of the complex as the ferric protein and

the activated superoxide ion. The appropriate strong field ligand in a hemoprotein could be a deprotonated histidyl imidazole or a histidyl imidazole activated by a strong hydrogen bond acceptor (*31*). These novel oxidation reactions are under further investigation in our laboratory.

Acknowledgment

The authors are indebted to A. L. Balch for numerous valuable discussions and to D. H. Chin for electrochemical data. This work was supported by a grant from the National Institutes of Health HL-16087.

Literature Cited

1. Harbury, H. A., Marks, R. H. L., in "Inorganic Biochemistry," G. L. Eichhorn, Ed., Chapter 26, Elsevier, Amsterdam, 1973.
2. Saunders, B. C., in "Inorganic Biochemistry," G. L. Eichhorn, Ed., Chapter 28, Elsevier, Amsterdam, 1973.
3. Feigelson, P., Brady, F. O., in "Molecular Mechanisms of Oxygen Activation," O. Hayaishi, Ed., p. 87, Academic, New York, 1974.
4. Takano, T., Kallai, O. B., Swanson, R., Dickerson, R. E., *J. Biol. Chem.* (1973) **248**, 5234.
5. Winfield, M. E., *J. Mol. Biol.* (1965) **12**, 600.
6. Hayaishi, O., *Mol. Mech. Oxygen Act.* (1974) 13.
7. Dolphin, D., Felton, R. H., *Acc. Chem. Res.* (1974) **7**, 26.
8. Piette, L. H., Yamazaki, I., Mason, H. S., *Free Radicals Biol. Syst. Proc. Symp.* (1961) Chapter 14.
9. Shiga, T., Imaizumi, K., *Arch. Biochem. Biophys.* (1975) **167**, 469.
10. Straub, D. K., Connor, W. M., *Ann. N. Y. Acad. Sci.* (1973) **206**, 383.
11. Swan, C. J, Trimm, D. L., ADV CHEM. SER. (1968) **76**, 182.
12. Weightman, J. A., Hoyle, N. J., Williams, R. J. P., *Biochim. Biophys. Acta* (1971) **244**, 567.
13. Epstein, L. M., Straub, D. K., Maricondi, C., *Inorg. Chem.* (1967) **6**, 1720.
14. Del Gaudio, J., La Mar, G. N., *J. Am. Chem. Soc.* (1976) **98**, 3014.
15. Cohen, I. A., Caughey, W. S., *Biochemistry* (1968) **7**, 636.
16. Adler, A. D., Longo, F. R., Finarelli, J. D., Goldmacher, J., Assour, J., Korsakoff, L., *J. Org. Chem.* (1967) **32**, 476.
17. Adler, A. D., Longo, F. R., Kampas, F., Kim, J., *J. Inorg. Nucl. Chem.* (1970) **32**, 2443.
18. Norris, A. R., *Can. J. Chem.* (1967) **45**, 2703.
19. Wuthrich, K., Baumann, R., *Helv. Chim. Acta* (1973) **56**, 585.
20. Fuhrhop, J. H., *Struct. Bonding* (1974) **18**, 1.
21. Goff, H., La Mar, G. N., *J. Am. Chem. Soc.*, in press.
22. Andreades, S., Zahnow, E. W., *J. Am. Chem. Soc.* (1969) **91**, 4181.
23. Frye, J. S., La Mar, G. N., *J. Am. Chem. Soc.* (1975) **97**, 3561.
24. Shriver, D. F., *Struct. Bonding* (1966) **1**, 32.
25. Chin, D. H., unpublished results.
26. Pratt, J. M., "Inorganic Chemistry of Vitamin B_{12}," Chapter 13, Academic, London, 1972.
27. Collman, J. P., Gagne, R. R., Reed, C. A., Halbert, T. R., Lang, G., Robinson, W. T., *J. Am. Chem. Soc.* (1975) **97**, 1427.
28. Baldwin, J. E., Huff, J., *J. Am. Chem. Soc.* (1973) **95**, 5757.
29. Atkins, P. W., Symons, M. C. R., in "The Structure of Inorganic Radicals," p. 113, Elsevier, Amsterdam, 1967.

30. Carrington, A., Fabris, A. R., Howard, B. J., Lucas, N. J. D., *Molec. Phys.*
    (1971) **20**, 961.
31. Del Gaudio, J., La Mar, G. N., unpublished results.
32. Bennett, L. E., *Prog. Inorg. Chem.* (1973) **18**, 1.

RECEIVED July 26, 1976.

Iron–Sulfur Proteins and Superoxide Dismutases in the Biology and Evolution of Electron Transport

D. O. HALL

Department of Plant Sciences, University of London King's College, 68 Half Moon Lane, London SE24 9JF, U.K.

Iron–sulfur proteins contain non-heme iron and sulfide in their active centers as 4Fe–4S or 2Fe–4S or with rubredoxin, as 1Fe alone. The iron is always bonded to cysteine sulfur. They catalyze redox reactions between +350 and −600 mV, (hydrogen electrode = −420 mV) and are usually small but can form complex enzymes with molybdenum and flavin. They occur as soluble or membrane-bound proteins, catalyze key reactions in carbon, hydrogen, sulfur, and nitrogen metabolism, and occur in all organisms. Synthetic analogs of iron–sulfur centers have provided evidence for the structure of active centers. Superoxide dismutases are enzymes which contain iron, manganese, or copper plus zinc in their active centers. They occur in anaerobes and aerobes and are involved in the dismutation of superoxide (O_2^-), which is thought to be very toxic and produced by free radical interaction with oxygen.

Two classes of enzymes, the iron–sulfur proteins and the superoxide dismutases, are metal-containing proteins which are probably closely involved in the evolution of electron transport from the origin of life onwards and in the interaction of oxygen with all living organisms. They thus afford specific examples in studying evolution, electron transport, and oxygen-associated reactions in biology and also probably in chemistry itself.

The Iron–Sulfur Proteins (1, 2, 3, 4, 5)

These proteins contain non-heme iron and inorganic (acid-labile) sulfur in the active centers as 4Fe–4S or 2Fe–2S or, in the case of rubredoxin, as 1Fe alone (Figure 1). The iron is always bonded to cysteine sulfur. Iron–sulfur proteins catalyze oxidation–reduction reactions between $+350$ and -600 mV (hydrogen electrode $= -420$ mV, pH 7). An important property of iron–sulfur proteins is the ability to catalyze electron transfer at different redox potentials depending on the state of the protein. For example, the so-called high potential iron proteins (HiPIP) from the red photosynthetic bacterium *Chromatium* transfer electrons at $+350$ mV or at about -500 mV, depending on the presence or absence of dimethylsulfoxide, a protein chaotropic agent (*see* below).

Iron–sulfur proteins are usually grouped into four different types as shown in Table I:

  (a) Containing 4Fe–4S centers
  (b) Containing 2Fe–2S centers
  (c) Containing a single Fe
  (d) Complex iron–sulfur proteins

Table I also shows the great diversity of organisms in which iron–sulfur proteins have been detected. Thus far there is no organism which when appropriately examined has not contained an iron–sulfur protein, either in the soluble or membrane-bound form. Iron–sulfur proteins catalyze reactions of physiological importance in obligate anaerobic bacteria, such as hydrogen uptake and evolution, ATP formation, pyruvate metabolism, nitrogen fixation, and photosynthetic electron transport. These properties and reactions can be considered "primitive" and thus make iron–sulfur proteins a good place to start the study of evolution. These key reactions are also important in higher organisms. Other reactions catalyzed by iron–sulfur proteins can be added such as hydroxylation, nitrate and nitrite reduction, sulfite reduction, NADH oxidation, xanthine oxidation, and many other reactions (Table II).

One of the best known groups of iron–sulfur proteins is the ferredoxins. These are small proteins consisting of relatively few amino acids,

Table I.   Properties of

4Fe + 4S center
  8Fe ferredoxins
    Clostridium (obligate anaerobic bacterium)
    Chlorobium (green photosynthetic bacterium)
    Chromatium (red sulfur photosynthetic bacterium)
    Rhodospirillum rubrum I (red non-sulfur photosynthetic bacterium)
    Azotobacter III (aerobic N_2 fixing bacterium)

  Ferredoxins with (8Fe + 8S have also been reported in: *C. pasteuri-*
    C. thermosaccharolyticum, Peptococcus aerogenes, Peptostrepto-

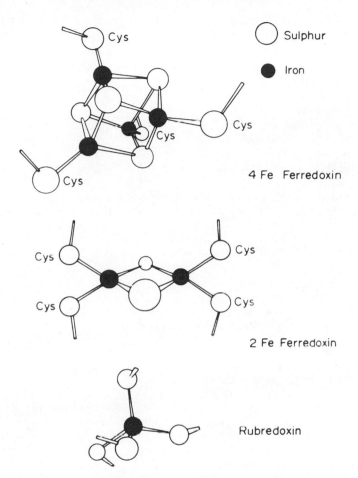

Figure 1. Structures of the 4Fe–4S centers and 2Fe–2S centers in ferredoxins and the 1Fe center in rubredoxin (2, 4)

Representative Iron–Sulfur Proteins[a]

Active Group (atoms/molecule)	Molecular Weight	No. of Amino Acids	Redox Potential (E_o' in mV)
8Fe, 8S	6,000	55	−395
8Fe, 8S	6,000	60	—
8Fe, 8S	10,000	81	−490
8Fe, 8S	13,000	—	—
8Fe, 8S	15,000	130	−420

anum, C. acidi urici, C. butyricum, C. tartarivorum, C. tetanomorphum, coccus elsdenii, Veillonella alcalescens.

Table I.

4Fe ferredoxins
  Desulfovibrio (anaerobic SO_4 reducing bacterium)
  Bacillus (facultative N_2 fixing bacterium)

4Fe HiPIP
  Chromatium
  Ferredoxins with (4Fe + 4S) have also been reported in *B. poly-*
  membrane, *D. gigas, D. desulfuricans, Rhodospirillum rubrum.*
  monas.

2Fe + 2S center
  2Fe ferredoxins
    Spinach (higher plant)
    Microcystis (blue-green alga)
    Scenedesmus (green alga)
    Azotobacter I (aerobic N_2 fixing bacterium)
    Pseudomonas putida (aerobic bacterium)
    E. coli (aerobic bacterium)
    Pig adrenals (mammal)
    Mitochondria, complex III (mammalian)

    Ferredoxins with (2Fe + 2S have also been reported in: *Aethusa,*
    sica, Botrydiopsis, Bumilleriopsis, Chenopodium, Chlamydomonas,
    locasia Cyanidium, Cyperus, Datura, Equisetum, Euglena, Gossy-
    Phaseolus, Phormidium, Pinus, Pisum, Polystichum, Porphyrid-
    Zea, pig adrenals, pig testes.

1Fe center
  1Fe rubredoxin
    Clostridium

    Rubredoxins have also been reported in: *Chloropseudomonas ethyl-*
    sulfuricans, Desulfovibrio gigas, Peptococcus aerogenes, Peptoco-

Complex Fe–S proteins
  Mitochondrial succinate dehydrogenase (mammalian)

  Mitochondrial NADH dehydrogenase (mammalian)
  Xanthine oxidase (milk, bacteria)
  Nitrogenase (*Clostridium* or *Klebsiella*): molybdenum iron protein

---

a For original references *see* Ref. *3.*

Continued

Active Group (atoms/molecule)	Molecular Weight	No. of Amino Acids	Redox Potential (E_0' in mV)
4Fe, 4S	6,000	56	−330
4Fe, 4S	8,000	78	−380
4Fe, 4S	9,650	86	−350

myxa, B. stearothermophilus, Spirochaeta aurantia, spinach chloroplast
High-potential iron–sulfur protein has been reported in *Rhodopseudo-*

2Fe, 2S	10,600	97	−420
2Fe, 2S	10,300	98	—
2Fe, 2S	10,600	96	—
2Fe, 2S	21,000	181	−350
2Fe, 2S	12,500	114	−240
2Fe, 2S	12,600	—	−360
2Fe, 2S	12,500	115	−270
2Fe, 2S	30,000	—	+280

Agrobacterium, Amaranthus, Anabaena, Anacystis, Aphanothece, Bras-
Chlorella, Cladophora, Clostridium (azoferredoxin and EPR protein), *Co-*
pium, Laminum, Leucaena, Medicago, Navicula, Nostoc, Petroselenium,
ium, Porphyra, Rhizobium, Sambucus, Spirulina, Stelluria, Tolypothrix,

1Fe	6,000	54	−60

ica, Clostridium butyricum, Clostridium stricklandii, Desulfovibrio de-
ccus glycinophilus, Peptostreptococcus elsdenii, Pseudomonas oleovorans.

8Fe, 8S, 1FAD	97,000 (dimer)	—	—
28Fe, 28S, 1FMN	—	—	—
8Fe, 8S, 2FAD, 2Mo	275,000	—	−343 and −303
18 or 24Fe, 18 or 24S, 2Mo	220,000	—	−60 and −280

Table II.   Electron Transfer

Type of Reaction

1. Phosphoroclastic reactions:
$$\text{pyruvate} + P_i \xrightarrow{\text{CoA}} \text{acetyl phosphate} + CO_2$$

2. Synthesis of α-keto acids (CO_2 fixation):
   e.g.: $\text{acetyl-CoA} + CO_2 \rightarrow \text{pyruvate} + \text{CoA}$
   $\text{succinyl-CoA} + CO_2 \rightarrow \alpha\text{-oxoglutarate} + \text{CoA}$
   $\text{propionyl-CoA} + CO_2 \rightarrow \alpha\text{-oxobutyrate} + \text{CoA}$

3. One-carbon metabolism:
   $CO_2 \rightleftarrows HCO_3$

4. Hydrogen metabolism:
$$2H^+ + 2e^- \xrightarrow{\text{hydrogenase}} H_2$$

5. Nitrogen fixation:
   $N_2 + 3H_2 \rightleftarrows 2NH_3$

6. Nicotinamide nucleotide oxidoreduction:
   $\text{NADH} + \text{NADP}^+ \rightleftarrows \text{NAD}^+ + \text{NADPH}$

7. Photosynthetic electron transfer in bacteria

8. α-Hydroxylation of hydrocarbons:
   $RCH_3 + \text{NADH} + H^+ + O_2 \rightarrow RCH_2OH + \text{NAD}^+ + H_2O$

9. Sulfite reduction:
   $SO_3^{2-} \rightarrow S^{2-}$

10. Nitrate reduction
    $NO_3^- \rightarrow NO_2^-$

11. Nitrite reduction:
    $NO_2^- \rightarrow NH_3$

12. Nicotinamide nucleotide reduction (chloroplasts):
    $\text{NADP}^+ + H_2O \rightarrow \text{NADPH} + \tfrac{1}{2}O_2$
    Photophosphorylation:
    $\text{ADP} + P_i \rightarrow \text{ATP}$

13. (a) Oxidation of NADH by mitochondria:
    $$\text{NADH} + \tfrac{1}{2}O_2 \longrightarrow \text{NAD} + H_2O$$
    ATP

    (b) Oxidation of succinate:
    $$\text{succinate} \longrightarrow \text{fumarate}$$
    ATP

14. Oxidation of xanthine and aldehydes:
    $R\text{—}H + H_2O + O_2 \rightarrow R\text{—}OH + H_2O_2$

15. Hydroxylation:
    $R\text{—}H + O_2 + \text{NAD(P)H} \rightarrow ROH + H_2O + \text{NAD(P)}^+$

[a] For original references *see* Ref. *3*.

Reactions Involving Iron–Sulfur Proteins[a]

Representative Organism	*Type of Fe Protein*
Clostridium	8Fe
fermentative and photosynthetic bacteria, e.g., *Clostridium, Chromatium*	8Fe
Clostridium	8Fe
certain algae; certain bacteria	4 or 12Fe
fermentative and photosynthetic bacteria	(a) 18–24 Fe/mol (b) 4Fe/mol
anaerobic bacteria	8Fe
Chromatium, Rhodopseudomonas	4Fe (ferredoxin and HiPIP)
Ps. oleovorans	1Fe (rubredoxin)
algae, plants, *D. gigas, E. coli, C. pasteurianum*	—
fungi, algae, bacteria	8 or 20Fe/mol
algae, plants, *Micrococcus denitrificans*	—
Pseudomonas putida, plant and algae	2Fe
eukaryotes	28Fe/mol
plants; mammals; bacteria	two Fe–S subunits
bacteria; mammals	8Fe/mol
mammalian (adrenal mitochondria)	2Fe

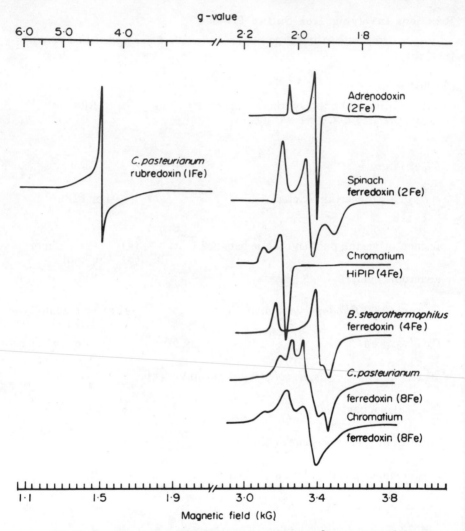

Figure 2.   Representative EPR spectra of iron–sulfur proteins (2)

usually between 55 and about 120 (mol wt 6,000–14,000). They contain either 4Fe–4S or 2Fe–2S active centers which can be extracted intact from the protein in the presence of appropriate ligands and solvents and studied by the so-called core extrusion techniques (4). Another important aspect of removing the active center is that the apoprotein so formed can be reconstituted under anaerobic, nonenzymatic conditions by the simple addition of ferrous ammonium sulfate and sodium sulfide (6).

The amino acid sequence in the ferredoxins is quite distinctive; the position of the cysteines which bind the irons in a specific environment do not vary in different classes of ferredoxins; this is shown in Figure 6, below. This invariance in the positions of key amino acids has been very useful in studying the evolution of ferredoxins.

The study of iron–sulfur proteins has been greatly helped by the recognition of the so-called $g_\perp \cong 1.94$ EPR signal which is generally seen in the reduced form of iron–sulfur proteins. Without the use of the EPR technique the study of iron–sulfur proteins is made very difficult by their low extinction coefficient and broad absorption in the visible region. Work using Mössbauer spectroscopy, circular dichroism, proton magnetic resonance, magnetic susceptibility, etc., has helped in the study of specific parts of the protein or the active center itself. Figure 2 shows characteristic EPR spectra of representative iron–sulfur proteins, including the high potential iron protein, which has an EPR signal of $g_\perp = 2.02$ in the oxidized state.

The existence of this HiPIP-type structure was the starting point of an interesting development in the study of the iron–sulfur active center. X-ray crystallography showed that there is little difference between the 4Fe–4S cluster in a ferredoxin ($E_o' = -400$ mV) and in HiPIP ($E_o' = +350$ mV). This anomaly was elucidated by the so-called "C" state hypothesis of Carter et al. (7) (Figure 3), in which the existence of a super-reduced HiPIP and a super-oxidized ferredoxin was postulated. The super-reduced HiPIP was shown to exist by Cammack (8) utilizing 80% DMSO (dimethylsulfoxide) to distort the protein environment of HiPIP. In this case a super-reduced HiPIP with EPR signal similar to a

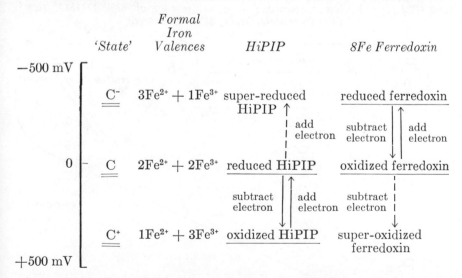

Figure 3. The three-state hypothesis for the redox potentials of electron transfer in the 4Fe–4S active centers of ferredoxins and HiPIP (7). The redox potential differs over a 1-V range (see also Ref. 2 for discussion).

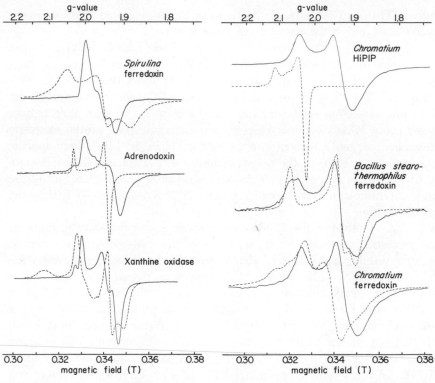

Figure 4.   Effect of DMSO (dimethylsulfoxide) on reduced iron–sulfur proteins (63). (· · ·), native proteins; (——), in presence of 80% DMSO. (left) 2Fe–2S centers; (right) 4Fe–4S centers.

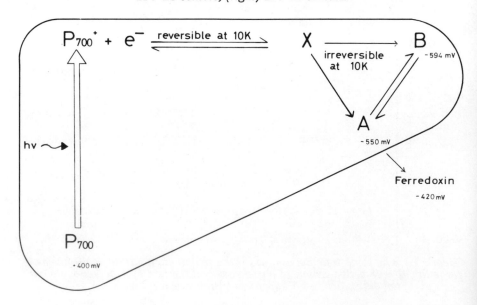

4Fe ferredoxin was detected. A superoxidized ferredoxin has been reported by Sweeney et al. (9).

The use of chaotropic agents such as DMSO has also facilitated studies of iron–sulfur proteins whose active centers have not yet been determined. In this case the protein in question is examined using EPR in the presence and absence of a given concentration of DMSO (Figure 4). The temperature dependence of these signals is also a diagnostic criterion, as is the shape of the EPR signals. In this way a two- or a four-iron center can be readily distinguished. This can also be resolved by the core extrusion experiments of Holm (4) and Orme-Johnson (10). The advantage of the chaotropic (DMSO) technique is that the type

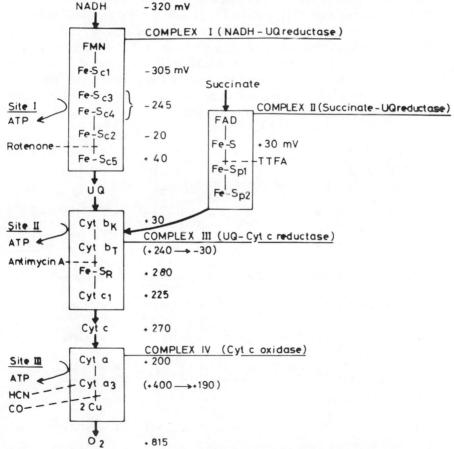

Figure 5. (a) (left) Functional relationships of the electron transport components of photosystem I reaction center from chloroplasts. A and B are iron–sulfur centers of the 4Fe–4S type, and X may be an Fe–S center (11). (b) (above) The respiratory chain of mitochondria, showing EPR-detectable Fe–S centers (2).

of center in a membrane-bound protein can also be distinguished, as has been the case with chloroplast membranes, where a 4Fe–4S center was detected as the primary electron transport acceptor (*11, 12*).

The existence of membrane-bound iron–sulfur proteins has been known for some time since their discovery by Rieske in complex III of animal mitochondria (*5*). However, it has only recently been recognized that membrane-bound iron–sulfur proteins probably play important and key roles in electron transport in, for example, mitochondria, chloroplasts, and chromatophores from photosynthetic bacteria. Figure 5(a) shows the sequence of electron transfer at the primary electron acceptor site of chloroplast membranes. A potential of approximately −600 mV can be measured at the primary electron acceptor in photosynthesis (*11, 12, 13*). A more oxidizing iron–sulfur protein with some properties of a HiPIP has also been detected in the membranes of chloroplasts (*14*). Figure 5b shows an interesting proposed scheme for iron–sulfur centers in mitochondria; it is shown that there might be 12–14 different centers (*15*), i.e., twice as many as have been recognized for cytochromes. It is possible that such a wide occurrence of iron–sulfur centers in mitochondria with potentials varying between −350 and +280 mV may be important

```
         1
A   Ala -Phe -Val -Ile  -       Asn-Asp -Ser  Cys Val -Ser  Cys Gly -Ala  Cys
B   Ala -Leu -Tyr -Ile  -       Thr-Glu -Glu  Cys Thr -Tyr  Cys Gly -Ala  Cys
C   Ala -Leu -Met -Ile  -       Thr-Asp -Gln  Cys Ala -Asn  Cys Asn -Val  Cys
D   ---- -Pro -Ile  -Gln -       Val -Asp -Asn  Cys Met-Ala  Cys Gln -Ala  Cys

         29
A   Gln -Phe -Val -Ile  -Asp-Ala -Asp -Thr  Cys Ile  -Asp  Cys Gly -Asn  Cys
B   Ile -Tyr -Val -Ile  -Asp-Ala -Asn -Thr  Cys Asn  -Giu  Cys Ala  Ala  Cys
                                              Gly         Leu -Asp
C   Thr -Tyr -Val -Ile  -Glu -Pro -Ser  -Leu Cys Thr -Glu  Cys Val  Asp  Cys

                                    Gly -His -Tyr -Glu -Thr-
D   Gyl -Asp -Lys -Ala -Val -Asn -Ile  -Pro -Asn -Ser  -Asn -Leu -Asp -Asp -Glu -

         1
E   Ala -Ser -Tyr -Lys -Val -Thr -Leu -Lys -Thr -Pro -Asp -Gly -Asp -Asn -Val -
F   Ala -Thr -Tyr -Lys -Val -Thr -Leu -Lys -Thr -Pro -Ser -Gly -Asp -Gln -Thr-
G   ---- ---- -Tyr -Lys -Thr -Val -Leu -Lys -Thr -Pro -Ser -Gly -Glu -Phe -Thr-
H   Ala -Ser -Tyr -Lys -Val -Lys -Leu -Val -Thr -Pro -Glu -Gly -Thr -Gln -Glu -

         31
E   Glu -Gly -Leu -Asp -Leu -Pro -Tyr -Ser  Cys Arg -Ala -Gly -Ala  Cys Ser -
F   Ala -Gly -Leu -Asp -Leu -Pro -Tyr -Ser  Cys Arg -Ala -Gly -Ala  Cys Ser -
G                      -Ser  Cys Arg -Ala -Gly -Ala  Cys Ser -
H   Glu -Gly -Ile  -Val -Leu -Pro -Tyr -Ser  Cys Arg -Ala -Gly -Ser  Cys Ser -

         61
H   Gly -Ser -Phe -Leu -Asp -Asp -Asp -Gln -Ile  -Glu -Glu -Gly -Trp -Val -Leu-

         91
H   Lys -Glu -Glu -Glu -Leu -Thr -Ala  (97)
```

Figure 6.   Amino acid sequences of representative ferredoxins.
A, Clostridium butyricum *(8Fe) obligate anaerobic fermenting bacterium; B,* *(8Fe) red (sulfur) photosynthetic bacterium; D,* Desulfovibrio gigas *(4Fe) sulfate* *green alga; G,* Equisetum *(2Fe) primitive*

both in electron transport and in phosphorylation itself, because of the proposed ability of iron–sulfur proteins to take on and give off protons. This invokes the proposal of Bruice et al. (*16*), who have suggested that the reduction of an iron–sulfur protein involves the making and breaking of two iron–sulfur bonds simultaneously with the taking on of protons. This proposal still seems speculative, but it is an interesting one in view of the necessity to transport electrons and protons in membrane electron transport schemes coupled to phosphorylation.

Much of our recent knowledge of the active centers of iron–sulfur proteins has come from the synthesis work of Holm (*4*) on analog compounds with 4Fe–4S, 2Fe–2S, and 1Fe centers. The additional ability to extract the iron–sulfur clusters from the proteins themselves and reinsertion of these clusters into other proteins has led to some interesting experiments which, among others, have shown that the 4Fe–4S configuration is more stable than the 2Fe–2S configuration.

In studying the evolution of iron–sulfur proteins, the requirements of the ferredoxins for specifically placed cysteines to bind the irons in the active center have been most useful. Figure 6 shows sequences of the ferredoxins from several obligate fermenting anaerobic bacteria, green

```
                                         28 -
-Ala -Gly -Glu Cys Pro-Val -Ser -Ala -Ile -Thr -Gln Gly -Asp-Thr-
-Glu-Pro -Glu Cys Pro-Val -Thr-Ala -Ile -Ser -Ala -Gly -Asp-Asp-
-Gln-Pro -Glu Cys Pro-Asn -Gly-Ala -Ile -Ser -Gln -Gly -Asp-Glu-
-Ile -Asn -Glu Cys Pro-Val -Asp-Val -Phe-Gln -Met-Asp -Glu-Gln

                                         55
-Ala -Asn -Val Cys Pro-Val -Gly-Ala -Pro-Asn -Gln -Glu
-Val -Ala -Val Cys Pro-Ala -Glu-Cys -Ile -Val -Gln -Gly (60)
-Glu-Gln                                                (81)
-Val -Glu -Val Cys Pro -Ile -Lys-Asp -Pro -Ser -His -Glu - - - - Gly
-Val -Cys
-Ser -Glu
-Cys-Val -Glu-Ala -Ile -Gln -Ser -Cys -Pro-Ala -Ala -Ile -Arg-Ser (56)

                                         30
-Ile -Thr -Val -Pro -Asp-Asp -Glu-Tyr -Ile -Leu -Asp -Val -Ala -Glu-Glu-
-Ile -Glu -Cys-Pro -Asp-Asp -Thr-Tyr -Ile -Leu -Asp -Ala -Ala -Glu-Glu-
-Leu-Asp -Val -Pro -Glu-                                      -Glu-
-Phe-Glu -Cys-Pro -Asp-Asp -Val -Tyr -Ile -Leu -Asp -His -Ala -Glu

              50
-Thr Cys Ala -Gly -Lys-
-Ser Cys Ala -Gly -Lys-
-Ser Cys Leu-Gly -Lys-                                 60
-Ser Cys Ala -Gly -Lys-Val -Ala -Ala -Gly-Glu -Val -Asn -Gln-Ser -Asp-

              90
-Thr Cys Val -Ala -Tyr-Ala -Lys-Ser -Asp-Val -Thr -Ile -Glu-Thr-His-
```

See Refs. 3 and 5 for original references and the text.
Chlorobium limicola (8Fe), green photosynthetic bacterium; C, Chromatium reducing bacterium; E, Aphanothece (2Fe) blue-greeen alga; F, Scenedesmus (2Fe) plant; H, Medicago (2Fe) alfala, higher plant.

and red photosynthetic (anaerobic) bacteria, sulfate reducers, and also from algae and plants. In the low molecular weight ferredoxins from *Clostridia* and photosynthetic bacteria (all of which contain two 4Fe–4S clusters per molecule), the first and second halves of the molecule are very similar. One iron–sulfur cluster is held by cysteine residues 9, 11, 14, and 43, and the other is held by residues 18, 35, 37, and 41. When one looks at the four-iron ferredoxins (e.g., *D. gigas*), there is a great homology between the first half of the *Clostridial* ferredoxins and the first half of the *D. gigas* ferredoxin. We have recently completed the sequence of the four-iron ferredoxin from *Bacillus stearothermophilus* and have shown that it has only four cysteines. Thus we now know that it is necessary to have three cysteines close together in a sequence Cys-X-X-Cys-X-X-Cys, followed by a fairly long portion of peptide, which can wrap around the "back" of the active center, containing a Cys-Pro to hold the iron–sulfur cluster in position. In the case of the algal and plant ferredoxins we also know that they require four invariant cysteines; these are positions 37, 44, 47, and 75, again emulating the requirement of the four-iron type ferredoxins, where one needs three cysteines fairly close together and another cysteine some distance along the peptide chain.

As yet we don't have the amino acid sequence of a ferredoxin from a red non-sulfur bacterium, e.g., *Rhodospirillum* or *Rhodopseudomonas*. This is unfortunate, since it is highly likely that these red non-sulfur bacteria form an intermediate type organism between the obligate anaerobic bacteria and aerobic organisms, with or without mitochondria. This evolutionary development has been proposed from studies of the sequences of cytochromes and also from physiological and metabolic criteria (*3, 18, 19, 20, 21*). It is likely that organisms like red non-sulfur bacteria and nitrate and sulfate reducers provide the transition between the anaerobic and the aerobic mode of life during the development of higher organisms (*22–27*). One way in which one can approach this is by looking at the antibodies to specific ferredoxins and seeing how they interact with ferredoxins from other types of organisms. We have done this with five antibodies against 8Fe, 4Fe, and 2Fe ferredoxins and interacted them with 32 different ferredoxins containing various types of active centers (*28*). This is giving us an idea of the evolution of the protein moiety during development of ferredoxins and is also giving us useful information on the way in which specific ferredoxin reactions may be inhibited in various organisms.

Figure 7 shows an evolutionary sequence based on the properties of iron–sulfur proteins from chemical, biological, and amino acid sequence information. It obviously is an over-simplification, but it shows how one may construct an evolutionary sequence from the properties of a single protein. The transition between an anaerobic and aerobic phase, as stated

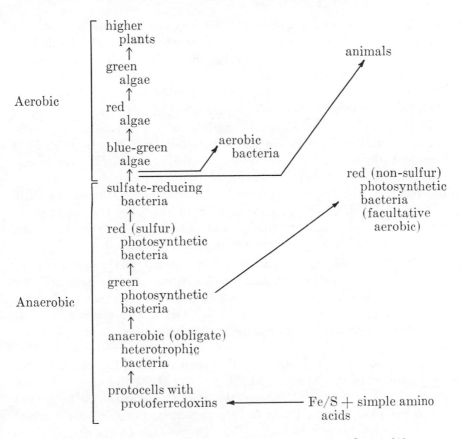

Figure 7.   Possible evolutionary development of ferredoxins (2)

before, is a difficult problem, and it is likely that the scheme as written is incorrect, since it does not take enough note of the cytochrome studies (*20*) of red non-sulfur bacteria—organisms likely to play a key role in this transition period.

Superoxide Dismutases (*24, 29–34*)

Oxygen in the atmosphere is a by-product of the evolution of oxygen-evolving photosynthesis first developed by the blue-green algae. Although it is a prerequisite for the development of life as we know it, oxygen itself is in fact quite toxic, and organisms have had to develop specific protection mechanisms against high concentrations of the gas. A key enzyme involved in this protection mechanism seems to be the fairly recently recognized superoxide dismutase (SOD), which contains iron or manganese or copper + zinc in its active centers. Even though nearly all the oxygen in the earth's atmosphere is derived from oxygen-evolving photosynthesis (from the splitting of water), there was undoubtedly oxygen present in small quantities during the early evolution of life as a result of the uv-induced photolysis of water (the so-called Urey effect) (35, 36, 37). Thus it is quite probable that even obligate anaerobic bacteria (nonphotosynthetic and photosynthetic) had to acquire some way of protecting themselves against the toxic effects of oxygen at a very early stage, before the development of oxygen-evolving photosynthesis. SOD was originally thought to occur only in aerobic organisms, but recent discoveries have shown that this enzyme also is present in obligate anaerobic bacteria, e.g., *Clostridium, Desulfovibrio, Chlorobium, Chromatium,* etc. (38, 39, 40, 41, 42). This provides us with an interesting enzyme to study the evolution of oxygen protection and production, both of which are key phenomena in the development of life.

The interaction of oxygen with free radicals produces superoxide (O_2^-). Figure 8 shows the reactions of oxygen with electrons and the enzymes involved in breaking down superoxide to peroxide (O_2^{2-}) and subsequent reaction to hydroxyl radicals $(OH^.)$. The toxicities of peroxide and superoxide are considered to be high, with superoxide much more so than peroxide. However, hydroxyl radicals are thought to be even more toxic. Protection by SOD against the toxic O_2^- radicals have been studied in some detail (30, 31, 32, 43, 44). For example, in rats living at different oxygen tensions, the amount of superoxide dismutase in the cytoplasm is considered to be proportional to the oxygen content of the atmosphere in which they live. In the case of blue-green algae (*Anacystis*) grown at different oxygen concentrations, the level of SOD is again proportional to the oxygen concentration. The SOD in mitochondria is thought to protect the NADH dehydrogenase, which is associated with complex I, from the O_2^- produced by the flavoproteins in

$$O_2$$
$$\downarrow e^-$$
$$O_2^-(HO_2) \qquad\qquad\qquad\qquad\quad SOD$$
$$\qquad\qquad\qquad\qquad 2O_2^- \longrightarrow O_2^{2-}(H_2O_2) + O_2$$
$$\downarrow e^- \qquad\qquad\qquad\qquad\qquad 2H^+$$
$$\qquad\qquad\qquad\qquad\qquad\quad CAT$$
$$O_2^{2-}(H_2O_2) \qquad\qquad 2H_2O_2 \longrightarrow 2H_2O + O_2$$
$$\downarrow e^-$$
$$O^-(OH) \qquad\qquad O_2^- + H_2O_2 \longrightarrow OH^. + H_2O + O_2$$
$$\downarrow e^-$$
$$O^{2-}(H_2O)$$

Figure 8.   Oxygen, superoxide, and catalase reactions. SOD, superoxide dismutase; CAT, catalase (24, 33)

Table III. Properties of Superoxide Dismutases[a]

Enzyme Type	Metal Ion Content (g-atom mol⁻¹)	Molecular Weight	Subunit Structure
Cu–Zn			
Many eukaryotes including mammalian tissues, yeast, *Neurospora* and green plants	2 Cu 2 Zn	32,000	α_2
Photobacterium leiognathi	1 Cu 2 Zn	33,100	$\alpha\beta$
Mn			
Escherichia coli	1.4	39,500	α_2
Streptococcus mutans	1.2	40,250	α_2
Rhodopseudomonas spheroides	1.1	37,400	α_2
Bacillus stearothermophilus	—	40,000	α_2
Thermus aquaticus	2	80,000	α_4
Chicken liver mitochondria	2.1	80,000	α_4
Yeast mitochondria	4	96,000	α_4
Pleurotus olearius (fungus)			
SODc	2.1	76,000	$\alpha_2\beta_2$
SODm	2	78,000	α_4
Fe			
Escherichia coli	1.0	38,700	α_2
Photobacterium leiognathi	1.6	40,600	α_2
Plectonema boryanum	0.9	36,500	α_2
Spirulina platensis	1.0	37,400	α_2

[a] For original references *see* Refs. *24, 30, 31, 32.*

that complex (*45*). No doubt more reactions will be found which require oxygen protection.

The enzymes SOD and catalase must have been present to take care of the oxygen radicals produced from an early stage in the development of life. It is interesting that catalase, a heme enzyme, does not seem to occur in the obligate anaerobic fermenters like *Clostridia*, but SOD does occur in these organisms, much to the surprise of many biologists (*42*). The primary electron acceptor(s) in both bacterial and plant-type photosynthesis seem to be able to interact with oxygen to produce superoxide, thus again requiring the presence of SOD for protection against inactivation (*46, 47, 48, 49, 50*).

There are at least three different types of SOD which we can distinguish (Table III). The iron- and manganese-containing SODs are both cyanide-insensitive and, until recently, were thought to occur only in prokaryotic organisms. The copper + zinc-containing SOD is cyanide-sensitive and again until recently was only thought to occur in eukaryotes. This hypothesis was very neat but unfortunately does not seem to be quite correct. It has been discovered recently that algae of the eukaryotic type (reds, browns, and greens) contain the iron or manganese type of SOD

Table IV. Occurrence of SOD in Some Representative Organisms[a]

Clostridium	fermentative bacterium (anaerobe)
Chlorobium	green photosynthetic bacterium (anaerobe)
Chromatium	red (sulfur) photosynthetic bacterium (anaerobe)
Rhodopseudomonas	red (non-sulfur) photosynthetic bacterium (facultative anaerobe)
Desulfovibrio	sulfate reducer (anaerobe)
Paracoccus	facultative nitrate reducer
Photobacterium	luminescent bacterium (aerobe)
Spirulina	blue-green alga
Scenedesmus	green alga (unicellular)
Codium	green alga (siphonaceous)—cytoplasm and chloroplast
Saccharomyces	yeast—cytoplasm and mitochondria
Neurospora	fungus—cytoplasm and mitochondria
Pleurotus	fungus
Spinacia	spinach—cytoplasm and chloroplast
Bos	ox—erythrocytes, cytoplasm, mitochondria, and rod outer segments

[a] Refs. *24, 30–32, 38, 41.*

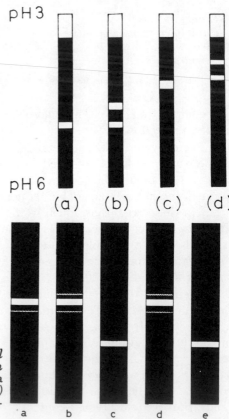

Figure 9. Polyacrylamide disc gel electrophoresis of soluble SOD from (A) Spinacia (spinach) (a, b), Spirulina (blue-green alga) (c,d), and from (B) rod outer segments from retina of cattle (a,b,d) and frogs (c,e) (39, 64)

(*41, 48*) and that fungi (*52*), protozoa, yeast, and chicken liver (which are eukaryotic) contain a cyanide-insensitive, manganese-containing SOD (*30, 31, 32*). Finally a copper + zinc SOD has been found in a bacterium (*53*). Thus the hypothesized distinction between eukaryotes and pro-karyotes does not seem to be universal. From our point of view this is in fact very fortunate since we now have two distinct types of enzymes (SOD and iron–sulfur proteins) which span the prokaryote/eukaryote transition and also the anaerobic/aerobic transition. Table IV shows the occurrence of SOD in some representative organisms from anaerobic bacteria to aerobic higher organisms.

Figure 9 shows the occurrence of SOD in a higher plant (spinach), a blue-green alga (*Spirulina*), and the rod outer segments from the frog and the cattle retina. This is just an example of a method for detecting SOD in the soluble components of various organisms. The assay on the gels is relatively easy since the stain is for the blue formazan which is precipitated from nitro-blue tetrazolium dye. This is the standard assay and has been very useful in detecting SOD in all kinds of organisms (*30, 31, 32*). Another technique for characterizing and detecting SOD is by use of their EPR characteristics. Figure 10 shows the purified iron-containing SOD from the alga *Spirulina* having an interesting spectrum in the $g = 5$ and $g = 9$ regions, which is as yet unexplained. The manganese-containing SOD and the copper + zinc-containing SOD also have characteristic EPR spectra which can be used as diagnostic criteria in relatively impure preparations.

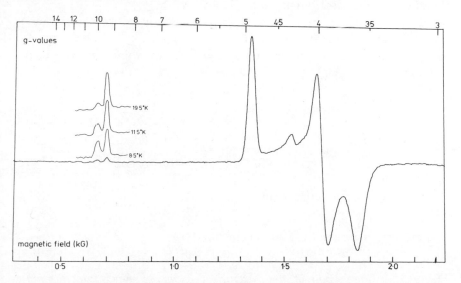

Figure 10.   EPR spectrum of Spirulina platensis *Fe-containing SOD at 8.5°K*
(62)

Mitochondria from animals contain two different SODs of the cyanide-insensitive iron and manganese types. The manganese SOD has an amino acid sequence (54) very similar to that from bacteria (see Figure 11). We have recently studied the mitochondria from Neurospora and from higher plants (55). In both cases we have detected the cyanide-insensitive SOD in the matrix, probably containing iron, and the cyanide-sensitive SOD in the inter-membrane space, which is probably a copper + zinc SOD. Thus SOD is distributed uniformly in animals, plants, and fungi. This provides interesting support for the theory of the symbiotic origin of mitochondria from bacteria (23, 36, 56, cf. 57).

Our recent survey of SOD in photosynthetic bacteria and various algae shows that cyanide-insensitive SOD occurs in all photosynthetic organisms from the red and green photosynthetic bacteria up to the green algae (39, 40, 41). The ability to isolate intact chloroplasts from the coenocytic green alga Codium allowed us to show the localization of a cyanide-insensitive SOD in the chloroplast itself. The higher plant chloroplasts do not contain the same type of SOD as has so far been found in the eukaryotic green and red algae (39, 40, 41, 47, 48). This interesting transition is now being investigated to study the evolution from algae to higher plant chloroplasts. It should be possible to find an alga or lower plant which has both the cyanide-insensitive and cyanide-sensitive types of SOD. Our preliminary data indicate that such algae do in fact exist (58). This biochemical data will be useful in complementing the structure and other characteristics used for propounding theories of evolution from algae to plants.

Higher plant chloroplasts contain a cyanide-insensitive, membrane-bound SOD which may also occur in blue-green algae (39–41, 47, 48, 59). This is rather a complicated system, but we think that the mem-

B. stearothermophilus	Mn	Pro–Phe–Glu–
E. coli	Mn	Ser –Tyr–Thr
E. coli	Fe	Ser
Chicken liver mitochondria	Mn	Lys–His–Thr
Bovine erythrocyte	Cu/Zn	Ac–Ala–Thr–Lys–

		16
B. stearothermophilus	Mn	Pro–His–Ile –
E. coli	Mn	Phe
E. coli	Fe	
Chicken liver mitochondria	Mn	
Bovine erythrocyte	Cu/Zn	Gly–Thr–Ile –

Figure 11. Comparison of the N-terminal region of superoxide dismutase and one mitochondrial dismutase. Residues identical to those in the B. region of the bovine

brane-bound SOD may be a manganese-containing enzyme which, during evolution, would have been simultaneously involved in the development of the water-splitting reaction (giving oxygen evolution) and the protection of the membrane against oxygen toxicity. This possibility is now being investigated. Asada (48) has solubilized this membrane-bound SOD from chloroplasts of red kidney bean. We have not yet been able to separate the SOD from membranes in our studies of the correlation between oxygen evolution and SOD activity.

The reason why less developed photosynthetic organisms and also the *Clostridial*-type bacteria probably used iron or manganese in the active center of SOD, instead of copper as higher organisms do, probably results from the fact that under anaerobic conditions copper is highly insoluble and would not have been available to form such enzymes (60, 61). With the development of oxygen in the atmosphere, which then became more oxidizing, copper probably became more soluble and would thus be available for complexing into various types of protein. Copper proteins are widespread in aerobic organisms and probably function just as efficiently in SOD as the cyanide-insensitive iron and manganese-containing SOD, which, to a large extent, they seem to have replaced in higher aerobic organisms (24).

The occurrence of SOD in photosynthetic and nonphotosynthetic and anaerobic and aerobic organisms can again provide a useful criterion for evolution studies. Figure 12 shows a number of hypotheses which can be postulated for such an evolution (24, 27). The second scheme has been favored (24). As with the iron–sulfur proteins, there seems to be an important transition between the anaerobic and the aerobic phase of life; it is possible that the red non-sulfur bacteria provide key transition-type organisms, probably in parallel, with the sulfate and nitrate

```
       5                    10                   15
–Leu–Pro–Ala–Leu–Pro –Tyr–Pro –Tyr–Asp–Ala –Leu–Glu
          Ser                   Ala
                                Ala–Lys                 Ala
          Asp                   Asp      Gly
–Ala –Val –Cys–Val–Leu –Lys–Gly–Asp–Gly–Pro –Val –Gln

       20                   25
–Asp–Lys–Glu–Thr–Met–Asn–Ile –His –His –Thr–Lys
          Gln                   Glu–Leu   ?        ?
 Ser –Ala        ?   Ile –Glu–Tyr      Tyr–Gly
 Ser –Ala       Ile      Gln–Leu         ?    ?
–His –Phe–Glu–Ala –Lys –Gly–Asp–Thr–Val –Val –Val
```

from B. stearothermophilus *with the N-terminal regions of two other bacterial* stearothermophilus *sequence have not been included. The corresponding erythrocyte enzyme is also shown (54).*

reducers. The similarity between the type of metabolism in red non-sulfur bacteria and that of mitochondria is quite striking, as we noted previously, and certainly the proposal here would tend to back up previous work. The sequence of SOD from *Rhodopseudomonas* and from mitochondria is very similar, as it is for other types of aerobic bacteria (*24, 62*).

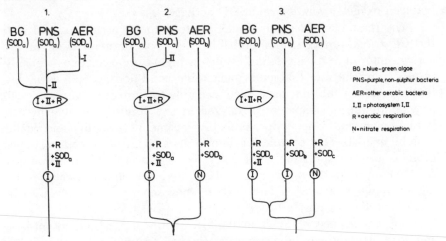

Figure 12. Possible evolutionary interrelationships of aerobic prokaryotes evolving from photosynthetic bacateria (24).

$SOD_{(a, b, c)}$ = *superoxide dismutases of these different types. The first scheme is favored (27) on the principle of "minimizing the amount of evolution inferred to have taken place." Nevertheless the second scheme seems probable, giving rise to two groups of SOD with somewhat different sequence homologies.*

Thus both iron–sulfur proteins and SOD will continue to provide us with two metal proteins which span the anaerobic/aerobic and the prokaryote/eukaryote transitions in the development of plants and animals from bacteria and algae.

Summary

Iron–sulfur proteins contain non-heme iron and inorganic (acid-labile) sulfur in their active centers as 4Fe–4S or 2Fe–2S or, in the case of rubredoxin, as one iron alone. The iron is always bonded to cysteine sulfur. They catalyze redox reactions between +350 and −600 mV (hydrogen electrode = −420 mV). They are usually of low molecular weight (6000–15,000 Daltons) but can form complex enzymes with molybdenum and flavin. They occur as soluble or membrane-bound proteins and catalyze key reactions in photosynthesis, oxidative phosphorylation, nitrogen fixation, H_2 metabolism, steroid hydroxylation, carbon and sulfur metabolism, etc. They occur in all organisms so far investigated and may have been among the first proteins sequenced which occur in all organisms. Synthetic analogs of iron–sulfur proteins have provided evidence for the structure of the active centers and may also have practical significance as catalysts. Superoxide dismutases are enzymes which contain iron, manganese, or copper + zinc in their active centers. They are involved in the dismutation of the superoxide anion (O_2^-) which is thought to be very toxic and which is produced by free radical interaction with oxygen. Surprisingly they seem to occur in all organisms, even those living in the absence of oxygen. The iron and manganese enzymes were thought to be associated only with prokaryotes and the copper/zinc enzyme with eukaryotes; this division, however, is being revised. They are usually soluble, but the manganese enzyme may possibly be membrane-bound and be involved in photosynthetic oxygen evolution from water.

Literature Cited

1. Hall, D. O., *Nature* (1975) **255**, 578–579.
2. Hall, D. O., Rao, K. K., Cammack, R., *Sci. Prog. (Oxford)* (1975) **62**, 285–317.
3. Hall, D. O., Rao, K. K., Mullinger, R. N., *Biochem. Soc. Trans* (1975) **3**, 472–479; 361 000.
4. Holm, R. H., *Endeavour* (1975) **34**, 38–43.
5. Lovenberg, W., "Iron–Sulphur Proteins," Vol. I and II. Academic, N.Y., 1973.
6. Hong, J. S., Rabinowitz, J. C., *J. Biol. Chem.* (1970) **245**, 6574–6581.
7. Carter, C. W., Kraut, J., Freer, S. T., Alden, R. A., Sieker, L. C., Adman, E., Jensen, L. H., *Proc. Natl. Acad. Sci. U.S.A.* (1972) **69**, 3526–3531.
8. Cammack, R., *Biochem. Biophys. Res. Commun.* (1973) **54**, 548–553.
9. Sweeney, W. V., Bearden, A. J., Rabinowitz, J. C., *Biochem. Biophys. Res. Commun.* (1974) **59**, 188–194.
10. Erbes, D. L., Burris, R. H., Orme-Johnson, W. H., *Proc. Natl. Acad. Sci. U.S.A.* (1975) **72**, 475–479.
11. Evans, M. C. W., Sihra, C. K., Cammack, R., *Biochem. J.* (1976) **158**, 71–77.
12. Cammack, R., Evans, M. C. W., *Biochem. Biophys. Res. Commun.* (1975) **67**, 544–549.
13. Malkin, R., Bearden, A. J., *Proc. Natl. Acad. Sci. U.S.A.* (1971) **68**, 16–20.
14. Malkin, R., Aparicio, P. J., *Biochem. Biophys. Res. Commun.* (1975) **63**, 1157–1160.
15. Ohnishi, T., *Eur. J. Biochem.* (1976) **64**, 91–103.
16. Bruice, T. C., Maskiewiez, R., Job, R., *Proc. Natl. Acad. Sci. U.S.A.* (1975) **72**, 231–234.
17. Hose, T., Ohmiya, N., Matsubara, H., Mullinger, R. N., Rao, K. K., Hall, D. O., *Biochem. J.* (1976) **159**, 55–63.
18. Broda, E., "Evolution of the Bioenergetic Processes," Pergamon, Oxford, 1975.
19. Cohen, Y., Jorgenson, B. B., Padan, E., Shilo, M., *Nature* (1975) **257**, 489–491.
20. Dickerson, R. E., Timkovich, R., Almassy, R, J., *J. Mol. Biol.* (1976) **100**, 473–491.
21. Dutton, P. L., Wilson, D. F., *Biochim. Biophys. Acta* (1974) **346**, 165–212.

22. Hall, J. B., *J. Theor. Biol.* (1973) **30**, 429–454.
23. John, P., Whatley, F. R., *Nature* (1975) **254**, 495–498.
24. Lumsden, J., Ph.D. Thesis, University of London King's College, 1975.
25. Olson, J. M., *Science* (1970) **168**, 438–446.
26. Stanier, R. Y., *Biochem. Soc. Trans.* (1975) **3**, 352–357.
27. Uzzell, T., Spolsky, C., *Am. Sci.* (1974) **62**, 334–343.
28. Tel-Or, E., Cammack, R., Rao, K. K., Roger, L. J., Hall, D. O., *Biochim. Biophys. Acta* (1976) in press.
29. Bors, S., Saran, M., Lengfelder, E., Spöttle, R., Michel, C., *Curr. Top. Radiat. Res. Q.* (1974) **9**, 247–309.
30. Fridovich, I., "Free Radicals in Biology," W. A. Pryor, Ed., p. 239, Academic, N.Y. (1976).
31. Fridovich, I., *Adv. Enzymol.* **41**, 35–97.
32. Fridovich, I., *Ann. Rev. Biochem.* (1975) **44**, 147–159.
33. Halliwell, B., *New Phytol.* (1974) **73**, 1075–1085.
34. Weser, U., *Struct. Bonding* (1973) **17**, 1–65.
35. Berkner, L. V., Marshall, L. C., *J. Atmos. Sci.* (1965) **22**, 225–261.
36. Margulis, L., "Origin of Eukaryotic Cells," Yale University, 1970.
37. Margulis, L., Walker, J. C. G., Rambler, M., *Nature* (1976) **264**, 620–624.
38. Hewitt, J., Morris, J. G., *FEBS Lett.* (1975) **50**, 315–318.
39. Lumsden, J., Hall, D. O., *Biochem. Biophys. Res. Commun.* (1974) **58**, 35–41.
40. *Ibid.*, **64**, 595–602.
41. Lumsden, J., Hall, D. O., *Nature* (1975) **257**, 670–672.
42. Morris, J. G., *Adv. Microb. Physiol.* (1975) **12**, 169–246.
43. Gregory, E. M., Fridovich, I., *J. Bacteriol.* (1973) **114**, 543–548.
44. *Ibid.*, **114**, 1193–1199.
45. Tyler, D. D., *Biochim. Biophys. Acta* (1975) **396**, 335–346.
46. Allen, J. F., Hall, D. O., *Biochem. Biophys. Res. Commun.* (1973) **52**, 856–862.
47. Asada, K., Kiso, K., *Eur. J. Biochem.* (1973) **33**, 253–257.
48. Asada, K., Yoshikawa, K., Takashi, M-A., Maeda, Y., Emmanji, K., *J. Biol. Chem.* (1975) **250**, 2801–2807.
49. Boucher, F., Gingras, G., *Biochem. Biophys. Res. Commun.* (1975) **67**, 421–426.
50. Elstner, E. F., Heupel, A., *Biochim. Biophys. Acta* (1973) **325**, 182–188.
51. Miller, R. W., MacDowell, F. D. H., *Biochim. Biophys. Acta* (1975) **387**, 176–187.
52. Lavelle, F., Dursay, P., Michelson, A. M., *Biochimie* (1974) **56**, 451–458.
53. Puget, K., Michelson, A. M., *Biochimie* (1974) **56**, 1255–1267.
54. Bridgen, J., Harris, J. I., Northrop, F., *FEBS Lett.* (1975) **49**, 392–395.
55. Arron, G. P., Palmer, J. M., Henry, L., Hall, D. O., *Biochem. Soc. Trans.* (1976) in press.
56. Margulis, L., *Evol. Biol.* (1974) **7**, 45–78.
57. Raff, R. A., Mahler, H. R., *Science* (1972) **177**, 575–582.
58. Henry, L., Hall, D. O., *Plant Cell Physiol.* (1976) in press.
59. Misra, H. P., Keele, B. B., *Biochim. Biophys. Acta* (1975) **379**, 418–425.
60. Egami, F., *J. Biochem. (Tokyo)* (1975) **77**, 1165–1169.
61. Osterberg, R., *Nature* (1974) **249**, 382–383.
62. Lumsden, J., Cammack, R., Hall, D. O., *Biochim. Biophys. Acta* (1976) **438**, 380–392.
63. Cammack, R., *Biochem. Soc. Trans.* (1975) **3**, 482–488.
64. Hall, M. O., Hall, D. O., *Biochem. Biophys. Res. Commun.* (1975) **67**, 1119–1204.

RECEIVED July 26, 1976.

Physical and Chemical Studies of Bovine Erythrocyte Superoxide Dismutase

STEPHEN J. LIPPARD, ALLAN R. BURGER, and KAMIL UGURBIL

Department of Chemistry, Columbia University, New York, N. Y. 10027

JOAN S. VALENTINE and MICHAEL W. PANTOLIANO

Department of Chemistry, Douglass College, Rutgers, The State University, New Brunswick, N.J. 08903

The structure and enzyme kinetics of bovine erythrocyte superoxide dismutase are reviewed. The protein has a novel imidazolate-bridged copper(II)–zinc(II) catalytic center in each of two identical subunits. Since a Cu^{II}/Cu^{I} redox couple is responsible for the dismutase activity of the enzyme, the role of zinc is of interest. Both 220-MHz NMR measurements of the exchangeable histidine protons and chemical modifications using diethylpyrocarbonate demonstrate that zinc alone can fold the protein chain in the region of the active site into a conformation resembling that of the native enzyme. Other possible roles for zinc are discussed. Synthetic, magnetic, and structural studies of soluble, imidazolate-bridged copper complexes of relevance to the 4 Cu(II) form of the enzyme have been made.

There is great interest in the biochemistry and relevant coordination chemistry of copper-containing proteins (1, 2, 3, 4, 5). They are widely distributed in both plants and animals and are often involved in oxygen metabolism, transport, and use. One of the most actively studied copper proteins is bovine erythrocyte superoxide dismutase (SOD) (6, 7, 8). This enzyme catalyzes the dismutation of superoxide ion, Reaction 1.

$$2O_2^- + 2H^+ \rightarrow O_2 + H_2O_2 \tag{1}$$

By thus scavenging O_2^-, SOD is proposed to protect respiring cells against the deleterious reactivities of this ion (9, 10). Because of its known anti-inflammatory properties, the enzyme is commercially available as

Table I.   Properties of Bovine Erythrocyte Superoxide Dismutase[a]

Molcular weight	31,400	two identical subunits
Amino acids/subunit	151	isoelectric point (pI)—4.95
Metals/subunit	1 Cu, 1 Zn	color—blue-green
Absorption spectrum	λ_{max} (nm)	ϵ_{max} (M^{-1} cm^{-1})
	258	10,300
	270, 282, 289	shoulders
	680	300
EPR spectrum (77°K)	g_m 2.080	$g_{\parallel}$ 2.265
		$A_{\parallel}$ 0.015 cm^{-1}
Redox potential (pH dependent)	$E^{\circ\prime} = 0.42$ V	

[a] Data taken from Refs. *12, 13, 14, 15, 16.*

Table II.   Amino Acid Sequence of Bovine Erythrocyte Superoxide Dismutase[a] (*13*)

1
                                                                                  10
Ac-Ala -Thr -Lys -Ala -Val -Cmc-Val -Leu -Lys -Gly -Asp -Gly -Pro-

                                                     20
Val -Gln -Gly -Thr -Ile -His¯-Phe -Glu -Ala -Lys -Gly -Asp -Thr-

              30
Val -Val -Val -Thr -Gly -Ser -Ile -Thr -Gly -Leu -Thr -Glu -Gly-

40   50
Asp -(His) -Gly -Phe -(His) -Val -(His) -Gln -Phe -Gly -Asp -Asn -Thr-

                                               60
Gln -Gly -Cmc-Thr -Ser -Ala -Gly -Pro -(His) -Phe -Asn -Pro -Leu-

                        70
Ser -Lys -Lys -[His] -Gly -Gly -Pro -Lys -Asp -Glu -Glu -Arg -[His]-

80   90
Val -Gly -[Asp] -Leu -Gly -Asn -Val -Thr -Ala -Asp -Lys -Asn -Gly-

                                          100
Val -Ala -Ile -Val -Asp -Ile -Val -Asp -Pro -Leu -Ile -Ser -Leu-

                                          100
Ser -Gly -Glu -Tyr -Ser -Ile -Ile -Gly -Arg -Thr -Met -Val -Val-

          120   130
(His) -Glu -Lys -Pro -Asp -Asp -Leu -Gly -Arg -Gly -Gly -Asn -Glu-

                                          140
Glu -Ser -Thr -Lys -Thr -Gly -Asn -Ala -Gly -Ser -Arg -Leu -Ala-

                        150
Cmc-Gly -Val -Ile -Gly -Ile -Ala -Lys

[a] Cmc denotes *S*-carboxymethylcysteine. Histidyl groups encircled are bound to copper, those in boxes bind to zinc, the shaded one is buried, and His 19 is accessible to solvent.

the drug, orgotein, used for the treatment of orthopedic disorders in horses and dogs. It is currently being investigated for possible human use under the trade name Ontosein (*11*).

Probably more is known about bovine erythrocyte SOD than about any copper protein (Table I), and it is the only one for which x-ray structural information is available. This chapter reviews the salient structural and mechanistic features of the enzyme, focusing on the active site. There are additional reviews of the literature in several excellent articles (*6, 7, 8, 17*).

Active Site Structure

The complete amino acid sequence (Table II) (*13*), subunit structure (*18*), and x-ray crystal structure at 3.0-Å resolution (*19, 20*) are known for bovine erythrocyte SOD. The enzyme consists of two identical subunits of molecular weight 15,700 daltons, each of which contains one copper and one zinc atom. A sketch of the peptide chain in the region of the active site is shown schematically in Figure 1 (*19, 20, 21*). The copper coordination geometry is distorted square-planar, with nitrogen donor atoms contributed from the imidazole side chains of four histidine residues. The peptide chain His–Val–His, with His 44 and His 46 in trans positions, completely blocks one side of the coordination plane from ligand attack. The opposite side is available to bind a fifth ligand, and there is evidence for water and anion coordination at that site in the native enzyme (*22*). The zinc ion has a tetrahedral environment, with four donor atoms of three histidine nitrogen atoms and the oxygen atom of Asp 81. Both copper and zinc can be reversibly removed by dialysis against EDTA at low pH to form the apoprotein (*12, 14, 15, 23*).

A remarkable feature of the active site structure in bovine SOD is the presumably deprotonated imidazole ring of His 61 that bridges the two metal ions:

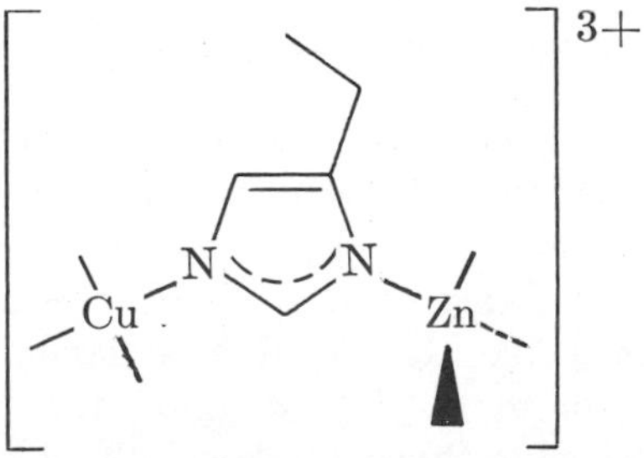

The imidazole rings of His 44, His 46, and His 118 are approximately normal to the copper coordination plane, whereas the imidazolate moiety of His 61 is rotated approximately parallel to that plane and may not be precisely coplanar with it (21). Additional x-ray data at higher resolution should provide further details about the active site structure.

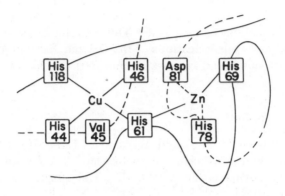

Figure 1. Active site of bovine superoxide dismutase showing the coordinated amino acids and partial tracing of the peptide chain (19, 20, 21)

Enzyme Kinetics and Mechanism: Role of Copper

Copper has a direct role in the catalytic mechanism. Complete removal of both copper and zinc destroys the activity of the enzyme which can be restored by addition of cupric ion but not by other metals (12). Pulse radiolytic methods have been used to generate superoxide ion and to follow the enzyme kinetics (24, 25). The reaction is second order with a rate constant of 2.37×10^9 M^{-1} sec^{-1} at 25°, independent of pH over the range 4.8–9.5. The following two-step mechanism has been proposed to account for the enzyme activity:

$$ECu(II) + O_2^- \rightarrow ECu(I) + O_2 \qquad (2)$$

$$ECu(I) + O_2^- + 2H^+ \rightarrow ECu(II) + H_2O_2 \qquad (3)$$

In this mechanism, divalent copper is reduced by superoxide ion in the first step and then reoxidized in the second step with concomitant reduction of superoxide ion to hydrogen peroxide. Protonation of the O_2^{2-} ion generated in Reaction 3 provides the thermodynamic driving force. As shown in Table III, metal ions having a reduction potential between -0.36 V (for the O_2/O_2^- couple) and $+0.90$ V (for the O_2^-/H_2O_2 couple) are capable thermodynamically of catalyzing the dismutation of superoxide ion. Both free cupric ion and enzyme bound copper fall in this range, and discussions of the factors that affect the reduction

Table III. Reduction Potentials of Copper and Superoxide Ions[a]

Half Reaction	$E^{\circ\prime}$ (V)	Reference
$O_2 + e^- \rightleftarrows O_2^-$	-0.36	26
Cu^{2+} (aq) $+ e^- \rightleftarrows Cu^+$ (aq)	0.17	27
$ECu(II)$[b] $+ e^- \rightleftarrows ECu(I)$	0.42	16
$O_2^- + 2H^+ + e^- \rightleftarrows H_2O_2$	0.90	26

[a] At pH 7.0.
[b] In bovine erythrocyte SOD.

potentials of copper complexes are available (*1, 28*). Hydrogen peroxide produced in Reaction 3 would be harmful to living cells, but it is scavenged by the enzyme catalase. Catalase and superoxide dismutase are thought to act in concert to protect respiring cells from the toxic byproducts of oxygen metabolism (*10*).

Role of the Zinc

Since only copper is invoked in the enzyme mechanism, the role of the zinc is of interest. One possibility is that it has mainly a structural role (*29*), organizing the polypeptide conformation in the region of the active site. This possibility is supported by thermal stability studies (*30*) and by recent chemical and NMR work in our laboratories (*31*). The reagent diethylpyrocarbonate (DEP) reacts with histidine in proteins to produce a strong uv absorbance at 242 nm (Figure 2) (*32, 33*). In the case of histidine bound to copper or zinc, it is likely that the metal–imidazole linkage will prevent this reaction. Folding of the peptide chain could also inhibit the reaction of DEP with histidine. Since six of the eight histidine residues per subunit of SOD are at the active site (Table II and Figure 1), the DEP reagent offers a way to probe the active site structure chemically. As shown in Figure 3, all eight histidines per subunit are ethoxyformylated when DEP is allowed to react with the apo enzyme. In the native enzyme, only one histidine residue per subunit can be modified, as reported previously (*34*). The modification most likely occurs at His 19, which the x-ray structural study

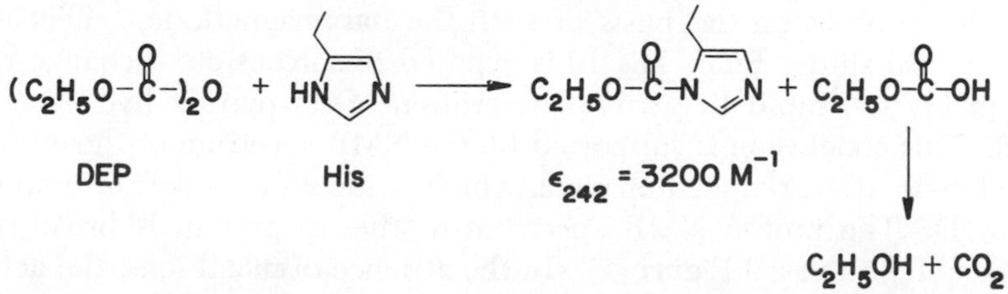

Figure 2. Diethylpyrocarbonate (DEP) and its reaction with histidine

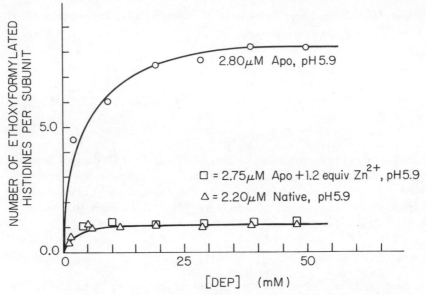

Figure 3.   The ethoxyformylation of histidyl residues in bovine SOD as a function of initial DEP concentration, pH 5.9. The concentrations of apo and native proteins were 2.8 μM and 2.2 μM, respectively. Experimental details are given in Ref. 31.

showed to be exposed to solvent (20). Addition of one or more equivalents per subunit of divalent zinc to the apoprotein reduces the number of ethoxyformylated histidine residues to one. This result suggests that zinc alone can restore much of the native structure of the peptide chain in the active site region, blocking access to the DEP reagent.

The major basis for this interpretation, however, derives from 220-MHz proton NMR studies of bovine SOD in H_2O (31). As shown in Figure 4, well resolved resonances are observed between 7 and 11 ppm downfield from the solvent. The NMR spectrum of the native enzyme consists of two broad, featureless absorptions owing to the presence of the paramagnetic copper(II) ion. By contrast, the spectrum of the reduced, copper(I) form of the native enzyme has several sharp, well resolved resonances in this region. These are assigned to histidine N–H (pyrrole) protons on the basis of both the paramagnetic ion effect and the chemical shifts. Since His 19 is exposed to solvent, its exchange rate is probably too rapid to permit observation of its pyrrole hydrogen by NMR. This conclusion is supported by the NMR spectrum of the ethoxyformylated native, reduced enzyme, which is identical to that of unmodified SOD. The proton NMR spectrum of the apoprotein is broad and relatively featureless (Figure 4). In the absence of metal ions, the active site structure is apparently loosely organized, which would permit rapid exchange of histidine protons and hence the broad lines. Addition of only one mole of zinc(II) per subunit, however, results in a series of narrow lines somewhat similar to that of the native reduced enzyme, and further addition of copper to the solution restores the NMR spectrum of oxidized SOD. It therefore appears that zinc alone can fold the polypeptide chain around the active site into a conformation resembling that of the native protein. Moreover, the quality of the spectra displayed in Figure 4 suggests that proton NMR will be useful in future studies designed to probe the details of the enzyme mechanism.

Another role for the zinc is suggested by the known ability of metal ions to attenuate the pK_a value of the pyrrole hydrogen of the imidazole group (35). The pK_a of this hydrogen is 14.4 in histidine, 11.7 in $Cu(L\text{-}His)_2$, and 10.8 in $[(en)Pd(L\text{-}His)]^+$. Substitution of the pyrrole

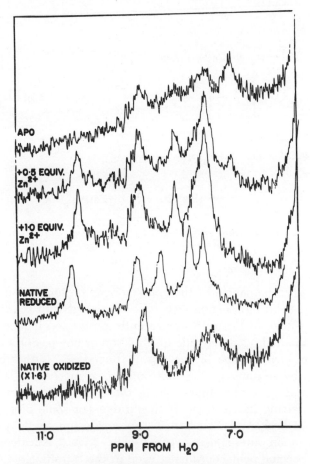

Figure 4.   220-MHz proton magnetic resonance spectra of bovine SOD in water, pH 6 phosphate buffer. Experimental details are given in Ref. 31.

hydrogen by a metal ion (36) occurs near pH 9.6 in solutions of neutral Ni(II), Cu(II), and Pd(II) complexes of glycyl-L-histidine and at even lower pH values for cationic $[Cu_2(bpim)(im)]_2^{4+}$ as discussed below. Thus in the enzyme, the zinc-substituted imidazole ring of His 61 might prefer to bind the [(His)$_3$Cu(II)] moiety instead of to a proton but might bind a proton in preference to [(His)$_3$Cu(I)], which is the product of the reduction step, Reaction 2. This possibility requires that the imidazolate bridge be broken during enzyme turnover, leading to the mechanism shown in Figure 5. There is not yet any direct evidence for this mechanism, but it has been suggested that the imidazolate bridge in the oxidized enzyme can be broken at low pH (37, 38). Chemical reduction of

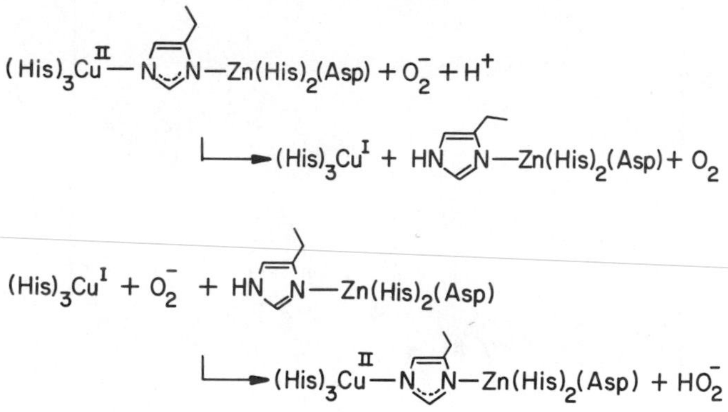

Figure 5.   Possible bridge-splitting mechanism for bovine superoxide dismutase activity

the enzyme is accompanied by the uptake of one proton per subunit (16), and a bridge-splitting reaction would be consistent with the result. The mechanism shown in Figure 5 was recently postulated independently in a report of the hydrogen peroxide inactivation of bovine SOD (39). One advantage in cleaving the bridge during Reaction 2 is that Reaction 3 can proceed by an inner-sphere electron transfer mechanism, with superoxide ion binding to the empty site in the copper coordination sphere before transferring its electron. If the bridge (or some other) bond to copper is not broken in Reaction 2, superoxide ion would presumably be reduced by an outer-sphere process in Reaction 3, since copper(I) is unlikely to become pentacoordinate, even in the transition state. Kinetic studies of the oxidation of related cuprous compounds by oxygen show the reaction to occur by an inner-sphere electron transfer mechanism (40).

Imidazolate-Bridged Copper(II) Complexes

Kinetic studies have shown that free cupric ion in aqueous solution can catalyze the dismutation of superoxide ion at a rate in excess of that of the enzyme (41). The rate can be substantially diminished when the copper is attached to chelating amino acid ligands (42, 43), however, suggesting that some feature(s) of the enzyme active site preserves the inherent ability of the cupric ion to catalyze Reaction 1. Although full treatment of this point is beyond the scope of this chapter, the presence of a bridging imidazolate ligand (I) is one such feature that might be tested in a model system. Recent synthetic and structural studies (44) of the imidazolate-bridged dicopper(II) species II suggest that it dimerizes reversibly in the presence of imidazole (imH) to form III, Reaction 4.

$$2II + 2imH_2^+ \rightleftharpoons III + 4H^+ \tag{4}$$

$[Cu_2(bpim)]^{3+}$, II

$[Cu_2(bpim)(im)]_2^{4+}$, III

X-ray structural studies of $[Cu_2(bpim)(im)]_2(NO_3)_4 \cdot 4H_2O$ establish the presence of imidazolate bridges in III, Figure 6. Solutions of this compound reversibly take up four protons between the end points at pH 9.75 and 4.25, consistent with Reaction 4. Magnetic susceptibility studies show that the copper(II) centers in III are antiferromagnetically

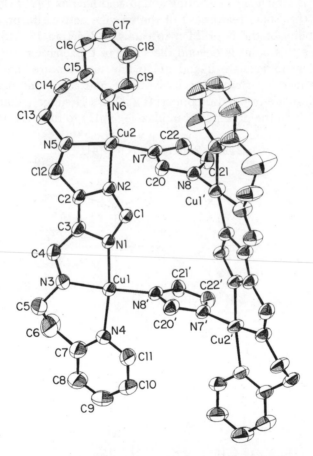

Figure 6. Structure of the $[Cu_2(bpim)(im)]_2{}^{4+}$ dimer as determined in an x-ray diffraction study. Primed and unprimed atoms are related by a crystallographically required twofold axis of symmetry. Further details are given in Ref. 44.

coupled, a result analogous to that found in the 4 Cu(II) form of bovine SOD (45). Continued studies of these and related soluble, imidazolate bridged complexes should help to provide insight into the catalytic mechanism of the histidine linked copper–zinc center at the active site of bovine SOD.

Acknowledgment

The work cited here was supported by NIH research grant no. GM-16449 from the National Institute of General Medical Science and by National Science Foundation Grants MPS75-05243 and MPS75-14463. We thank D. C. Richardson for providing additional, unpublished details about the bovine SOD x-ray structure.

Literature Cited

1. Brill, A. S., Martin, R. B., Williams, R. J. P., *Electron. Aspects Biochem. Proc. Int. Symp.* (1964) 519.
2. Frieden, E., Osaki, S., Kobayashi, H., *J. Gen. Physiol.* (1965) 49, 213.
3. Peisach, J., Aisen, P., Blumberg, W. E., Eds., "The Biochemistry of Copper," Academic, New York, 1966.
4. Malkin, R., Malmström, B. G., *Adv. Enzymol.* (1970) 33, 177.
5. Malkin, R., "Inorganic Biochemistry," G. L. Eichhorn, Ed., Vol. 2, p. 689, Elsevier, Amsterdam, 1973.
6. Fridovich, I., *Acc. Chem. Res.* (1972) 5, 321.
7. Fridovich, I., *Annu. Rev. Biochem.* (1975) 44, 147.
8. Weser, U., *Struct. Bonding* (1973) 17, 1.
9. Fridovich, I., *N. Engl. J. Med.* (1974) 290, 624.
10. Fridovich, I., *Am. Sci.* (1975) 63, 54.
11. Lund-Olesen, K., Menander, K. B., *Curr. Ther. Res. Clin. Exp.* (1974) 16, 706.
12. McCord, J. M., Fridovich, I., *J. Biol. Chem.* (1969) 244, 6049.
13. Steinman, H. M., Naik, V. R., Abernathy, J. L., Hill, R. L., *J. Biol. Chem.* (1974) 249, 7326.
14. Bannister, J., Bannister, W., Wood, E., *Eur. J. Biochem.* (1971) 18, 178.
15. Keele, B. B., Jr., McCord, J. M., Fridovich, I., *J. Biol. Chem.* (1971) 246, 2875.
16. Fee, J. A., DiCorleto, P. E., *Biochemistry* (1973) 12, 4893.
17. Fridovich, I., *Adv. Enzymol. Relat. Areas Mol. Biol.* (1974) 41, 35.
18. Abernethy, J. L., Steinman, H. M., Hill, R. L., *J. Biol. Chem.* (1974) 249, 7339.
19. Richardson, J. S., Thomas, K. A., Rubin, B. H., Richardson, D. C., *Proc. Natl. Acad. Sci. USA* (1975) 72, 1349.
20. Richardson, J. S., Thomas, K. A., Richardson, D. C., *Biochem. Biophys. Res. Commun.* (1975) 63, 986.
21. Richardson, D. C., private communication, 1976.
22. Fee, J. A., Gaber, B. P., *Oxidases Relat. Redox Syst. Proc. Symp.* (1973) 77.
23. Fee, J. A., *Biochim. Biophys. Acta* (1973) 295, 87.
24. Klug-Roth, D., Fridovich, I., Rabani, J., *J. Am. Chem. Soc.* (1973) 95, 2786.
25. Fielden, E. M., Roberts, P. R., Bray, R. C., Lowe, D. J., Mautner, G. N., Rotilio, G., Calabrese, L., *Biochem. J.* (1974) 139, 49.
26. Bennett, L. E., *Prog. Inorg. Chem.* (1973) 18, 1.
27. Lattimer, W. M., "Oxidation Potentials," Prentice-Hall, Englewood Cliffs, N.J., 1952.
28. Patterson, G. S., Holm, R. H., *Bioinorg. Chem.* (1975) 4, 257.
29. Rotilio, G., Calabrese, L., Bossa, F., Barra, D., Agró, A. F., Mondovi, B., *Biochemistry* (1972) 11, 2182.
30. Forman, H. J., Fridovich, I., *J. Biol. Chem.* (1973) 28, 2645.

31. Lippard, S. J., Burger, A., Ugurbil, K., Pantoliano, M., Valentine, J. S.,
    Biochemistry (1977) **16**, 1136.
32. Ovadi, J., Liber, S., Elödi, P., *Acta Biochim. Biophys. Acad. Sci. Hung.*
    (1967) **2**, 455.
33. Pradel, L. A., Kassab, R., *Biochim. Biophys. Acta* (1968) **167**, 317.
34. Stokes, A. M., Hill, H. A. O., Bannister, W. H., Bannister, J. V., *Biochem.
    Soc. Trans.* (1974) **2**, 489.
35. Martin, R. B., *Proc. Natl. Acad. Sci. USA* (1974) **7**, 4346.
36. Morris, P. J., Martin, R. B., *J. Inorg. Nucl. Chem.* (1971) **33**, 2913.
37. Fee, J. A., Phillips, W. D., *Biochim. Biophys. Acta* (1975) **412**, 26.
38. Calabrese, L., Cocco, D., Morpurgo, L., Mondovi, B., Rotilio, G., *FEBS
    Lett.* (1975) **59**, 29.
39. Hodgson, E. K., Fridovich, I., *Biochemistry* (1975) **14**, 5294.
40. Zuberbühler, A., *Helv. Chim. Acta* (1967) **50**, 466.
41. Rabani, J., Klug-Roth, D., Lilie, J., *J. Phys. Chem.* (1973) **77**, 1169.
42. Brigelius, R., Hartmann, H-J., Bors, W., Saran, M., Lengfelder, E., Weser,
    U., *Z. Physiol. Chem.* (1975) **356**, 739.
43. Klug-Roth, D., Rabani, J., *J. Phys. Chem.* (1976) **80**, 588.
44. Kolks, G., Frihart, C. R., Rabinowitz, H. N., Lippard, S. J., *J. Am. Chem.
    Soc.* (1976) **98**, 5720.
45. Fee, J. A., Briggs, R. G., *Biochim. Biophys. Acta* (1975) **400**, 439.

RECEIVED July 26, 1976.

Characterization of the Copper(II) Site in Galactose Oxidase

ROBERT D. BEREMAN, MURRAY J. ETTINGER,
DANIEL J. KOSMAN, and ROBERT J. KURLAND

The Bioinorganic Graduate Research Group, Departments of Chemistry and
Biochemistry, State University of New York at Buffalo, Buffalo, NY 14214

Spectral and model studies of the Cu(II)-containing metallo-protein galactose oxidase suggest that the enzymic metal coordination center is a square planar system involving two nitrogenous ligands. Model studies suggest that exogenous ligands coordinate in an equitorial position. Difference absorbance spectra with exogenous anions establish that only the 314-nm transition exhibited by the enzyme is of charge-transfer character. Difference absorbance, EPR, and fluoride relaxation are consistent with formation of a complex with the anion ferricyanide near the copper site. In the enzyme, a histidine imidazole and tryptophan indole contribute directly to catalysis. Spectral results imply that enzyme activity is associated with a relatively unique geometry of the active site Cu(II) complex.

Copper is an essential element to most life forms. In humans it is the third most abundant trace element; only iron and zinc are present in higher quantity. Utilization of copper usually involves a protein active site which catalyzes a critical oxidation reaction, e.g., cytochrome oxidase, amine oxidases, superoxide dismutase, ferroxidases, dopamine-β-hydroxylase, and tyrosinase. Accordingly, animals exhibit unique homeostatic mechanisms for the absorption, distribution, utilization, and excretion of copper (1). Moreover, at least two potentially lethal inherited diseases of copper metabolism are known: Wilson's Disease and Menkes's Kinky Hair Syndrome (1).

Cu(II) sites in proteins have been classified into three types based on their spectral properties (2). Type I Cu(II) sites are characterized by very high molar absorbtivity values for the visible band near 600 nm

(16.5 kK). Accordingly, proteins which contain Type I sites are often referred to as blue copper proteins since their solutions are blue at typical enzyme concentrations in a research laboratory (10^{-5}–$10^{-4}M$). The 600-nm region band has been suggested to arise from $n \rightarrow \pi^*$ (σ^*) charge transfer involving a copper–cysteine bond. Metal replacement studies (3, 4) and recent model studies (5) tend to substantiate this assignment. Electron spin resonance (ESR) parameters for these Type I sites are also different than for simple copper complexes, particularly in the unusually low value for the spin Hamiltonian parameter, A_{zz}, which is typically less than 100 G (2). The Cu(II) atoms in these sites are believed to be in a trigonal environment (4 or 5 coordinate) (6, 7), and functional activity is associated with a change in oxidation state of the copper (8).

Type II Cu(II), or low-blue copper, is less colored at common research concentrations. These systems have received less attention than Type I copper. However, even low-blue cupric copper can possess high molar absorbtivities when compared with simple coordination complexes of Cu(II). The Cu(II) sites in such proteins also yield A_{zz} values normally greater than 140 G, i.e., more like that of low molecular weight square planar Cu(II) complexes (2, 8). The only available crystal structure of a copper protein is that of a low blue protein bovine erythrocyte superoxide dismtuase (9). The two copper atoms in this protein are each coordinated to four histidine nitrogens in an approximate square planar array.

Few prototypes are available for Type III Cu(II). These systems are ESR inactive, i.e., although Cu(II) is present, no ESR spectrum can be obtained. Recent magnetic susceptibility results indicate that the Type III Cu(II) in Rhus laccase is an antiferromagnetic-coupled Cu(II) dimer (10). Little, however, is known about the copper ligands or the nature of the dimeric interaction.

The entire metal active site in any metalloprotein consists of the metal chelate plus all of the protein groups which contribute to its spectral and catalytic properties. A copper protein has both very special and specific catalytic and spectral properties. The premise of our research is that both the catalytic and spectral properties must reflect the same unique characteristics of the ligands to the metal, the geometry of the metal complex, the properties of the active site protein groups, and the general protein environment of the active site. Our ultimate objective is to elucidate the pertinent interrelationships among these parameters. Variations in any or all of these factors might be the underlying basis for the distinguishing spectral and chemical properties of the two classes of paramagnetic copper in copper proteins.

Galactose Oxidase—Background

Our recent interest has centered on the fungal enzyme, galactose oxidase, which may be the only copper protein with a single non-blue Cu(II) (in the non-blue family) and which contains no other prosthetic groups. (We take the view that ESR parameters establish the families of the cupric sites. Operationally, one goal of the copper–protein research is to determine precisely what is constant and what varies among the various examples of each type of Cu(II) site.) We can summarize a few pertinent properties of this enzyme. Galactose oxidase was first isolated by Cooper in 1959 (*11*). Early available literature about the enzyme was contributed primarily by Horecker and co-workers (*12*). The enzyme is elaborated by *Dactylium dendroides* and is an extracellular protein (*13*). Its molecular weight has been recently established in these laboratories to be 68,000 ± 3,000 daltons (*14*). An amino acid analysis has been reported (*14*). The protein has an isoelectric point around pH = 12 (*15*). The single Cu(II) atom can be readily removed by diethyl dithiocarbamate coordination or by H_2S. The apoenzyme is stable, and reconstitution of the enzyme results in total reactivation (*12, 16*). As the name implies, the enzyme catalyzes the oxidation of galactose by molecular oxygen as indicated below:

Nearly any primary alcohol serves as a substrate with the exception of methanol and ethanol. Ferricyanide (*17, 18*), porphyrexide (*18*), and hexachloroiridate(IV) (*18*) can replace oxygen as oxidant. Hexachloroiridate(IV) is consumed to the exclusion of oxygen in aerobic mixtures. When hexachloroiridate(IV) and H_2O_2 serve as oxidant and reductant respectively, the normal reaction, vis-a-vis H_2O_2, is reversed, and oxygen is produced (*18*).

The first spectral study of galactose oxidase was the report of the electron spin resonance spectrum by Blumberg et al. (*19*). More recently, Cleveland et al. (*20*) reported a further ESR study which was based on a computer fit to the spectrum. They concluded that four nitrogens were bound to the Cu(II) atom.

Ligands to Metal in Galactose Oxidase: ESR and Model Studies

Model systems may be very useful in elucidating the atoms liganded to the copper. We earlier proposed a pseudo-square planar N_2O_2 Schiff base complex of copper as a model for the equitorial coordination of Cu(II) in galactose oxidase (21). That model complex mimics some of the spectral properties of the enzymic Cu(II). In support of that model, better-resolved ESR spectra of the enzyme have been obtained. Figure 1 shows the time-averaged ESR spectrum of galactose oxidase at 100°K. A five-line superhyperfine splitting on the parallel lines (A_{zz} component) caused by two nitrogens is clearly indicated. This conclusively demonstrates that at least two nitrogen atoms must be liganded to the copper atom; the presence of four nitrogens (20) appears unlikely. (Recent spin-echo data are consistent with coordination of the Cu(II) by two histidine imidazole nitrogen atoms. (Work in collaboration with J. Peisach and W. Mims).)

However, the Schiff base complex lacks the stability towards reduction by CN⁻ that characterizes the Cu(II) in galactose oxidase. While the enzyme binds a single CN⁻ even at large CN⁻ excess (22), the Cu(II) in the model is reduced by the ligand. To assess the underlying structural components which stabilize the enzymic Cu(II) towards reduction by CN⁻, a five-coordinate model (Figure 2) having square bipyramidal symmetry was prepared (23). (The conditions and system procedures

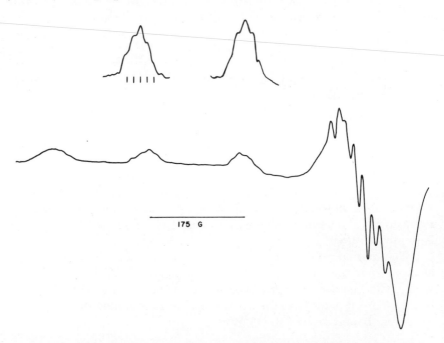

175 G

Figure 1. ESR spectrum of galactose oxidase at ~ 5 × 10⁻⁴M. T = 100°K, average of six scans. Obtained with Nicolet Lab-80 CAT on a Varian E-9 spectrometer. Insert shows the second and third parallel lines and the five-line superhyperfine splitting.

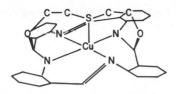

Cu (TAAB) + S (CH₂CH₂ONa)₂   *Figure 2.   Cu[N₄S] model (see text)*

reported by Busch et al. (*23*), for the N₄S complex do not yield the complex reported. More likely, the complex they obtained was one in which only one side of the bridging -O-C-C-S-C-C-O- group is attached to the TAAB backbone and the other end is free, probably as the alcohol. A detailed synthetic procedure for the complex discussed here is in preparation (*24*)). In a qualitative way, nitrogen and oxygen atom ligation yields ionic bonds to Cu(III). Thus an N₄S system, which is synethetically more easily approached, will probably serve the same purpose as an N₂O₂S system. The utility of the model lies in its encompassing two possibly important structural features—axial ligation by sulfur (*14, 25*) and a "cage" shielding one axial position from exogenous ligands. Table I summarizes the pertinent ESR and CN⁻ data. In fact, none of these macrocyclic Cu(II) complexes bind CN⁻ at all as evidenced by both optical absorbance and ESR studies. Thus, although they are stable to CN⁻, their stability is caused by a total lack of reactivity towards CN⁻, quite unlike the enzyme. Also, as indicated in Table I, there is little indication of axial ligation by either -S- or -O- as evidenced by the ESR parameters. We conclude from these data that binding of exogenous ligands requires access to an essentially equitorial coordination site and that the rigid macrocycle is simply too rigid to allow this substitution. The apparent lack of axial coordination by -S- may be caused by the thioether's somewhat weaker affinity for Cu(II) (cf. RS⁻) and does not rule out mercaptide ligation in the protein (*14, 25*). However, both the magnitude of the spin Hamiltonian parameters when compared with other systems (*26*) and the lack of a substantial linear electric field effect (LEFE) on the g-values make it unlikely that sulfur ligation, if present at all, is equitorial. In this limited way, the N₄S system is geometrically, if not chemically, appropriate.

Galactose Oxidase: Optical Transitions and Anion Binding

Given some notion of the nature of the endogenous ligands, we can next ask how exogenous ligands perturb the copper atom and what this can tell us about the electronic transitions exhibited by the Cu(II) atom in galactose oxidase.

Table I. Stoichiometry of CN⁻ Ligation to (and Redox Stability

Complex	Ligand Type	Symmetry
$Cu(TAAB)[S(CH_2CH_2O)_2]$[b]	N_4S	C_{4v} (pseudo)
$Cu(TAAB)[O(CH_2CH_2O)_2]$	N_4O	C_{4v} (pseudo)
$Cu(TAAB)[CH_3N(CH_2CJ_2O)]_2$	N_4N	C_{4v} (pseudo)
$Cu(TAAB)[CH_2(CH_2CH_2O)_2]$	N_4	C_{4v} (pseudo)
$Cu(tren-OH)BPh_4$[c,d]	N_4O	C_{3v} (pseudo?)
$Cu(F_3Ac)_2en$[e]	N_2O_2	rhombic planar (C_{2v})
$Cu(Ac)_2en$[f]	N_2O_2	rhombic planar (C_{2v})
$Cu(tren-NH_2Ph)BPh_4$[g]	N_4N	C_{3v} (pseudo)
Galactose oxidase	N_2O_2 (?)	rhombic planar (C_{2v})

[a] Values in Gauss.
[b] See Figure 2 and Refs. 23 and 24 for explanation of terminology.
[c] tren-NH₂Ph-(2,2′,2″-triaminotriethylamine-phenylamine)copper(II).
[d] Ph = $-C_6H_5$.

At least five electronic transitions can be detected by absorbance, difference absorbance, and circular dichroic (CD) spectra for galactose oxidase prior to resolution by computer analysis or magnetic CD studies. These occur at energies which correspond to 314, 395, 500, 630, and about 775 nm (Table II) (27). Noteworthy among these bands are the transitions near 650, 450, and 800 nm since they are exhibited by all copper proteins (2). In any event, since only a maximum of four d–d transitions are permitted, at least one of the transitions exhibited by galactose oxidase must be charge transfer in nature. Moreover, although the pseudo-square planar system which is likely in galactose oxidase would not be expected to exhibit lower energy transitions, CD and magnetic CD spectra in the near-infrared region should be obtained as have been recently reported in Type I Cu(II) systems (28).

It is interesting to consider the effect of exogenous ligands (which have previously been shown to bind to the Cu(II) atom inner sphere by ESR studies (22)) on the optical spectrum of galactose oxidase. (While

Table II. Electronic Transitions Exhibited by Galactose Oxidase[a]

Absorbance			Circular Dichroism		
$\lambda(nm)$	$\nu(cm^{-1})$	$\epsilon(M^{-1}\,cm^{-1})$	$\lambda(nm)$	$\nu(cm^{-1})$	$[\theta]$ $[(deg\,cm^2)/dMol]$
314	31,800	1,370	314	31,800	+18,900
445	22,500	1,155	395	25,300	+3,000
630	15,900	1,015	500	20,000	+1,500
775	12,900	905	610	16,400	−8,200

[a] Data from Ref. 27.

of) Some 4- and 5-Coordinate Copper(II) Complexes[a] (24)

CN^- Effects

$A_{zz}{}^a - g_{zz}$	1:1 Complex	Excess
144/2.160[g]	no	stable
145/2.159[g]	no	stable
179/2.171[g]	no	stable
144/2.160[g]	no	stable
163[h]	?	reduces $Cu(II) \to Cu(I)$
187.2/2.220[h]	no	reduces $Cu(II) \to Cu(I)$
211/2.186[h]	no	reduces $Cu(II) \to Cu(I)$
163/2.244[h]	?	reduces $Cu(II) \to Cu(I)$
173/2.273	yes	stable

[e] $(F_3Ac)_2en = N,N'$-ethylenebis(trifluoroacetylacetoniminato)copper(II).
[f] $(Ac)_2en = N,N'$-ethylenebis(acetylacetoniminato)copper(II).
[g] Data in DMF solvent.
[h] Data in C_2H_5OH solvent.

it was not indicated in our previously reported anion binding studies that inner coordination sphere binding occurs, more recent superhyperfine detection of exogenous ligand binding certainly substantiates this fact. Further changes in the A_{zz} component are quite similar to the fine studies of Coleman et al. (29, 30) on artificial Cu(II) proteins where exogenous ligands gave superhyperfine structure to the ESR spectrum.) Azide, for example, at approximately 100:1 molar excess causes a very large blue shift of the 775-nm peak not shown here and the 630- and 445-nm absorbance bands (Figure 3). Absorbance maxima near 380 nm with azide have been attributed to charge transfer complexes in other proteins (31), but since the shift from 445 is the same as for the 630-nm band in energy terms, we suggest that this 380-nm band is related to the 450-nm band. Cyanide also uniformly blue shifts the three low energy transitions. Most importantly, common anion exogenous ligands affect mainly the intensity of the 314-nm transition, but not its energy. Generally, transition intensity increases are two to three fold with the binding of such anions. The simple fact that anions affect the energy of the lowest energy transitions similarly suggests that these are primarily *d–d* in character while the 314-nm transition may be a transition with unique charge transfer character.

All anions which bind to the Cu(II) in galactose oxidase lower the g_{zz} and A_{zz} values (22). This is consistent with (but not required for) a blue shift in the "*d–d*" transitions (32, 33, 34). $Fe(CN)_6{}^{3-}$ is the only anion among the limited ones we have studied which produces a red shift in the optical bands (Figure 4). At 1:1, 5:1, or 100:1 molar ratios of $Fe(CN)_6{}^{3-}$ to enzyme the same difference absorbance spectrum is obtained, and it is consistent with complex formation between galactose oxidase and the anion. Namely, the positive difference peaks at 455, 830, and near-650 nm indicate red shifts of the copper transitions. Similar to the other anions studied, ferricyanide also leads to relatively large increases in absorbance. Analogous to the effects of azide, for example, the intensity effects of ferricyanide are more pronounced on the 445- and 775-nm transitions than on the 630-nm band. Further evidence for a $Fe(CN)_6^{3-}$–protein complex is the observation that removal of the anion is very difficult, e.g., by treatment with an anion exchange resin.

$Fe(CN)_6^{4-}$ also apparently binds to the protein; the difference absorbance spectrum (Figure 4) indicates a small red shift. Again, increasing the molar ratio of ferrocyanide causes no further absorbance changes. The two anions added at identical 5:1 molar ratios to enzyme cause a difference spectrum consistent with a competitive binding to the protein; the resultant difference absorbance is that expected for an equal partitioning of the enzyme between the two anions (Figure 4).

Similar results are obtained by monitoring the Cu(II)-mediated relaxation of added $^{19}F^-$ or of bulk water (35, 36). For example, $Fe(CN)_6^{3-}$ decreases the relaxivity of the Cu(II) towards water protons by 60% while $Fe(CN)_6^{4-}$ has little effect. Addition of equal molar amounts of these anions produces a 30% reduction. Effects such as these occur at as little as 1:1 molar ratios (37).

These results could be attributed to a change in the redox state of the enzyme copper (17, 38). However, ESR spectra at 1:1 and 6:1 molar

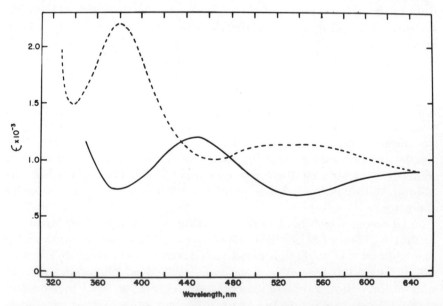

Figure 3. Copper absorbance spectrum of galactose oxidase (775 nm not recorded) (——) and the enzyme in the presence of sodium azide (– – –). Spectra were recorded in 5-cm cells, 0.1M sodium phosphate buffer, pH 7.0.

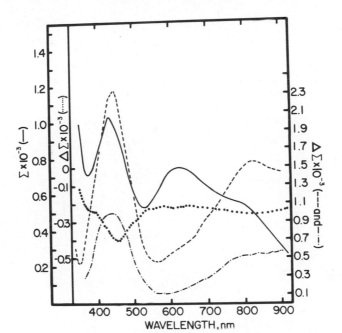

Figure 4. Visible absorbance spectrum of galactose oxidase (——) and the difference absorbance spectra with 5:1 molar ratios of ferricyanide (– – –), ferrocyanide (· · ·), and the two together (– · –). The scale on the right is for both solutions containing Fe(CN)$_6^{3-}$. Spectra were recorded in 4.5-cm double compartment cells; 0.1M sodium phosphate buffer, pH 7.0.

ratios of Fe(CN)$_6^{3-}$-to-enzyme indicate little change in the spectral intensity (Figure 5). Thus, the effects seen in absorbance and nuclear spin relaxation at these ratios cannot obviously be attributed to an oxidation of Cu(II) to Cu(III), for example (*17, 38*). The fact that the difference absorbance does not change even up to 100:1 molar ratio suggests that as far as the ligand field energy of the enzymic copper and the transition probabilities are concerned, the addition of further Fe(CN)$_6^{3-}$ is unimportant. This is of interest since the Cu(II) ESR signal (as in Figure 5) disappears at these high ratios. Thus, the effects of ferricyanide on the optical absorbance and spin characteristics of the Cu(II) are apparently distinct. Whether or not the differences are related to differences in the observation temperature used, while a possible explanation (*39*), remains an intriguing but unresolved question. Further information on the effects of Fe(CN)$_6^{3-}$ may be provided by x-ray spectroscopic experiments now in progress.

A time-dependent, apparent reduction of Fe(CN)$_6^{3-}$ by both the holo- and apoenzymes is indicated by difference absorbance measure-

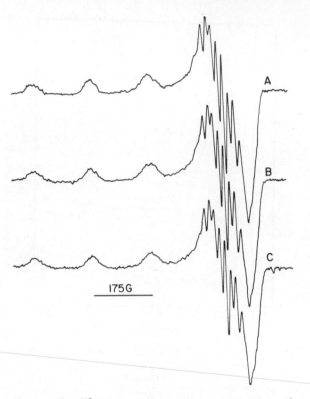

Figure 5. Electron spin resonance spectrum of galactose oxidase (A) and galactose oxidase and $Fe(CN)_6^{3-}$ *at a 1:1 (B) and 1:6 (C) molar ratio spectra were recorded at 100°K, with a power of 20 mw (9.115 GHz) and a modulation amplitude of 2G; [galactose oxidase] = 0.5 mM.*

ments, as in Figure 4. At a 1:1 molar ratio of ferricyanide-to-enzyme, the difference spectrum (Figure 4) slowly ($\tau_{1/2} \sim 7$ hr) changes to yield a single negative difference peak at 420 nm, the λ_{max} for $Fe(CN)_6^{3-}$. The solution, too, loses its characteristic yellow color caused by $Fe(CN)_6^{3-}$ during this period. The fact that this reduction is independent of the presence of the Cu(II) in the enzyme and that, with the holoenzyme, the only difference peak is that attributable to the reduction of $Fe(CN)_6^{3-}$ (i.e., no copper difference peaks present) indicates that the protein, and not the Cu(II), is involved in this redox reaction. This protein product has not been characterized.

Non-Ligand Active Site Groups

Other protein groups which contribute to the molecular dynamics of the entire copper active site must be complementary to inner-coordination sphere ligands to the copper atom. One of the most dramatic results which first suggested that a tryptophan residue might be within the active site locus was the near-uv–CD spectrum of galactose oxidase in the presence of dihydroxyacetone (which is an excellent substrate) or the aldehyde product of the galactose reaction, each in the absence of oxygen. Binding of a substrate or product has enormous effects on the tryptophan optical activity in the 285–300-nm region (*40*). Furthermore, incorporation of copper into the apoenzyme causes a 29% reduction in tryptophan fluorescence (*41*). A holoenzyme–apoenzyme difference absorbance spectrum also clearly shows perturbation of a typtophan environment by the copper atom (*27*). While these results could reflect very indirect interactions, selective oxidation of the tryptophans in galactose oxidase with *n*-bromosuccinimide (NBS) revealed a critical structure-function role for at least one residue (*41, 42*). Galactose oxidase is inactivated as exactly two of its 18 tryptophans are oxidized. Moreover, the inactivation profile implies that just one of the most reactive residues in the enzyme is probably associated with the inactivation (*41*). One manifestation of the specificity of the reaction is the observation that tryptophan optical activity associated with only the 290-nm extremum is affected. The 295-nm peak is unaffected (Figure 6). A new extremum

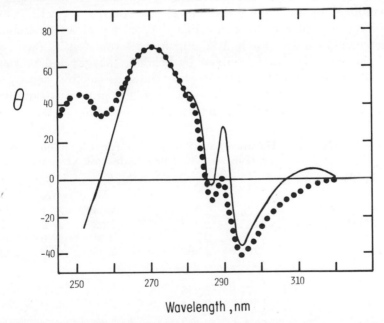

Figure 6. Near-uv CD spectrum of galactose oxidase in 0.1M sodium phosphate buffer, pH 7.0 (——) and the NBS-modified enzyme; 2 trp equivalents oxidized (· · ·). Spectra were recorded in a 1-cm cell.

near 250 nm reflects optical activity of the oxindole product of the oxidation reaction. Fluorescence is also markedly affected; the first two residues oxidized account for 48% of the total fluorescence of the enzyme which further indicates that the most reactive tryptophan residues have unique properties (41).

What is most interesting about this modification is what it suggests about the molecular interactions within the copper active site. Prior experiments established that with the native enzyme, galactose in the absence of oxygen markedly reduces its copper optical activity, but oxygen in the absence of galactose has no significant effect (40). The fact that galactose also markedly reduces copper optical activity in the NBS-inactivated enzyme suggests that the inactivation involves an effect on catalysis specifically rather than binding. The same influence can be drawn from fluorescence data (41). What is particularly interesting here is that while binding of common exogenous ligands invariably leads to a uniform reduction in each of the optical activity transitions, the inactivation by selective tryptophan oxidation is associated with a decrease at 314 nm, but an increase near 600 nm.

While several alternate inferences are possible for the increase near 600 nm, one possibility is that the change in chemistry of the copper site as exemplified by abolition of catalysis in this case may reflect a conversion of the normal pseudo-square planar geometry to an environment characterized by a decrease in axial unpaired electron density (see Table III). This might be brought about by bonding changes and/or stereochemical effects, both of which can effect the axial electron density.

Interestingly, the cupric ion in this modified protein is rapidly reduced by cyanide. At a 10:1 molar ratio, CN^- effects complete reduction

Table III.   Spin Hamiltonian Parameters for Liganded and
Non-Liganded Native and Modified Galactose Oxidase[a]

Enzyme Form/Ligand (Ratio)	A_{zz}	g_{zz}	A_{xx}	g_{xx}	A_{yy}	g_{yy}
Native[b]	176.5	2.273	28.8	2.048	30.1	2.058
CN^- (1:1)	155.8	2.226	41.6	2.035	45.2	2.048
N_3^- (100:1)	166.8	2.262	27.2	2.049	27.9	2.040
NBS oxidized[c]	166.0	2.267	38.0	2.055	43.0	2.065
CN^- reduces Cu(II)						
Iodoacetamide alkylated[d]	177.8	2.268	30.5	2.035	30.6	2.064
CN(1:1)	160.1	2.234	38.8	2.041	43.0	2.051

[a] Spectra were obtained on a Varian E-9 X-band spectrometer at 110°K with a 100-KH$_z$ modulation amplitude of 2 Gauss and a microwave power of 30 mw at 9.5 GH$_z$ with proton Gauss meter and frequency counter for spectral marking.
[b] (22).
[c] (42).
[d] (44).

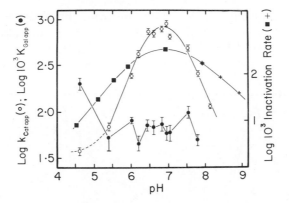

Figure 7. Plot of the pH-dependent values for the oxidation rate of β-methyl-D-galactopyranoside at 1.4 mM O_2 by galactose oxidase ($k_{cat_{app}}$) and the rate of inactivation of galactose oxidase by 1mM iodoacetamide.

The smooth curve for the former (○) is calculated using pK_a values 6.3 and 7.1 with $k_{cat} = 1350$ sec^{-1}. The curve for the latter process (●) corresponds to pK_a values 6.3 and 7.0 with a pH-independent rate of 2.2 min^{-1}.

(Table III). Atomic absorption analysis shows that the copper is still present in the enzyme. These two facts coupled with the 600-nm region intensity increase might be interpreted to suggest that a stereochemical distortion accompanies tryptophan oxidation in galactose oxidase. For example, a saddling of the planar environment would be expected to lower the reduction potential of the Cu(II)–Cu(I) couple.

The active site role of one other protein moiety has been examined in detail. In the first reported work with galactose oxidase, pH-rate data were reported that implicated an imidazole group (*11*). The pH-dependence of both the enzymic reaction as well as enzyme inactivation by iodoacetamide reflects the essential ionization of a conjugate acid, $pK_a = 6.3$ (Figure 7) (*43*). Inactivation is caused by the specific alkylation of a single histidine at its N-3 nitrogen (*43*). The alkaline pH-dependence may reflect the ionization of a copper-bound water molecule (vide supra) (*36, 37, 43, 44*). Like the NBS-modified protein, the alkylated enzyme still binds sugar substrate, as indicated by fluorescence experiments (*43*). Thus, catalysis is again uniquely affected. Moreover, CN⁻ also binds normally and does not reduce the Cu(II) (*43*). Correlated to this is the near identity of the spin Hamiltonian parameters of native and alkylated forms (Table III). Furthermore, the apoenzyme does not react with iodoacetamide (*43*). Therefore, the affected imidazole is most likely not a copper ligand. The apoenzyme result does suggest, however, that a critical relationship between histidine reactivity and the copper atom does exist. Moreover, the NBS-oxidized protein is not alkylated, which establishes a critical link between the reactive tryptophan and histidine residues (43) which are probably both within the active site locus.

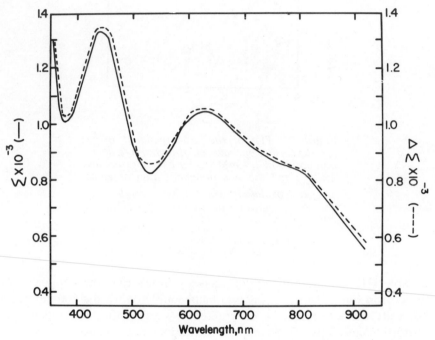

Figure 8. Copper absorbance spectrum of galactose oxidase (——) and difference absorbance spectrum with the alkylated enzyme as the reference and the unmodified enzyme as the sample (−−−). Spectra were recorded in 5-cm cells, 0.1M sodium phosphate buffer, pH 7.0.

Results with galactose oxidase illustrate that it is extremely dangerous to rely on any one spectral method to evaluate a perturbation of a metal system. ESR has been a sensitive probe for inner coordination sphere ligands but apparently a relatively poor one of metal chelate conformation in galactose oxidase. The difference spectrum which is recorded with the alkylated enzyme as the reference and the native enzyme as the sample is indistinguishable from the native enzyme's absorbance spectrum, i.e., alkylation virtually abolishes copper absorbance (Figure 8). By CD, some small magnetic and/or electric transitions are detected with the alkylated enzyme (Figure 9). The lack of shifts in energies of the copper absorbance transitions further argues against a change in ligation. The decrease in transition probability can perhaps be best rationalized by the copper chelate assuming a more centrosymmetric geometry. In any event, in addition to its role as a specific base

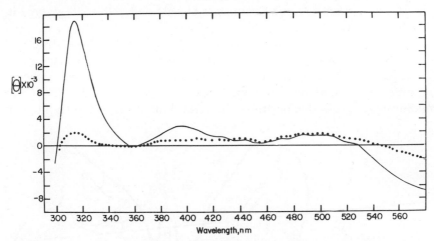

Figure 9. Copper CD spectrum of galactose oxidase (———) and the alkylated enzyme (· · ·) which were recorded in 5-cm cells in 0.1M sodium phosphate buffer, pH 7.0

catalyst, the affected histidine also probably influences activity through a critical role in maintaining the copper chelate conformation (43).

Galactose oxidase can illustrate how ligands, geometry, and active site groups together provide the basis for the structure-function properties of a metal active site. Figure 10 summarizes mutual interactions

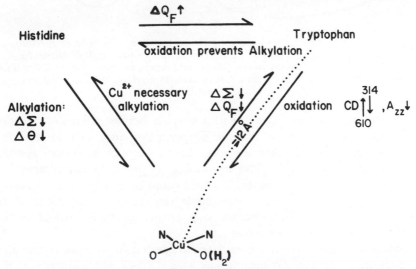

Figure 10. Summary of interdependent interactions between the three groups proposed to be at the active site. Q_f = fluorscence quantum yield, ΔE = difference absorbance, $\Delta\theta$ = change in ellipticity (CD). The distance estimation is derived from fluorescence energy-transfer methods.

which may pertain between the copper chelate and active site groups in galactose oxidase.

A representation of the active site which contains these structural elements is seen in Figure 11. Model and spectral studies have suggested the in-plane ligands. That the copper influences the tryptophan is shown by difference absorbance and fluorescence quantum yield. The NBS-modification in turn demonstrates the critical influence of the tryptophan

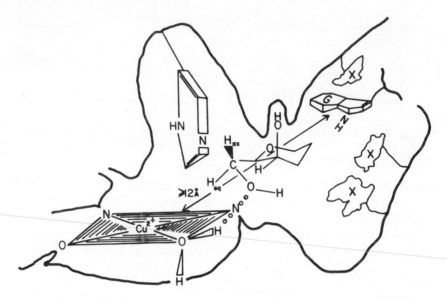

Figure 11. Diagrammatic representation of active site which includes a four-coordinate copper complex, sugar-substrate bound outer-sphere to the Cu(II) atom, imidazole and indole rings, and nonpolar side chains (×). The distance between the Cu(II) and indole is estimated by fluorescence energy-transfer methods.

on the copper chelate. The spectral overlap between the fluorescence spectrum of the tryptophan and the copper absorbance at 314 nm allows one to estimate the distance between these groups by a Förster energy-transfer calculation (*41*). The tryptophan is probably not closer than 12 Å, but even at this distance, it could come in contact with a bound sugar substrate. Most likely, the indole ring is one component of an active site cluster of hydrophobic side chains which is critical to the conformation of the entire active site.

Structure-function roles have been suggested for unique tryptophan residues in other copper proteins as well (*44, 45, 46*). Moreover, the single tryptophan that is quenched by including the copper atom in azurin is apparently not in contact with the indole ring, as evidenced by metal replacement and phosphorescence results (*45, 46*).

Oxidation of this tryptophan in galactose oxidase also prevents alkylation of the histidine residue. Alkylation of the histidine residue in turn markedly affects the fluorescence quantum yield of this tryptophan (*43*) and nearly abolishes the absorbance of the copper atom. The copper atom itself is also essential to the reactivity of this histidine. Thus, we appear to have a consistent set of highly interdependent components. Not unexpectedly, the copper site cannot be fully understood without considering its interactions with non-ligand protein groups.

Acknowledgments

The work outlined here is a product of the cooperative efforts of the several graduate and undergraduate students who are members of the Bioinorganic Graduate Research Group. The Group is grateful for the support of the National Science Foundation (B404662) and the Graduate School of this University. Time-averaged ESR spectra were obtained with the aid of a Nicolet Lab 80 CAT (NSF-MPS 7506183). R. D. B. is a recipient of a Camille and Henry Dreyfus Fellowship.

Literature Cited

1. Evans, G. W., *Physiol. Rev.* (1973) **54**, 535.
2. Vänngård, T., "Copper Proteins," in "Biological Applications of Electron Spin Resonance," H. M. Swartz, J., R. Bolton, D. C. Borg, Eds., p. 441, Wiley-Interscience, New York, 1972.
3. McMillin, D. R., Holwerda, R. A., Gray, H. B., *Proc. Natl. Acad. Sci. USA* (1974) **71**, 1339.
4. McMillin, D. R., Rosenberg, R. G., Gray, H. B., *Proc. Natl. Acad. Sci. USA* (1974) **71**, 4760.
5. Bereman, R. D., Wang, F. T., Najdzionek, J., Braitsch, D. M., *J. Am. Chem. Soc.* (1976) **98**, 7266.
6. Gray, H. B., ADV. CHEM. SER. (1977) **162**, 145.
7. Spiro, T. S., *Acc. Chem. Res.* (1974) **7**, 339.
8. "Biological and Biochemical Applications of Electron Spin Resonance," D. J. E. Ingram, Plenum, New York, 1969.
9. Richardson, J. S., Thomas, K. A., Rubin, P. H., Richardson, D. C., *Proc. Natl. Acad. Sci. USA* (1975) **72**, 1349.
10. Solomen, E. I., Dooley, D. M., Wang, R. H., Gray, H. B., Cerdonio, M., Mogno, F., Monani, G. L., *J. Am. Chem. Soc.* (1976) **98**, 1029.
11. Cooper, J. A., Smith, W., Bacila, M., Medina, H., *J. Biol. Chem.*, (1959) **234**, 445.
12. Amaral, D., Kelly-Falcoz, F., Horecker, B. L., *Methods Enzymol.* (1966) **9**, 87.
13. Nobles, M. K., Madhosingh, C., *Biochem. Biophys. Res. Commun.* (1963) **12**, 146.
14. Kosman, D. J., Ettinger, M. J., Weiner, R. E., Massaro, E. J., *Arch. Biochem. Biophys.* (1974) **165**, 456.
15. Bauer, S., Blauer, G., Avigad, G., *Isr. J. Chem. Proc.* (1967) **5**, 126.
16. Giblin, F., unpublished data.
17. Hamilton, G. A., Libby, R. D., Hartzell, G. R., *Biochem. Res. Commun.* (1973) **53**, 715.

18. Kosman, D., unpublished data.
19. Blumberg, W., Horecker, B. L., Kelly-Falcoz, F., Peisach, J., *Biochim. Biophys. Acta* (1965) **96**, 336.
20. Cleveland, L., Coffman, R. E., Coon, P., Davis, L., *Biochemistry* (1975) **14**, 1108.
21. Giordano, R. S., Bereman, R. D., *J. Am. Chem. Soc.* (1974) **96**, 1019.
22. Giordano, R. S., Bereman, R. D., Kosman, D. J., Ettinger, M. J., *J. Am. Chem. Soc.* (1974) **96**, 1023.
23. Katovic, V., Taylor, L. T., Busch, D. H., *Inorg. Chem.* (1971) **10**, 458.
24. Bereman, R. D., Shields, G., unpublished data.
25. Kelly-Falcoz, F., Greenberg, H., Horecker, B. L., *J. Biol. Chem.* (1965) **240**, 2966.
26. Peisach, J., Blumberg, W. E., *Arch. Biochem. Biophys.* (1974) **165**, 691.
27. Ettinger, M. J., *Biochemistry* (1974) **13**, 1242.
28. Solomon, E. J., Hare, L. W., Gray, H. B., *Proc. Natl. Acad. Sci. USA* (1976) **73**, 1389.
29. Taylor, J. S., Mushak, P., Coleman, J. E., *Proc. Natl. Acad. Sci. USA* (1970) **67**, 1410.
30. Taylor, J. S., Coleman, J. E., *J. Biol. Chem.* (1971) **246**, 7058.
31. Williams, R. J. P., in "The Biochemistry of Copper," J. Peisach, P. Aisen, W. E. Blumberg, Eds., p. 131, Academic, New York.
32. Neiman, R., Kievelson, D., *J. Chem. Phys.* (1961) **35**, 149.
33. *Ibid.* (1961) **35**, 156.
34. *Ibid.* (1961) **35**, 159.
35. Marwedel, B. J., Kurland, R. J., Kosman, D. J., Ettinger, M. J., *Biochem. Biophys. Res. Commun.* (1975) **63**, 773.
36. Fabry, T. L., Kim, J. P., Titcomb, L. M., *IBM Res. Rept.* (1969) RW108, No. 11415.
37. Marwedel, B. J., Kurland, R. J., unpublished data.
38. Dyrkacz, G. R., Libby, R. D., Hamilton, G. A., *J. Am. Chem. Soc.* (1976) **98**, 626.
39. Nickerson, K. W., Phelan, N. F., *Bioinorg. Chem.* (1974) **4**, 79.
40. Ettinger, M. J., Kosman, D .J., *Biochemistry* (1974) **13**, 1247.
41. Weiner, R. E., Ettinger, M. J., Kosman, D. J., *Biochemistry*, in press.
42. Kosman, D. J., Ettinger, M. J., Giordano, R. S., Bereman, R. D., *Biochemistry*, in press.
43. Kwiatkowski, L. D., Siconolfi, L., Ettinger, M. J., Weiner, R. E., Giordano, R. S., Bereman, R. D., Kosman, D. J., *Arch. Biochem. Biophys.*, in press.
44. Morpurgo, L., Finazzi-Agró, A., Rotilio, G., Mondovi, B., *Biochem. Biophys. Acta* (1974) **271**, 292.
45. Finazzi-Agró, A., Giovagnoli, C., Arigliano, L., Rotilio, G., Mondovi, B., *Eur. J. Biochem.* (1973) **34**, 20.
46. Finazzi-Agró, A., Rotilio, G., Avigliano, L., Guerrieri, P., Boffi, V., Mondovi, B., *Biochemistry* (1970) **9**, 2009.

RECEIVED July 26, 1976.

16

Copper(II)– and Copper(III)–Peptide Complexes

DALE W. MARGERUM, LOUIS F. WONG, FRANK P. BOSSU,
K. L. CHELLAPPA, JOHN J. CZARNECKI, SANFORD T. KIRKSEY, JR.,
and THOMAS A. NEUBECKER

Department of Chemistry, Purdue University, West Lafayette, IN 47907

Coordination of deprotonated-peptide nitrogens greatly affects the binding and kinetic reactivity of Cu(II) and the ease of oxidation to Cu(III). Substitution reactions of oligopeptide and serum albumin complexes of Cu(II) are characterized by three reaction pathways—(1) proton transfer to the peptide group, (2) nucleophilic attack on copper, and (3) a combination of proton transfer and nucleophilic attack. Depending upon the nature of the coordinated ligands, the electrode potentials of Cu(III)–Cu(II) couples vary from 0.5 to 1.0 V (vs. NHE). Copper(III)–peptide complexes are characterized in solution in terms of their electrochemical, spectral, and kinetic properties. At high pH amine groups coordinated to Cu(III) also will deprotonate.

The binding of Cu(II) to the backbone of polypeptides by coordination to deprotonated-peptide nitrogens has been known for some time (*1–7*), but only recently has it been discovered that this coordination facilitates the formation of Cu(III) (*8, 9*). This chapter concerns recent studies of Cu(II)–peptide complexes, in particular those reported since a general review of this topic by Margerum and Dukes (*10*), and summarizes the present knowledge of Cu(III)–peptide complexes.

The presence of L-histidine as the third amino acid residue in tripeptide complexes of Cu(II) drastically decreases their susceptibility to both nucleophilic attack and acid attack (*11, 12*). Thus, the doubly deprotonated complex of glycylglycyl-L-histidine (Cu(H$_{-2}$gly-gly-his)$^-$) shown in Structure **I** is relatively slow to react with the nucleophilic triethylenetetramine (trien) since this reaction is seven orders of magnitude slower than the corresponding reaction with Cu(H$_{-2}$gly-gly-gly)$^-$. The rate of acid dissociation of the gly-gly-his complex is also many orders of magnitude slower than for the gly-gly-gly complex. However, a previously unobserved pathway which is a combination of attack by H^+ and H_2trien^{2+} has been found. This proton-assisted nucleophilic mechanism provides a third major pathway for the transfer of Cu(II) from peptide complexes and is a major pathway for the removal of Cu(II) bound to serum albumin. The histidine-containing peptide complexes provide evidence of the importance of axial coordination to Cu(II) by carboxylate groups (13). In addition, for the reactions at low pH "outside" protonation occurs to give $[Cu(H_{-2}gly\text{-}gly\text{-}his)H]$ and $[Cu(H_{-2}gly\text{-}gly\text{-}his)H_2]^+$, where the peptide oxygens add protons (12, 13).

Copper(III) has been proposed as a highly reactive intermediate for a number of organic oxidations using copper (14, 15, 16). A few Cu(III) complexes have been identified in the solid state, e.g., $KCuO_2$ (17), K_3CuF_6 (18), $Na_3KH_3[Cu(IO_6)_2]\cdot14H_2O$ (19), and $CuBr_2(S_2CN\text{-}(C_4H_9)_2)$ (20). A detailed crystal structure is given for the latter compound with the dithiocarbamate (dtc) group and two Br^- forming a distorted square-planar geometry around Cu(III). In the periodate complex four of the periodate oxygens form a square plane about Cu(III) (Cu–O, 1.9Å), and a H_2O molecule (Cu–O, 2.7Å) forms a fifth bond. Except for CuF_6^{3-} the Cu(III) complexes appear to be low spin, although $CuBr_2(dtc)$ has a magnetic moment of 0.5 B.M.

Olson and Vasilekskis (21) electrochemically oxidized macrocyclic tetramine complexes in acetonitrile solutions to give Cu(III) species which are moderately stable but which undergo spontaneous reduction to Cu(II). Meyerstein (22) produced extremely reactive Cu(III) complexes of amines and amino acids in aqueous solution by pulse radiolysis. In the work of Levitski, Anbar, and Berger (23) $IrCl_6^{2-}$ was used to oxidize Cu(II)–peptides. The resulting Cu(III)–peptides were proposed as intermediate species prior to further oxidation and fragmentation of the peptides. Crystalline Cu(III) complexes of biuret and oxamide were isolated by Bour, Birker, and Steggerda (24) and provided some of the first evidence that Cu(III) could be stabilized by deprotonated amide groups. The compounds were all diamagnetic.

Table I summarizes some of the uv and visible spectral properties of Cu(III) complexes. Spectral bands of high intensity, attributed to charge-transfer transitions, have been observed for all Cu(III) complexes at 360 ± 60 nm. Regardless of the coordinating groups all the complexes listed have at least one absorption band in this spectral region.

Burce, Paniago, and Margerum (8) first observed the formation of Cu(III)–peptide complexes in the reactions of oxygen with Cu(II)–tetraglycine(G_4) in neutral solutions. Previously, Ni(II)–tetraglycine catalyzed oxygen uptake (25). The Cu(II)–tetraglycine reaction with oxygen is unusual in that it is inhibited by light (8). The exact nature of this photochemical inhibition remains to be established, but in the dark an intense yellow–green color forms as oxygen reacts. This yellow species passes through a Chelex ion exchange column, which quantitatively removes all forms of Cu(II). The yellow effluent contains copper and has oxidizing properties. On standing the solution generates Cu(II) and tetraglycine as well as some peptide oxidation products (26).

Table I. Ultraviolet and Visible Absorption Bands of Cu(III) Complexes

Cu^{III} Complex	λ_{max} nm (ϵ, M^{-1} cm^{-1})	Media	References
Cu(*trans*-tetramine)$^{3+}$	425 (15,000)	CH_3CN	21
	375 (12,000)		
	275 (6,700)		
Cu(*trans*-diene)$^{3+}$	395 (14,530)	CH_3CN	21
	335 (12,690)		
Cu(en)$_2$$^{3+}$	300 (2,500)	H_2O	22
Cu(gly)$_2$$^+$	310 (7,800)	H_2O	22
Cu(IO$_6$)$_2$$^{7-}$	414 (12,000)	H_2O	19
CuBr$_2$(dtc)	560 (2,400)	CH_2Cl_2	20
	370 (26,500)		
KCu(bi)$_2$	244 (strong)	BaSO$_4$ reflectance	24
	340 (v. strong)		
	461 (weak, sh)		
KCu(3-Rbi)$_2$ · 2H$_2$O	270 (5,000)	DMSO	24
	373 (8,500)		
	490 (weak, sh)		

Other oxidizing agents which can convert Cu(II)G$_4$ to Cu(III)G$_4$ include S$_2$O$_8$$^{2-}$ and IrCl$_6$$^{2-}$. The IrCl$_6$$^{2-}$ reaction is quantitative under suitable conditions and was used to help prove that Cu(III)–peptide complexes could be formed and characterized in aqueous solutions (9). Electrochemical oxidation is more efficient and avoids the need to remove the iridium complexes. The Cu(III)–peptide complexes have now been characterized by:

(1) Loss of Cu(II) spectral bands and the formation of intense absorption bands at 350–370 nm

(2) Loss of EPR signal

(3) Reversible Ir(IV)–Ir(III) redox reactions as a function of pH

(4) Sluggish reactions with acid and with Chelex ion exchanger resin

(5) 100% recovery of the original peptide after reaction with reducing agents

(6) Redox capabilities with a variety of substrates

(7) Molar absorptivity determinations

(8) Cyclic voltammetry

(9) Kinetics of decomposition in acid and in base

(10) Spectral shifts and pK_a determinations in strong base.

These studies show that Cu(III)–peptide complexes have relatively low electrode potentials and suggest that Cu(III) may be a far more common oxidation state than had previously been thought possible. Furthermore, the decomposition reactions of Cu(III)–peptides indicate that two-electron transitions to give Cu(I) species are possible. Two-electron redox reactions in biological systems are intriguing because high energy, free radical intermediates are avoided. However, as yet we know very little about possible Cu(I) complexes. This oxidation state is poorly characterized in aqueous solution, and studies with various model complexes are needed.

Hamilton and co-workers (27, 28) have suggested Cu(III) as a probable intermediate in the reaction catalyzed by galactose oxidase. Papers by Kosman and co-workers (29, 30) seem at variance with this interpretation. Regardless of the outcome of this dispute, we hope that our evidence for the existence and properties of Cu(III)–peptide complexes will encourage more investigations of the presence of trivalent copper in biological systems. Our work shows that this oxidation state is readily attained under biological conditions.

Copper(II)–Peptide Complexes

Stability Constants. The Cu(II)–polyglycine formation constants have been reviewed previously (10). Some of these constants have been redetermined (31, 32, 33) but are in substantial agreement with earlier work (34, 35, 36). New constants are available for peptide amides (37) and for tripeptides containing β-alanine and glycine (32). Table II summarizes the cumulative and stepwise constants (log stability constants and log K_a values for the formation of deprotonated species). A few constants for bis-peptide complexes have been determined (31, 37, 38). The β-alanyl-containing tripeptides form more stable complexes than the corresponding α-alanyl (39) or glycyl (31, 32, 33) species. Relative stabilities of the fused-ring systems in the deprotonated chelates are in the order, $5\text{-}6\text{-}5(G \cdot \beta\text{-}A \cdot G) \simeq 6\text{-}5\text{-}5(\beta\text{-}A \cdot G \cdot G) > 5\text{-}5\text{-}6(G \cdot G \cdot \beta\text{-}A) > 5\text{-}5\text{-}5\text{-}(G \cdot G \cdot G)$ (32). This effect is similar to that observed for polyamine complexes of Cu(II) and Ni(II) where the 5-6-5 chelates are much more stable than the 5-5-5 chelates (40). This effect can be attributed to bond strain in the complexes with 5-5-5 linked consecutive rings and to the difficulty of forming the most favorable square-planar geometry about the metal ion.

Included in Table II are the stability constants for gly-gly-L-his and its N-acetyl derivative. In acidic solutions both ligands use the imidazole group to initiate their coordination to copper. The only measured constants for the former correspond to the reaction in Reaction 1 where three

$$Cu(HGGhis)^{2+} \overset{\beta_3}{\rightleftarrows} Cu(H_{-2}GGhis)^- + 3H^+ \tag{1}$$

protons (one from the protonated amino group, two from peptide linkages) are lost simultaneously with $\beta_3 = 10^{-14.65} M^3$. Although stepwise constants were estimated (41) for the peptide ionization constants, these assignments were made on a statistical basis. The values for the N-acetylglycylglycyl-L-histidine constants suggest that these assignments are not valid. We estimate the cumulative stability constant for $Cu(H_{-2}GGhis)^-$ given in Equation 2 from the combination of the first

$$\frac{[Cu(H_{-2}GGhis)^-][H^+]^2}{[Cu^{2+}][GGhis^-]} = 10^{-2.2}M \tag{2}$$

complexation constant of the N-acetyl derivative ($10^{4.24}$), the protonation constant for the amino group of GGhis⁻ ($10^{8.22}$), and the constant for Reaction 1. Thus, the $Cu(H_{-2}GGhis)^-$ complex is $10^{4.5}$ times more stable than the $Cu(H_{-2}G_3)^-$ complex and is $10^{3.9}$ times more stable than the $Cu(H_{-2}CG-\beta A)^-$ complex (which also has a 5-5-6 membered ring system) because the imidazole nitrogen forms a stronger bond to copper than does the carboxylate group.

Copper(II) *d-d* Absorption Band. Billo (42) correlated spectral data for the Cu(II)–peptide complexes with the type of coordinated groups. The wavelength of the absorbance band observed between 500 and 740 nm varies with the number of deprotonated peptide groups and amine groups. The $\nu_{max}(kK)$ of the *d-d* band (aqueous solution spectra) can be expressed as the sum of the individual ligand field contribution of the four donor atoms which define the square plane with copper. In Equation 3 the *n*'s refer to the number of each type of donor atom

$$\nu_{obsd} = n_a\nu_{N(peptide)} + n_b\nu_{N(amino)} + n_c\nu_{N(imidazole)} + n_d\nu_{0(carboxylate)} +$$
$$n_e\nu_{0(peptide, H_2O \ or \ OH^-)} \tag{3}$$

($\Sigma n = 4$), and $\nu_{N(peptide)} = 4.85$, $\nu_{N(amino)} = 4.53$, $\nu_{N(imidazole)} = 4.3$, $\nu_{0(carboxylate)} = 3.42$, and $\nu_{0(peptide, H_2O, or OH^-)} = 3.01$. The effect of axial coordination by hydroxide, carboxylate, or amino groups is to shift ν_{max} to lower energy by 1kK.

Axial Coordination to Copper(II)–Peptide Complexes. Although in-plane coordination dominates the thermodynamic, kinetic, and spectral properties of the Cu(II)–peptides, axial coordination also is important. While the carboxylate groups in $Cu(H_{-3}G_4)^{2-}$ and $Cu(H_{-2}GGhis)^-$

Table II. Cumulative and Stepwise Stability and Deprotonation Constants for Cu(II)–Peptide Complexes[a]

Species	$\log \beta$ (or $\beta\ddagger$)	$\log K$ (or K_a)	Reference
$CuGa^{2+}$	5.29	5.29	37[b]
$Cu(Ga)_2^{2+}$	9.45	4.16	
$Cu(H_{-1}Ga)(Ga)^+$	2.54	−6.91	
$Cu(H_{-1}Ga)_2$	−5.58	−8.12	
$Cu(H_{-1}Ga)^+$	−1.63	−6.92	
CuG_2a^{2+}	4.88	4.88	37[b]
$Cu(H_{-1}G_2a)^+$	−0.19	−5.07	
$Cu(H_{-2}G_2a)$	−8.20	−8.01	
$Cu(H_{-2}G_2a)(OH)^-$	−18.02	−9.82	
CuG_3a^{2+}	4.77	4.77	37[b]
$Cu(H_{-1}G_3a)^+$	−0.51	−5.28	
$Cu(H_{-2}G_3a)$	−7.50	−6.99	
$Cu(H_{-3}G_3a)^-$	−16.19	−8.69	
CuG_2^{2+}	5.56, 5.68, 5.50	5.56, 5.68, 5.50	31,[c] 32,[c] 34[b]
$Cu(H_{-1}G_2)^+$	1.50, 1.47, 1.40	−4.06, −4.21, −4.10	
$Cu(H_{-1}G_2)(OH)$	−7.79, −7.77	−9.29, −9.24	31, 34
$Cu(H_{-1}G_2)(G_2)^+$	4.34	2.84*	31
$[Cu(H_{-1}G_2)]_2(OH)^-$	−4.14	2.15**	31
CuG_3^{2+}	5.12, 5.08, 5.25	5.12, 5.08, 5.25	31,[c] 32,[c] 33[b]
$Cu(H_{-1}G_3)^+$	0.01, −0.08, 0.02	−5.11, −5.16, −5.23	
$Cu(H_{-2}G_3)$	−6.67, −6.82, −6.71	−6.68, −6.74, −6.73	
$Cu(H_{-2}G_3)(OH)^-$	−18.68, −18.32	−12.0, −11.5	
$Cu(H_{-1}G_3)_2^{2-}$	−4.43	2.24‡	38[d]
CuG_4^+	5.13, 5.16	5.13, 5.16	31,[c] 35[b]
$Cu(H_{-1}G_4)$	−0.28, −0.36	−5.41, −5.52	
$Cu(H_{-2}G_4)^-$	−7.09, −7.14	−6.81, −6.78	
$Cu(H_{-3}G_4)^{2-}$	−16.24, −16.30	−9.15, −9.16	
CuG_5^+	5.32	5.32	36[e]
$Cu(H_{-1}G_5)$	−0.68	−6.00	
$Cu(H_{-2}G_5)^-$	−7.58	−6.90	
$Cu(H_{-3}G_5)^{2-}$	−15.62	−8.04	
$CuGG\beta A^+$	5.25	5.25	32[c]
$Cu(H_{-1}GG\beta A)$	−0.02	−5.27	
$Cu(H_{-2}GG\beta A)^-$	−6.10	−6.08	
$CuG\beta AG^+$	5.60	5.60	32[c]
$Cu(H_{-1}G\beta AG)$	0.24	−5.36	
$Cu(H_{-2}G\beta AG)^-$	−5.50	−5.74	
$Cu\beta AGG^+$	5.28	5.28	32[c]
$Cu(H_{-1}\beta AGG)$	−0.04	−5.32	
$Cu(H_{-2}\beta AGG)^-$	−5.58	−5.54	
$Cu\beta AG^+$	5.50	5.50	
$Cu(H_{-1}\beta AG)$	1.40	−4.10	
$CuGGA^+$	5.08	5.08	39[e]

Table II.   Continued

Species	log β (or β‡)	log K (or K_a)	References
Cu(H$_{-1}$GGA)	−0.02	−5.10	
Cu(H$_{-2}$GGA)$^-$	−6.91	−6.89	
CuGAG$^+$	5.18	5.18	39^e
Cu(H$_{-1}$GAG)	−0.14	−5.32	
Cu(H$_{-2}$GAG)$^-$	−6.76	−6.62	
Cu(H$_{-2}$GAG)(OH)$^{2-}$	−18.16	−11.4	
CuAGG$^+$	4.81	4.81	39^e
Cu(H$_{-1}$AGG)	−0.17	−4.98	
Cu(H$_{-2}$AGG)$^-$	−7.01	−6.84	
Cu(H$_{-2}$AGG)(OH)$^{2-}$	−18.2	−11.2	
Cu(HGGhis)$^{2+}$	−14.65 (β$_3$) ^{f}	—	
Cu(H$_{-2}$GGhis)$^-$	−2.2 (est.) ^{g}	—	12, 41^e
Cu(N-acetyl-GGhis)	4.24	4.24	41^e
CuH$_{-1}$(N-acetyl-GGhis)	−2.26	−6.50	
CuH$_{-2}$(N-acetyl-GGhis)	−9.61	−7.35	
CuH$_{-3}$(N-acetyl-GGhis)	−18.86	−9.25	

$$* \ Cu(H_{-1}G_2) + G_2^- \overset{K}{\rightleftarrows} Cu(H_{-1}G_2)(G_2)$$

$$** \ Cu(H_{-1}G_2)OH^- + Cu(H_{-1}G_2) \overset{K}{\rightleftarrows} [Cu(H_{-1}G_2)]_2OH^-$$

$$\ddagger \ Cu(H_{-2}G_3)^- + G_3^- \overset{K}{\rightleftarrows} Cu(H_{-1}G_3)_2^{2-}$$

^{a} 25.0° and 0.10M NaClO$_4$ or KNO$_3$
^{b} 0.10M (NaClO$_4$)
^{c} 0.10M (KNO$_3$)
^{d} 0.10–0.17M (NaClO$_4$)
^{e} 0.16M KCl
^{f} See Reaction 1
^{g} See Equation 2

(Structure I) cannot reach an axial coordination site, this is not the case for gly-gly-his-gly (Structure II) or for asp-ala-his-lys (Structure III) where axial coordination changes the chemical reactivity of the complexes. The equilibrium constant for the ratio of bound-to-free carboxylate, K_{free}^{bound}, is ∼ 100 for the terminal COO$^-$ in these tetrapeptides. The aspartyl side chain COO$^-$ group has a K_{free}^{bound} value of ∼ 30, but it is reduced to about 4 when another COO$^-$ group is trans to it, as in Structure III. These constants were determined from kinetic studies (13).

"Outside" Protonation. When metal–peptide complexes are placed in acidic solutions, the complexes dissociate. Metal ions which are sluggish in their substitution reactions, such as Ni(II) (43), Pd(II) (44), and Co(III) (45), add protons to the peptide oxygens prior to the metal–N(peptide) bond dissociation. The Cu^{II}(H$_{-2}$GGhis)$^-$ complex is sufficiently sluggish in its reaction with acid to permit outside protonation to be observed kinetically (12). Protonation constants of $10^{4.2}$ and

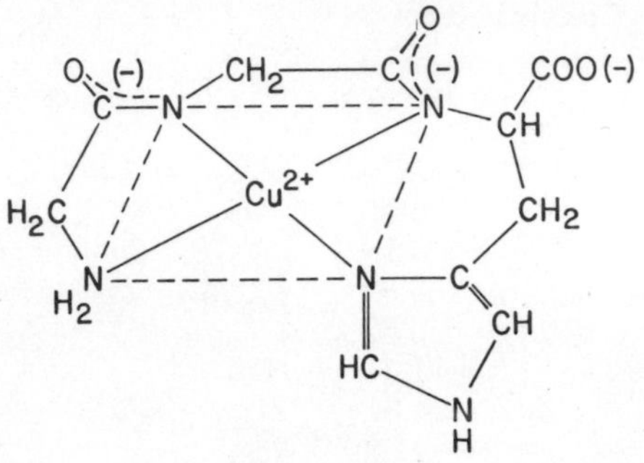

I

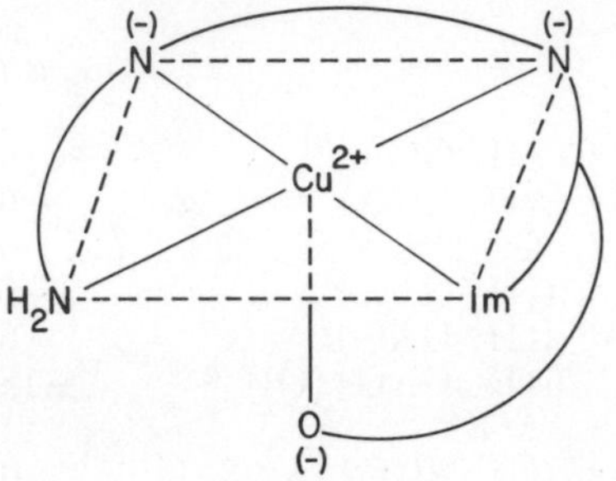

II

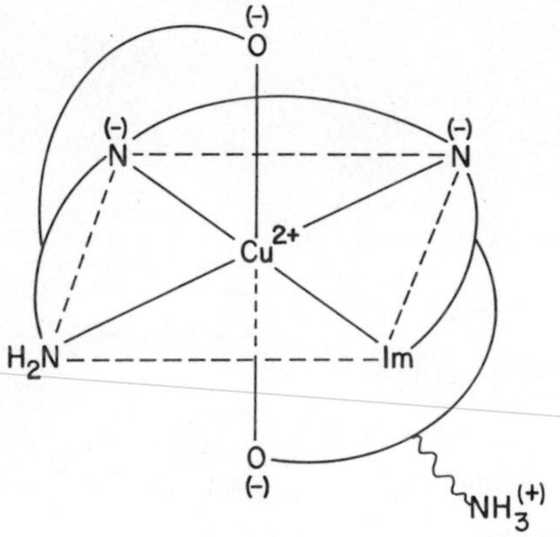

III

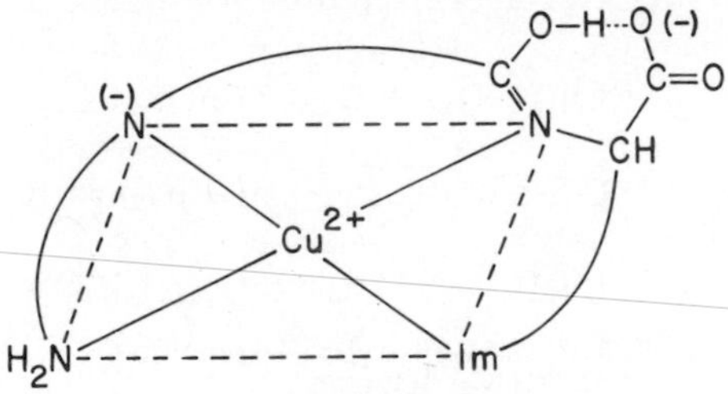

IV

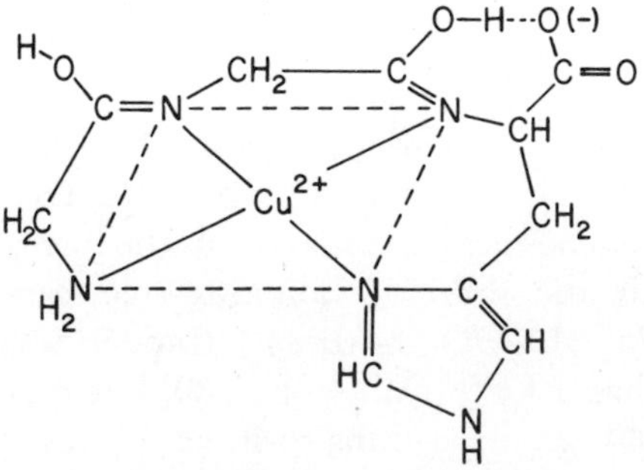

V

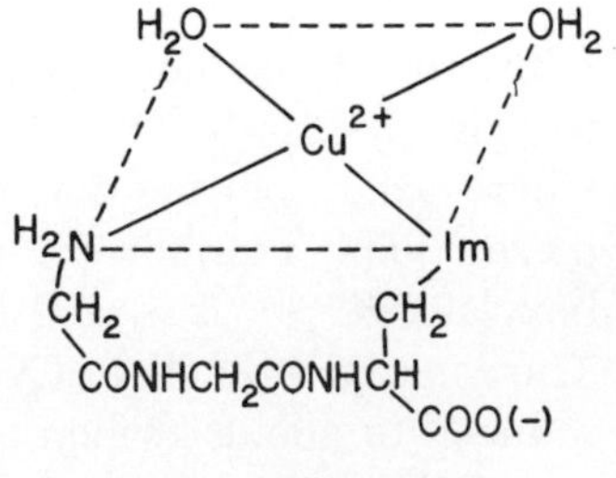

VI

$10^{2.3}$ have been measured and are believed to correspond to Structures **IV** and **V**. Internal hydrogen bonding from the carboxylate group of the histidyl residue helps to stabilize the outside protonated form.

Substitution Kinetics of Copper(II)–Peptide Complexes. Three main reaction pathways have been found for the displacement of copper from peptide complexes—(1) proton transfer to the peptide group, (2) nucleophilic attack on copper, and (3) a combination of proton transfer and nucleophilic attack (*11, 12*). In each case there is a concurrent or subsequent metal–N(peptide) bond cleavage. The reaction rates for the three mechanisms differ in their pH dependence, general acid dependence, and nucleophile dependence.

PROTON TRANSFER MECHANISM. Acids can react with the metal–N-(peptide) group by adding rapidly to the peptide oxygen, giving an outside protonated species which then rearranges by metal–nitrogen bond cleavage. Alternatively, acids can transfer a proton directly to the deprotonated nitrogen atom accompanied by metal–nitrogen bond cleavage (*10, 46, 47*). The latter reaction is slower than normal proton transfer rates but may nevertheless be the preferred kinetic pathway because of the rapid metal–N(peptide) bond cleavage. Different Cu(II)–peptide complexes display different mechanisms. When the direct proton transfer reaction is the rate-determining step, general acid catalysis also is observed. This is the case for the reactions of $Cu(H_{-2}G_3)^-$ with acids (*46*).

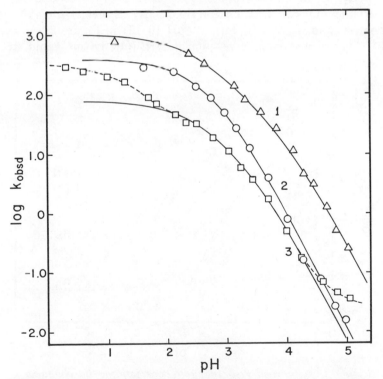

Figure 1. The observed first-order dissociation rate constants level off at low pH because of the addition of two "outside" protons. There is a $[H^+]^2$ dependence at higher pH. (1) Cu(II)–gly-gly-his; (2) Cu(II)–gly-gly-his-gly; (3) Cu(II)–asp·ala·his·lys.

On the other hand the reactions of Cu(H.₂GGhis)⁻, Cu(H.₂GGhisG)⁻ and Cu(H.₂asp-ala-his-lys)⁻ are not general acid catalyzed and at low pH show kinetic evidence of outside protonation as seen in Figure 1 (*12, 13*). The rate-determining step for these histidyl-containing peptide complexes requires two protons which add to the peptide-nitrogens while the amine and the imidazole ends of the oligopeptides are coordinated, so that in the dissociation, the Cu(II) can be pictured as "skipping" over the ligand to give Structure **VI** (*12*).

NUCLEOPHILIC ATTACK. These reactions are characterized by a dependence on the nucleophile concentration and by increasing rate with increasing pH (*48*). The pH dependence for the reaction of trien with Cu(H.₂G₃)⁻ is typical and is given by curve A in Figure 2 for the reaction of $3 \times 10^{-3}M$ trien. The availability of an equatorial site (i.e., of a group which is displaced easily from an equatorial site) is important in nucleophilic attack. Chelating amine ligands are particularly effective as nucleophiles, but these reactions are sensitive to steric effects blocking the availability of an equatorial site (*48, 49*). The corresponding direct nucleophilic attack by trien on Cu(H.₂GGhis)⁻ is seven orders of magnitude slower and is given by line A′ (pH 10–12) in Figure 2. As a result of the sluggish nucleophilic reaction with Cu(H.₂GGhis)⁻ and its slug-

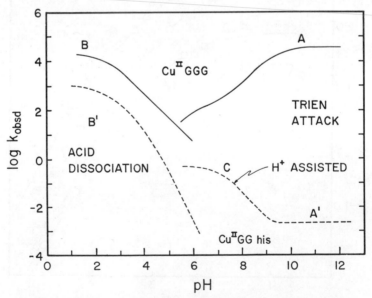

Figure 2. Observed first-order rate constants for the reactions of $Cu^{II}(H_{-2}GGG)^-$ (———) and of $Cu^{II}(H_{-2}GGhis)^-$ (– – –) with: (A,A′) trien (nucleophilic attack); (B,B′) H⁺ (acid dissociation); (C) H⁺ and H_2trien^{2+} (proton-assisted nucleophilic attack), where $[trien]_{total}$ is $3 \times 10^{-3}M$.

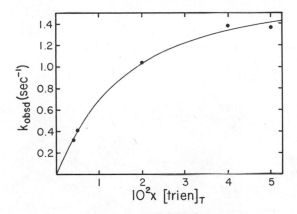

Figure 3.   Dependence of the observed first-order rate constant of $Cu^{II}(H_{-2}GGhis)^-$ on the trien concentration. $Cu^{II} = 2 \times 10^{-4}M$,   pH   6.9,   1M $NaClO_4$, 25.0°. At high $[trien]_T$ the rate constant is proton transfer rate limited.

gish acid dissociation reaction (curve B'), it is possible to observe another reaction pathway, the proton-assisted nucleophilic path, from pH 6 to 9 (curve C).

PROTON-ASSISTED NUCLEOPHILIC MECHANISM (*11, 12*). The reaction rate between trien and $Cu(H_{-2}GGhis)^-$ increases below pH 9 (curve C, Figure 2) in contrast to the behavior of $Cu(H_{-2}G_3)^-$ (curve A). The proton-assisted nucleophilic rate (M sec^{-1}) equals 1.7×10^9 [H$^+$]-[H_2trien^{2+}][$Cu(H_{-2}GGhis)^-$]. As the trien concentration increases, the rate dependence in trien falls off (Figure 3), and rate becomes proton-transfer limited. Below pH 7 H_3trien^{3+} begins to form, effectively removing the nucleophile, and in the vicinity of pH 5–6, the acid dissociation path (curve B', Figure 2) takes over from the proton-assisted nucleophilic path. The full mechanism is given in Figure 4. The reason that the proton-assisted nucleophilic path was not detected for the trien reaction with $Cu(H_{-2}G_3)^-$ can be seen from Figure 2. The direct nucleophilic path (curve A) and the acid dissociation path (curve B) are so favorable that the proton-assisted nucleophilic path makes very little contribution. However, for the histidine-containing peptides the proton-assisted pathway is very important for reactions with trien, EDTA, and histidine.

Copper(II) Transfer from Serum Albumin (*13*). The kinetics of transfer of Cu(II) from its complexes with human serum albumin, bovine serum albumin, gly-gly-his, gly-gly-his-gly, and asp·ala·his·lys to trien all exhibit parallel behavior. The histidine-containing peptides model the first Cu(II) binding site in the serum albumins, where Cu(II) is coordinated to the amine terminal, to two deprotonated peptide nitrogens,

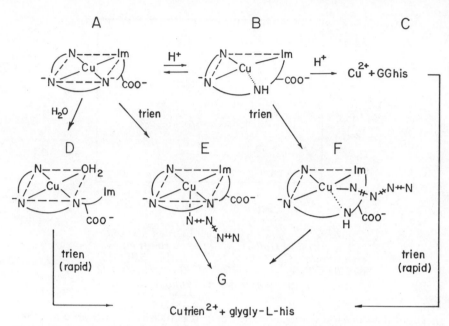

Figure 4. Proposed mechanism for the transfer of Cu(II) from gly-gly-L-his to trien. The predominant pathway in neutral solution is the combined proton and nucleophilic attack shown in path ABFG.

and to the imidazole nitrogen of the histidyl residue (50, 51, 52, 53). The coordination of imidazole alters greatly the kinetic behavior of the Cu(II) complexes. All the complexes react by a proton-assisted nucleophilic pathway at physiological pH, and the serum albumin complexes are faster to transfer Cu(II) than are the tetra-peptide complexes. Thus, in Figure 5 the relative reactivity at pH 7–8 is gly-gly-his > BSA > gly-gly-his-gly > asp·ala·his·lys. The Cu(II) binding site in BSA appears to be fully exposed to solution even when the protein undergoes the N to B conformational change between pH 7 and 9. The reactions with trien and with other nucleophiles involve displacements initiated at non-terminal peptide positions. These reactions are sensitive to acids, are relatively insensitive to peptide steric factors, and are influenced by axial coordination of available carboxylate groups.

Copper(III)–Peptide Complexes. Molecular oxygen reacts with Cu(II)tetraglycine (G_4) in neutral solution to produce a yellow species with an intense absorption band at 362 nm. As the oxygen in the solution is consumed, the amount of the yellow species decays (Figure 6). The uv-visible spectrum, molar absorptivity, dissociation kinetics in acid and in base, and the redox behavior of this yellow species are similar to those of $Cu^{III}(H_{-3}G_4)^-$, which is generated by $IrCl_6^{2-}$ or by electrolytic oxidation of the corresponding Cu(II) complex. The peptide products after the decay of the oxygen-generated yellow species also are similar to those formed when $Cu^{III}(H_{-3}G_4)^-$ decays in neutral solution. These products include a substantial recovery of unreacted G_4 (50–75% depending on the pH) as well as oxidized peptide fragments such as glycylglycinamide and glyoxylglycine.

The Cu(II)–tetraglycine complex is oxidized to Cu(III) by $IrCl_6^{2-}$ (Figure 7). The redox equilibrium is reversible with pH change. The pH dependence is a result of the variable degree of protonation of the Cu(II)–tetraglycine complexes, whereas the Cu(III) complex is present only as the triply deprotonated peptide complex. The curves in Figure 7 correspond to the redox equilibrium in Reaction 4 offset by the acid–

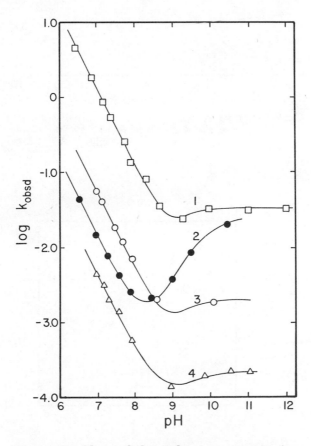

Figure 5. Observed first-order rate constant for the reaction of Cu(II) complexes with 0.07M trien (1.0M NaClO$_4$, 25.0°). (1) Cu(H$_{-2}$GGhis)$^-$; (2) Cu(H$_{-2}$GGhisG)$^-$; (3) Cu-BSA (bovine serum albumin); (4) Cu(H$_{-2}$asp·ala·his·lys)$^-$.

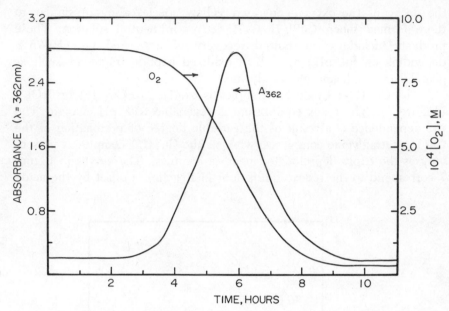

Figure 6.   Oxygen uptake compared with the formation and decay of the yellow species (Cu^{III}) upon the reaction of 8.7×10^{-4}M O_2 with 2×10^{-3}M $Cu^{II}G_4$ at pH 8, 25.0° (26)

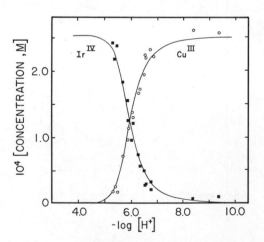

Figure 7.   Formation of $[Cu^{III}(H_{-3}G_4)]^-$ and corresponding loss of $Ir^{IV}Cl_6^{2-}$ as a function of $[H^+]$. The lines are calculated based on an E° value of 0.631 V for $Cu^{III,II}$-$(H_{-3}G_4)$ and 0.891 V for $Ir^{IV,III}Cl_6$.

$$Cu^{II}(H_{-3}G_4)^{2-} + IrCl_6^{2-} \overset{K_{ox}}{\rightleftharpoons} Cu^{III}(H_{-3}G_4)^- + IrCl_6^{3-} \qquad (4)$$

base and complexation equilibria between Cu^{2+} and HG_4, CuG_4^+, $CuH_{-1}G_4$, $CuH_{-2}G_4^-$ and $CuH_{-3}C_4^{2-}$ (9). The K_{ox} value is 2.7×10^4, and the resulting electrode potential for $Cu^{III, II}$ is given in Reaction 5. The

$$Cu^{III}(H_{-3}G_4)^- + e \rightleftharpoons Cu^{II}(H_{-3}G_4)^{2-} \qquad E° = 0.631 \text{ V} \qquad (5)$$

low potential and relatively high stability of this Cu(III) species in aqueous solution is of special interest.

In order to confirm that Cu(III) was present, EPR spectra were taken before and after oxidation (Figure 8). In this case the pentaglycine complex was oxidized electrolytically at pH 10 using a ground graphite

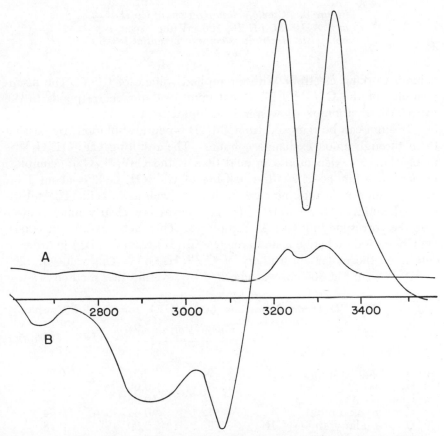

Figure 8. EPR spectra of copper-pentaglycine solutions liquid N_2 temperature and 9.075 GHz. (A) After electrolytic oxidation to give $Cu^{III}(H_{-3}G_5)^-$; (B) Cu^{II}-$(H_{-3}G_5)^{2-}$ before oxidation.

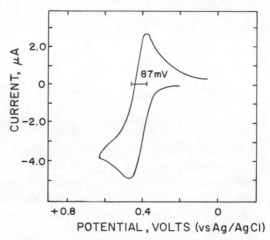

Figure 9.   Cyclic voltammogram of Cu(H₋₃G₄)
at 5 × 10⁻⁴M, pH 10, 100 mV sec⁻¹ scan, car-
bon paste electrode, electrode potential is 0.63
V vs. NHE

column working electrode with an applied voltage of 0.9 V. The disappearance of the Cu(II) EPR signal upon oxidation corresponds to the expected formation of a low spin d^8 complex for Cu(III).

Iridium can be removed from Cu(III)–peptide solutions by passing them through anion exchange columns. The resulting $Cu^{III}(H_{-3}G_4)^-$ is much slower to decompose in acid than is the $Cu^{II}(H_{-3}G_4)^{2-}$ complex. In neutral solutions at 25° the half-life of $Cu^{III}(H_{-3}G_4)^-$ is about 1 hr. The decomposition rate increases in base as well as in acid. The substitution kinetics of the Cu(III)(d^8) complexes are clearly much slower than the corresponding Cu(II) complexes. This fact was used in choosing Chelex ion exchange resin to remove Cu(II) from Cu(III) in order to determine the molar absorptivity of $Cu^{III}(H_{-3}G_4)^-$. This value of 7200 ± 300 M^{-1} cm⁻¹ at 365 nm was checked by several other methods (9).

Table III.   Electrode Potentials (vs. NHE) for $Cu^{III, II}$ Couples[a]

Cu(III)–Peptide	Cyclic Voltammetry[b]		Ir^{IV}–pH Equil.
	ΔmV	E° (V)	E° (V)
Cu(H₋₃hexaglycine)⁻	95	0.67	0.67
Cu(H₋₃pentaglycine)⁻	80	0.66	0.65
Cu(H₋₃tetraglycine)⁻	85	0.63	0.63
Cu(H₋₃triglycinamide)	72	0.64	0.64
Cu(H₋₂di-L-alanylamide)(OH)	165	0.80	0.78

[a] 25°C, $\mu = 0.1M$ NaClO₄.
[b] Carbon paste working electrode, scan rate = 100 mV sec⁻¹ scan, pH 10 [CuL]_T = 5 × 10⁻⁴M.

Table IV. Effect of Coordinating Groups on E^0 of $Cu^{III, II}$ Couples

$Cu(III)$–Peptide	pH	E^0 (V) (NHE)[a]
Cu(H₋₂glycylglycyl-L-histidine)	7.5	0.98
Cu(H₋₂triglycine)	7.7	0.92
Cu(H₋₂diglycinamide)(OH)	9.2	0.89
Cu(H₋₃triglycinamide)	9.5	0.64
Cu(H₋₃tetraglycine)⁻	9.3	0.63
Cu(H₋₄N-formyltetraglycine)²⁻	11.5	0.55

[a] Determined by cyclic voltammetry with carbon paste working electrode, $[CuL]_T$ $= 5 \times 10^{-4}M$, $\mu = 0.1M$ NaClO₄, 25°C.

Cyclic voltammetry is a convenient way to measure the E^0 values for $Cu^{III, II}$ couples. A carbon paste electrode gives quasi-reversible behavior as shown in Figure 9. In order to be certain that the E^0 from cyclic voltammetric experiments were valid, five peptide complexes were examined by both $Ir^{IV}Cl_6{}^{2-}$–pH profile methods and electrochemically (Table III). The agreement of the E^0 values determined by the two methods is excellent even when the peak-to-peak separation of the oxidation and reduction waves ($\triangle$mV) were significantly greater than 60 mV. The effect of varying the nature of the coordinating groups on the $Cu^{III, II}$ potential is seen in Table IV. As the number of deprotonated peptide groups increase, the E^0 values decrease. The N-formyl derivative with the equivalent of four deprotonated peptide groups coordinated to copper has an E^0 value as low as 0.55 V. The N-formyl derivatives consistently lower the E^0 values (Table V). The use of alanyl and valyl residues instead of glycyl residues in the peptide chain also lowers the E^0 value. As seen in Table VI, the relative effectiveness in giving lower potentials is $CH(CH_3)_2 > CH_3 > H$. The bulkiness of the R group favoring coordination to the smaller metal ion (i.e., Cu(III)) may be a factor.

The effect of high pH on solutions of Cu(III)–polyglycine complexes is unusual because there is a color change from yellow to red between pH

Table V. Effect of N-Formyl Derivatization on E^0 of $Cu^{III, II}$ Couples

$Cu(III)$–Peptide	E^0 (V)(NHE)[a]
Cu(H₋₃tetraglycine)⁻	0.63
Cu(H₋₄N-formyltetraglycine)²⁻	0.55[b]
Cu(H₋₃triglycinamide)⁻	0.64
Cu(H₋₄N-formyltriglycinamide)²⁻	0.49[b]
Cu(H₋₂triglycine)	0.92[c]
Cu(H₋₃N-formyltriglycine)⁻	0.75

[a] Determined by cyclic voltammetry with a carbon paste working electrode, $[CuL]_T = 5 \times 10^{-4}M$, pH $= 10$, 25°C, $\mu = 0.1M$ NaClO₄, scan rate 100 mV sec⁻¹.
[b] pH $= 11.5$.
[c] pH $= 7.7$.

11 and 12. This color change can be taken back and forth by alternatively adding acid and base. The spectrum of the red complex is difficult to obtain by conventional methods because of the rapid redox reactions of the Cu(III) species at high pH. Within 28 sec all the Cu(III) species have disappeared. However, Figure 10 shows spectra obtained by the stopped-flow vidicon technique (54). In Figure 10A the $Cu^{III}(H_{-3}G_3a)$ spectrum is shown with an absorption peak at 365 nm. The spectrum taken 0.5 sec after mixing with $1.0M$ OH$^-$ has new peaks at 310 and 525 nm, and the initial peak at 365 has disappeared. This spectral shift does not occur when $Cu^{III}(H_{-4}N\text{-formyl-}G_4)^{2-}$ reacts with $1M$ OH$^-$ (Figure 10B). Furthermore, the yellow $\rightleftarrows$ red shift does not occur with $Cu^{III}(H_{-3}(CH_3)_2N\text{-}G_4)^-$ where there are no hydrogens on the terminal amine. Therefore, the spectral shift at high pH is attributed to ionization of an amine hydrogen as shown in Figure 11 for $Cu^{III}(H_{-3}G_3a)$. A similar type of ionization and spectral shift has been reported (55) for $[Au^{III}(H_{-1}dien)X]^+$. Figure 12 shows that the amine hydrogen ionization constant can be measured from the absorbance changes at 525 nm which are obtained by extrapolation to the initial value after mixing Cu^{III}-(H_3G_3a) with various amounts of NaOH. The pK_a values for loss of amine hydrogens for four Cu(III)–peptide complexes (Table VII) are

Table VI.   Effect of R-Group Ligand Substitution on the $E°$ of $Cu^{III,II}$ Couples

Cu(III)–Peptide	E° (V) (NHE)[a]
Cu(H$_{-3}$tetraglycine)$^-$	0.63
Cu(H$_{-3}$tetra-L-alanine)$^-$	0.60
Cu(H$_{-3}$tetra-L-valine)$^-$	0.51
Cu(H$_{-3}$pentaglycine)$^-$	0.66
Cu(H$_{-3}$penta-L-alanine)$^-$	0.61

[a] Determined by cyclic voltammetry with carbon paste working electrode, $[CuL]_T$ $= 5 \times 10^{-4}M$, $25°C$, $0.1M$ NaClO$_4$, 100 mV sec^{-1}.

Table VII.   Ionization Constants for the Loss of Coordinated Amine Hydrogens from Cu(III)–Peptide Complexes[a]

$$Cu^{III}(H_{-3}L)^- \rightleftarrows Cu^{III}(H_{-4}L)^{2-} + H^+$$

L	pK_a
G$_3$a	12.6
G$_4$	12.1
G$_5$	11.6
G$_6$	11.3

[a] $25.0°C$, $\mu = 1M$ NaClO$_4$. The [H$^+$] is calculated from $pK_w = 13.8$. Ref. 57 after calibration of the pH readings in terms of NaOH.

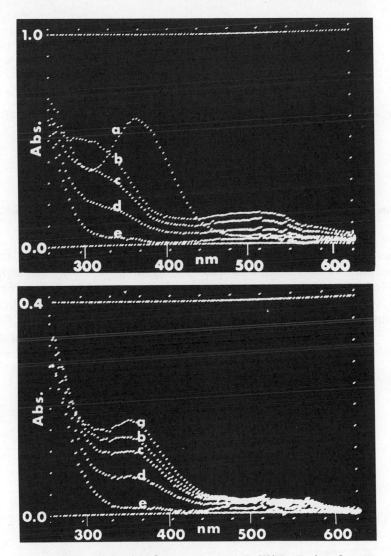

Figure 10.   (A) (top) Vidicon spectra of $Cu^{III}(H_{-3}G_3a)$ and Cu^{III}-$(H_{-4}G_3a)^-$.   a, $Cu^{III}(H_{-3}G_3a)$ at pH 8; b, 0.5 sec after mixing (0.5M OH^-); c, 2.5 sec; d, 7.5 sec; e, 28.0 sec.   (B) (bottom) Vidicon spectra of $Cu^{III}(H_{-4}N\text{-}formyl\text{-}G_4)^{2-}$.   a, at pH 9; b, 0.5 sec after mixing (0.5M OH^-); c, 2.5 sec; d, 7.5 sec; e, 28.0 sec.

as low as 11.3. It is interesting how great an effect coordination to Cu(III) has on this ionization reaction, which is seldom seen in aqueous solution. The reduction potentials for these $Cu^{III}(H_{-4}L)$ complexes decrease with increasing pH because the amine hydrogen is not ionized from the corresponding Cu(II) complexes. Unfortunately the strong

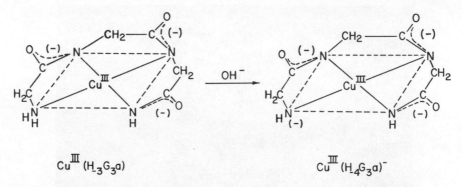

$$Cu^{III}(H_{-3}G_3a) \qquad\qquad Cu^{III}(H_{-4}G_3a)^-$$

Figure 11. Proposed ionization of the amine hydrogen causing the yellow-to-red color change when base is added to Cu(III)-peptide complexes

base needed to form the $Cu^{III}(H_{-4}L)$ complexes also causes their rapid decomposition because of base-catalyzed ligand oxidation by the copper. The reduction of Cu(III) is not caused by solvent oxidation because no oxygen or peroxide can be detected. Only 25% of the coordinated ligand is oxidized in the case of G_4, and 75% is recovered intact (57). These and other peculiarities about the nature of the products suggest that electron exchange reactions between Cu(II)– and Cu(III)–peptide complexes may be rapid.

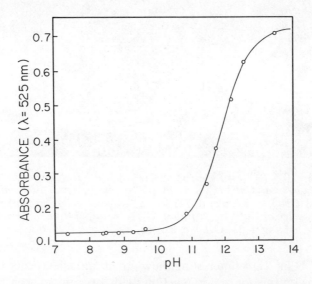

Figure 12. Absorbance (525 nm) vs. pH titration curve for amine hydrogen ionization from Cu^{III}-$(H_{-3}G_3a)$. Correction of the pH readings to $-\log [H^+]$ is needed to give the pK_a value in Table VII.

The electron transfer reaction in Reaction 6 was measured by circular dichroism stopped-flow methods at pH 7.7. Only the tetra-L-alanine

$$Cu^{III}(H_{-3}G_5)^- + Cu^{II}(H_{-3}A_4)^{2-} \xrightarrow{k_e} Cu^{II}(H_{-3}G_5)^{2-} + Cu^{III}(H_{-3}A_4)^-$$
$$\Updownarrow OH^- \qquad\qquad \Updownarrow OH^- \qquad\qquad (6)$$
$$Cu^{II}(H_{-2}A_4)^- \qquad Cu^{II}(H_{-2}G_5)^-$$

complexes are CD active, and the CD spectra change when the Cu(II) complex is oxidized to Cu(III) as shown in Figure 13. Under the conditions used (pH 7.7 and $2 \times 10^{-4}M$ Cu^{III}(H$_{-3}$G$_5$)$^-$), the overall rate was

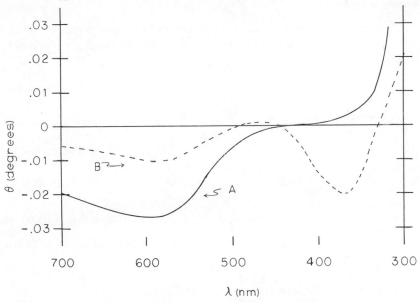

Figure 13. CD spectra showing the electron transfer reaction to form Cu^{III}(H$_{-3}$A$_4$)$^-$. (A) Cu^{II}A$_4$, $9.5 \times 10^{-4}M$, pH 7.2, 2 cm cell; (B) After mixing equal volumes of Cu^{III}(H$_{-3}$G$_5$)$^-$ and Cu^{II}A$_4$ at 7.0, both complexes approximately $5 \times 10^{-4}M$, 2 cm cell.

first order with a rate constant of 0.1 sec^{-1}, depending only on the conversion of Cu^{II}(H$_{-2}$A$_4$)$^-$ to Cu^{II}(H$_{-3}$A$_4$)$^{2-}$. Hence, the electron exchange rate constant k_e must be greater than $10^4 M^{-1}$ sec^{-1}. Additional studies in progress in our laboratory confirm that electron exchange reactions of this type are very fast.

The reduction of Cu^{III}(H$_{-3}$G$_4$)$^-$ has been examined with a number of substrates including I$^-$, Fe(CN)$_6$$^{4-}$, (tert-Bu)$_2$NO, ascorbic acid, cysteine, hydroquinone, and SO$_3$$^{2-}$. All of the reductions are rapid and appear to proceed by one-electron steps. Two-electron transfer reactions between Cu(III) and Cu(I), which would avoid high energy, free-radical intermediates, would be of considerable interest in biological oxidations, but as yet we have no clear-cut examples of this behavior.

In summary, Cu(III) is stabilized by deprotonated peptide bonding, and this trivalent oxidation state is much more accessible in aqueous solution than had been realized. Variations in E^0 values of more than 500 mV occurs for $Cu^{III, II}$ redox couples as the nature of the coordinating groups change. If Cu(III) exists in nature, it is likely to be coordinated at least in part to deprotonated peptides, which play an important role in stabilizing this trivalent oxidation state.

Acknowledgment

This investigation was supported by Public Health Service Grant Nos. GM-12152 and GM-19775 from the National Institute of General Medical Sciences.

Literature Cited

1. Rising, M. M., Parker, F. M., Gaston, D. P., *J. Am. Chem. Soc.* (1934) **56**, 1178.
2. Chouteau, J., Leonormant, H., *C. R. Acad. Sci., Paris* (1951) **232**, 1479.
3. Chouteau, J., *C. R. Acad. Sci., Paris* (1951) **232**, 2314.
4. Dobbie, H., Kermack, W. O., *Biochem. J.* (1955) **59**, 246, 257.
5. Manyak, A. R., Murphy, C. B., Martell, A. E., *Arch. Biochem. Biophys.* (1955) **59**, 373.
6. Murphy, C. B., Martell, A. E., *J. Biol. Chem.* (1957) **226**, 37.
7. Koltun, W. L., Fried, M., Gurd, F. R. N., *J. Am. Chem. Soc.* (1960) **82**, 233.
8. Burce, G. L., Paniago, E. B., Margerum, D. W., *Chem. Commun.* (1975) 261.
9. Margerum, D. W., Chellappa, K. L., Bossu, F. P., Burce, G. L., *J. Am. Chem. Soc.* (1975) **97**, 6894.
10. Margerum, D. W., Dukes, G. R., "Metal Ions in Biological Systems," H. Sigel, Ed., Vol. 1, p. 157, Marcel Dekker, New York, N.Y., 1974.
11. Cooper, J. C., Wong, L. F., Venezky, D. L., Margerum, D. W., *J. Am. Chem. Soc.* (1974) **96**, 7560.
12. Wong, L. F., Cooper, J. C., Margerum, D. W., *J. Am. Chem. Soc.* (1976) **98**, 7268.
13. Wong, L. F., Margerum, D. W., unpublished data.
14. Kochi, J. K., in "Free Radicals," J. K. Kochi, Ed., Vol. 1, Chapter 11, Wiley-Interscience, New York, 1973.
15. Cohen, T., Lewarchik, R. J., Tarino, J. Z., *J. Am. Chem. Soc.* (1974) **96**, 7753.
16. Cohen, T., Wood, J., Dietz, Jr., A. G., *Tetrahedron Lett.* (1974) **40**, 3555.
17. Wahl, K., Klemm, W., *Z. Anorg. Allgem. Chem.* (1952) **270**, 69.
18. Hoppe, R., *Angew. Chem.* (1950) **62**, 339.
19. Hadinec, I., Jenšovský, L., Línek, A., Syneček, V., *Naturwissenschaften* (1960) **47**, 377.
20. Beurskens, P. T., Cras, J. A., Steggerda, J. J., *Inorg. Chem.* (1968) **7**, 810.
21. Olson, D. C., Vasilekskis, J., *Inorg. Chem.* (1971) **10**, 463.
22. Meyerstein, D., *Inorg. Chem.* (1971) **10**, 638, 2244.

23. Levitzki, A., Anbar, M., Berger, A., *Biochemistry* (1967) **6**, 3757.
24. Bour, J. J., Birker, P. J. M. L., Steggerda, J. J., *Inorg. Chem.* (1971) **10**, 1202.
25. Paniago, E. B., Weatherburn, D. C., Margerum, D. W., *Chem. Commun.* (1971) 1427.
26. Burce, G. L., Ph.D. Thesis, Purdue University, 1975.
27. Hamilton, G. A., Libby, R. D., Hartzell, C. R., *Biochem. Biophys. Res. Commun.* (1973) **55**, 333.
28. Dyrkacz, G. R., Libby, R. D., Hamilton, G. A., *J. Am. Chem. Soc.* (1976) **98**, 626.
29. Kwiatkowski, L. D., Kosman, D. J., *Biochem. Biophys. Res. Commun.* (1973) **53**, 715.
30. Kosman, D. J., Bereman, R. D., Ettinger, M. J., Giordano, R. S., *Biochem. Biophys. Res. Commun.* (1973) **54**, 856.
31. Kaneda, A., Martell, A. E., *J. Coord. Chem.* (1975) **4**, 140.
32. Yamauchi, O., Nakao, Y., Nakahara, A., *Bull. Chem. Soc. Japan* (1973) **46**, 2119.
33. Hauer, H., Billo, E. J., Margerum, D. W., *J. Am. Chem. Soc.* (1971) **93**, 4173.
34. Brunetti, A. P., Lim, M. C., Nancollas, G. H., *J. Am. Chem. Soc.* (1968) **90**, 5120.
35. Nancollas, G. H., Poulton, D. J., *Inorg. Chem.* (1969) **8**, 680.
36. Hartzell, C. R., Gurd, F. R. N., *J. Biol. Chem.* (1969) **244**, 147.
37. Dorigatti, T. F., Billo, E. J., *J. Inorg. Nucl. Chem.* (1975) **37**, 1515.
38. Dukes, G. R., Margerum, D. W., *J. Am. Chem. Soc.* (1972) **94**, 8414.
39. Bryce, G. F., Gurd, F. R N., *J. Biol. Chem.* (1966) **241**, 1439.
40. Weatherburn, D. C., Billo, E. J., Jones, J. P., Margerum, D. W., *Inorg. Chem.* (1970) **9**, 1557.
41. Bryce, G. F., Roeske, R. W., Gurd, F. R. N., *J. Biol. Chem.* (1965) **240**, 3837.
42. Billo, E. J., *Inorg. Nucl. Chem. Lett.* (1974) **10**, 613.
43. Paniago, E. B., Margerum, D. W., *J. Am. Chem. Soc.* (1972) **94**, 6704.
44. Cooper, J. C., Wong, L. F., Margerum, D. W., unpublished data.
45. Barnet, M. T., Freeman, H. C., Buckingham, D. A., Hsu, I., Van der Helm, D., *Chem. Commun.* (1970) 367.
46. Pagenkopf, G. K., Margerum, D. W., *J. Am. Chem. Soc.* (1968) **90**, 6963.
47. Bannister, C. E., Margerum, D. W., Raycheba, J. M. T., Wong, L. F., *Faraday Symposia of the Chemical Society* (1975) **10**, 78.
48. Pagenkopf, G. K., Margerum, D. W., *J. Am. Chem. Soc.* (1970) **92**, 2683.
49. Hauer, H., Dukes, G. R., Margerum, D. W., *J. Am. Chem. Soc.* (1973) **95**, 3515.
50. Breslow, E., *J. Biol. Chem.* (1964) **239**, 3252.
51. Peters, Jr., T., Blumenstock, F. A., *J. Biol. Chem.* (1967) **242**, 1574.
52. Shearer, W. T., Bradshaw, R. A., Gurd, F. R. N., Peters, Jr., T., *J. Biol. Chem.* (1967) **242**, 5451.
53. Bradshaw, R. A., Shearer, W. T., Gurd, F. R. N., *J. Biol. Chem.* (1968) **243**, 3817.
54. Milano, M. J., Pardue, H. L., Cook, T., Santini, R. E., Margerum, D. W., Raycheba, J. M. T., *Anal. Chem.* (1974) **46**, 374.
55. Baddley, W. H., Basolo, F., Gray, H. B., Nölting, C., Pöe, A. J., *Inorg. Chem.* (1963) **2**, 921.
56. Fischer, R., Byé, J., *Bull. Soc. Chim. France* (1964) 2920.
57. Kurtz, J. L., Margerum, D. W., unpublished data.

RECEIVED July 26, 1976.

17

Recent Studies on the Effect of Copper on Nucleic Acid and Nucleoprotein Conformation

GUNTHER L. EICHHORN, JOSEPH M. RIFKIND, and YONG A. SHIN

Laboratory of Molecular Aging, Gerontology Research Center,
National Institute on Aging, National Institutes of Health,
Baltimore City Hospitals, Baltimore MD 21224

Copper ions cause cooperative disordering of the helical structure of nucleic acid strands, as demonstrated by ORD and uv changes as a function of Cu^{2+} activity. Equilibrium dialysis binding studies reveal that the cooperative disordering is paralleled by cooperative binding. The concentration and chain length dependence of these phenomena indicate that Cu^{2+} ions form both intermolecular and intramolecular crosslinks between and within polynucleotide chains. Copper ions also have dramatic effects on nucleoprotein structure. They invert the CD spectrum of the DNA–polylysine complex, which simulates the protein–DNA binding in nucleoprotein, indicating that metal binding significantly changes the way in which the DNA is packed in a superstructure. Thus copper ions influence the conformation of not only nucleic acids but also their complexes with proteins.

It is well known that copper ions are required for the biological function of some enzymes and are detrimental to the function of many others. These effects can often be ascribed to conformational changes. The impact of copper ions on the conformation of nucleic acids is equally dramatic and of potential significance in disturbing the function of the genetic material (1, 2).

We have previously shown that Cu^{2+} ions can reversibly unwind the double helix of DNA (3). Recent experiments have demonstrated that

Cu²⁺ ions also disorder single-stranded helical structures by forming cross-links between and within polynucleotide strands (*4*).

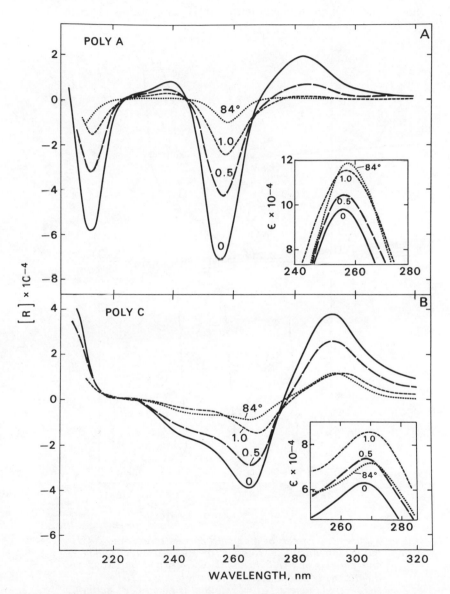

Figure 1.   Effect of Cu(II) on the ORD and uv spectra of single-stranded poly(A) and poly(C). Nucleotide residue concentration 1.5 × 10⁻⁴M, pH 7, 25°. Copper/nucleotide residue given in Figure, compared with spectrum without Cu²⁺ at 84°.

Intramolecular and Intermolecular Crosslinking of Polynucleotides

The disordering of single-stranded poly(A) and poly(C) helices by Cu(II) is demonstrated by the collapse of the ORD curves characteristic of these helices when increments of Cu(II) are added (Figure 1). The

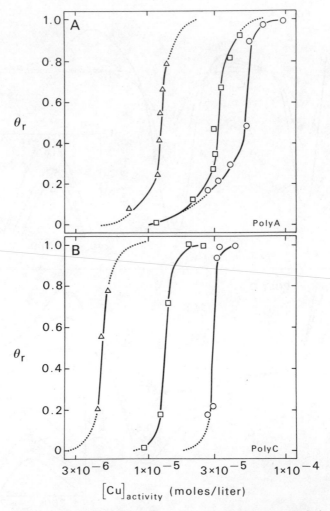

Figure 2. Effect of polymer concentration on Cu(II)-induced changes in rotation of single-stranded poly(A) and poly(C) at pH 6 in 0.1M NaClO$_4$ and 0.01M caco-dylate. θ_r is fractional change in rotation $= [\alpha] - [\alpha_o]/[\alpha]_d - [\alpha]_o$ where $[\alpha] =$ rotation at given Cu^{2+} activity, $[\alpha_o] =$ rotation in the absence of Cu^{2+}, and $[\alpha_d] =$ rota-tion of fully disordered structure. A: poly(A), B: poly(C). Concentrations: ($\triangle$) 5×10^{-4}M; ($\square$) 5×10^{-5}M; ($\bigcirc$) 5×10^{-6}M.

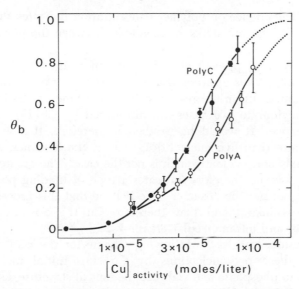

Figure 3.   Binding of Cu(II) to single-stranded poly(A) and poly(C) in 0.1M NaClO₄ and 0.01M sodium cacodylate, pH 6, 25° ± 0.1°. θ_b is fractional saturation of potential binding sites assuming Cu(II):nucleotide stoichiometry of 1:2.

simultaneous increase in the absorbance indicates a destacking of the nucleotide bases that must occur as a result of the disordering (Figure 1). The helical structures are, of course, also collapsed by heating, as the ORD curves at 84° illustrate. The curves produced by the incremental addition of Cu(II) pass through an isosbestic point, which however does not include the high temperature curve. Therefore, the disordering produced by Cu(II) must be different from thermally induced disordering.

Figure 2 provides some clues to the difference in the mechanism of Cu(II)-induced and thermal disordering. The latter is a rather noncooperative process as shown by the broad temperature range in which the disordering transition takes place (5, 6). The Cu(II)-induced transition, on the other hand, occurs over a very narrow concentration range, as shown in Figure 2, in which the fractional change in rotation is plotted as a function of copper activity. The cooperativity in the spectral transition is accompanied by cooperativity in the binding of Cu(II) to the polynucleotides (Figure 3). It is therefore apparent that the initial binding of Cu(II) to poly(A) and to poly(C) facilitates further binding and that increased binding promotes enhanced disordering.

The effect of polymer concentration on the disordering provides an additional clue to its mechanism. Figure 2 shows that the higher the polymer concentration, the lower the Cu^{2+} activity required for the transition. This dependence on polymer concentration indicates that Cu(II) forms intermolecular crosslinks, or crosslinks between the polynucleotide chains.

The disordering process also depends on chain length. Figure 4 reveals that poly(A) is disordered at much lower copper concentrations than the hexamer. From the results with the polymer it has been calculated (4) that four to five bases are disordered by Cu(II) in one continuous sequence. It would be predicted, therefore, that an oligomer containing more than five nucleotides, such as the hexamer, should behave like a polymer; evidently such is not the case. The greater cooperativity of the polymer therefore indicates a type of binding possible only in a polymer but not the hexamer. We believe that this process is intramolecular crosslinking and that therefore Cu(II) can produce both intermolecular and intramolecular crosslinks.

These results lead to the following scheme for the cooperative disordering of these polynucleotides by Cu(II). Initial increments of Cu(II) bind to phosphate and stabilize the helical structure, as the initial increase in rotation in Figure 4 demonstrates. As the concentration of polymer-bound Cu(II) increases, crosslinking to bases takes place. Such crosslinks may be within the same polymer strand or between different strands. Once such crosslinking has begun, additional cooperative cross-

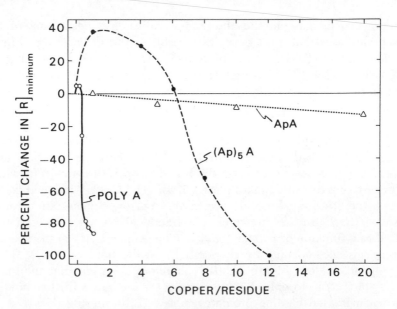

Figure 4. *Effect of chain length on Cu(II)-induced changes in rotation for single-stranded poly(A) and oligo(A) in water at 25°. Presented as % change in mean residue rotation at trough (see Figure 1).*

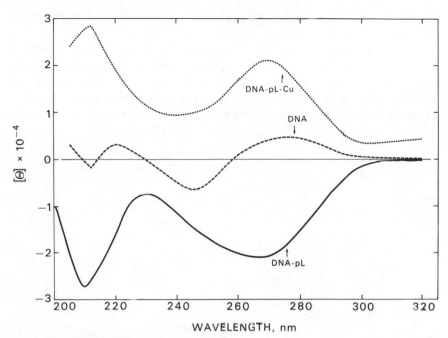

Figure 5. Effect of heating with CuCl₂ (1 × 10⁻⁴M) on CD spectrum of complex of DNA and poly(Lys) in 1:0.5 ratio (DNA concentration, 1 × 10⁻⁴M)

linking in the vicinity of the first crosslink is facilitated since the entropy lost in bringing strands or strand segments together is no longer required.

Intermolecular Interactions in Nucleoprotein Models

The disordering of polynucleotides by Cu(II) binding may be contrasted to the effects of Cu(II) on complexes of DNA with polylysine. Such complexes serve as models for the nucleoproteins found in the nuclei of eukaryotic cells. The nucleoproteins contain DNA bound to a variety of proteins, many of which contain a large amount of lysine. These proteins are believed to be involved in the regulation of genetic activity.

Complexing of DNA with polylysine results in a very large change in the ORD and CD spectra of DNA. The CD change is shown in Figure 5. The amplitude of the ellipticity is greatly magnified, and a high negative ellipticity is observed in the 270-nm region (7). This high ellipticity is caused by anisotropic packing of the DNA polylysine complex, i.e., an intermolecular association with a directional twist.

When DNA–polylysine is heated with Cu(II), a CD spectrum is obtained (8) that has ellipticity of positive sign in the 270-nm region (Figure 5). This CD inversion could be caused by a reversal in the

directional twist of the intermolecular arrangement, but it can also result
from a change in the distance between repeating units (9).

The CD spectra with positive and negative ellipticities are similar to
two forms of CD spectra that have been observed with complexes of DNA
and several of the proteins that are associated with DNA in the cell
nucleus (10). Apparently metal binding can bring about interconversions
between the intermolecular forms that produce these two types of CD
spectra.

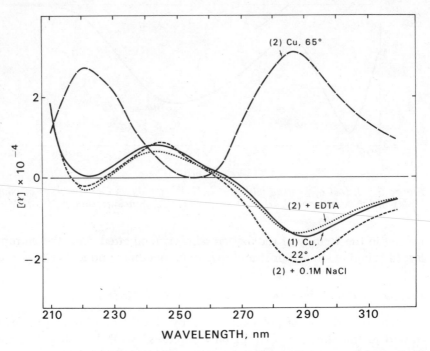

*Figure 6.   Reversibility of the effect of CuCl₂ on the ORD spectrum of
DNA–polylysine by addition of EDTA or 0.1M NaCl*

Figure 6 shows the Cu(II)-induced changes in the ORD spectrum
of DNA–polylysine at room temperature, after heating to 65°, and on
addition of EDTA or 0.1M NaCl to the heated solution. The ORD spectrum of the copper complex of DNA–polylysine at room temperature is
not very different from the spectrum of DNA–polylysine itself. This is
explained by the fact that at room temperature Cu(II) binds mainly to
the phosphate groups of DNA and therefore does not drastically alter
the DNA structure. The dramatic effect on the ORD spectrum is observed on heating when there is a large increase in the binding of Cu(II)
to the DNA bases (3). Both EDTA and high ionic strength remove

Cu(II) from DNA, and therefore the previously characterized ORD spectrum of DNA–polylysine is regenerated.

We believe that Cu(II) binding to the DNA bases alters the DNA conformation in such a way that a gross alteration occurs in the packing of the DNA molecules. The addition and subsequent removal of copper make this a reversible process.

These studies demonstrate that Cu(II) has substantial effects on the structure of polynucleotides as well as on the intermolecular packing of complexes of nucleic acids with proteins or polypeptides.

Literature Cited

1. Eichhorn, G. L., in "Inorganic Biochemistry," G. L. Eichhorn, Ed., Elsevier, Amsterdam, 1973, 2, 1210.
2. Marzilli, L. G., *Prog. Inorg. Chem.* (1977) 23, in press.
3. Eichhorn, G. L., Clark, P., *Proc. Natl. Acad. Sci. U.S.A.* (1965) 53, 586.
4. Rifkind, J. M., Shin, Y. A., Eichhorn, G. L., *Biopolymers* (1976) 15, 1879.
5. Fasman, G. D., Lindblow, C., Grossman, L., *Biochemistry* (1964) 3, 1015.
6. Leng, M., Felsenfeld, G., *J. Mol. Biol.* (1966) 15, 455.
7. Jordan, C. F., Lerman, L. S., Venable, J. H., *Nature New Biol.* (1972) 236, 67.
8. Shin, Y. A., Eichhorn, G. L., *Biopolymers* (1977) 16, 225.
9. Holzwarth, G., Holzwarth, N. A. N., *J. Opt. Soc. Am.* (1973) 63, 324.
10. Fasman, G. D., Schaffhausen, B., Goldsmith, L., Adler, A., *Biochemistry* (1970) 9, 2814.

RECEIVED July 26, 1976.

18

Bio(in)organic Views on Flavin-Dependent One-Electron Transfer

P. HEMMERICH

Universität Konstanz, Fachbereich Biologie, Postfach 7733,
D-7750 Konstanz, West Germany

Flavin-dependent 1e⁻-transfer in enzymes and chemical model systems can be differentiated from 2e⁻-transfer activities, i.e., (de)hydrogenation and oxygen activation, by chemical structure and dynamics. For 1e⁻-transfer, two types of contacts are discussed, namely outer sphere for interflavin and flavin–heme and inner sphere for flavin–ferredoxin contacts. Flavin is the indispensable mediator between 2e⁻- and 1e⁻-transfer in all biological redox chains, and there is a minimal requirement of three cooperating redox-active sites for this activity. The "switch" between 2e⁻- and 1e⁻-transfer is caused by apoprotein-dependent prototropy between flavin positions N(1)/O(2α) and N(5) or by N(5)-metal contact.

During the current symposium the state of chemical knowledge concerning all except one type of biological electron-transferring protein has been reviewed with expertise, such as e⁻-transferring copper by Gray et al. (*1*), heme iron by Sutin (*2*), and "sulfur iron" (ferredoxin) by Hall (*3*). In fact, however, and in spite of the diversity and number of iron–sulfur and heme proteins participating in biological electron transport, no single electron from any organic CH-substrate ever sees an iron center before it has been digested by a flavoprotein. The "flavin passage" is in fact mandatory for all one-electron equivalents passing to and from carbon–hydrogen bonds in animal and plant as well as microbial metabolism (Figure 1).

The wealth of seemingly random flavoproteins can be brought into a certain order by distinguishing them according to the number of redox equivalents operating at once in their input and output reactions (Table

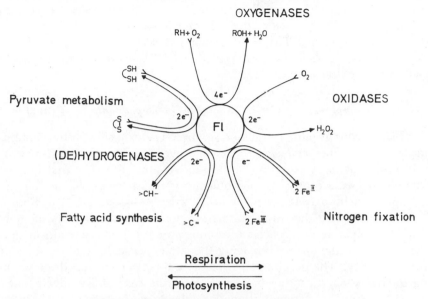

Figure 1. Functions of animal, plant, and microbial flavoproteins and redox stoichiometry

I). In the present chapter, which is devoted to flavin-dependent $1e^-$-transfers, we can differentiate two main classes of $1e^-$-transferring flavoproteins, namely the flavodoxins or "pure e^--transferases" and the $(2 \times 1e^- \rightleftarrows 1 \times 2e^-)$-"transformases" which constitute one essential part of

Table I. Prototypes of Flavoproteins and Number of Redox Equivalents Transferred in a Single Input and Output Step[a]

Redox-Equivalents Input/Output	Flavoprotein Function	Prototype
1 : 1	pure e^--transfer	flavodoxin
(2x) 1 : 2	e^--transfer/hydrogenation ("1/2-transformase")	NADP-reductase of chloroplasts
2 : 1 (2x)	dehydrogenation/e^--transfer ("2/1-transformase")	NADH-dehydrogenase of mitochondria
2 : 2	dehydrogenation/O_2-reduction	bacterial glucose oxidase
	dehydrogenation/oxygenation	bacterial phenol oxygenase
	pure transhydrogenation	bacterial NAD(P)-transhydrogenase

[a] Mitochondrial NADH-dehydrogenase contains $1e^-$-FeS-acceptor sites within the molecule.

flavin-dependent dehydrogenases. In this transformase activity flavin is unique and mandatory for all biological redox chains. With respect to the transformases, however, this chapter deals only with one "half-activity" of the flavin. The residual main types of flavin action—(de)-hydrogenation (4) and oxygen activation (4, 5)—have been extensively reviewed in the recent past.

Types of Complexes or Contacts in Flavin-Dependent 1e⁻-Transfer

Flavodoxins as pure e⁻-transferases are ferredoxin substitutes containing only one center while it appears that every biological "transformase" unit must contain at least three redox-active sites (6). The first one, or (de)hydrogenase center, is the obligatory transformase flavin, having 2e⁻-input and 1e⁻-output (or vice versa), such as, e.g., carrier I in Figure 2. The second and the third centers are one-electron collectors, namely either two separate FeS centers (ferredoxin units) or one flavodoxin plus one ferredoxin, the former replacing the latter, particularly in the case of iron deficiency. The requirements for those three carriers, all three working in the 1e⁻-shuttle (the lower shuttle in the case of flavodoxin) and only the first one working additionally in the 2e⁻-shuttle, becomes obvious from Figure 2. Nature aims at the maximum yield of single electrons with

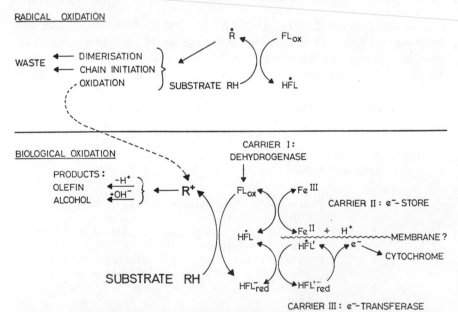

Figure 2. The three-center hypothesis for 1e⁻/2e⁻-"transformase" activity. Dehydrogenase-flavin (Fl) = carrier I; pure e⁻-transferase flavin (Fl', flavodoxin —replaceable by a second FeS-protein) = carrier II; e⁻ storage FeS-protein (Fe, ferredoxin) = carrier III.

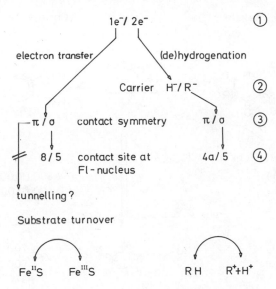

*Figure 3. The four questions in flavin-depend-
ent biological oxidoreduction mechanisms*

no loss by substrate radical dimerization or chain reactions. The low-potential "1e⁻-only" carrier II of either flavodoxin or ferredoxin type transfers the energy-rich electron from HFl$_{red}^-$ into the redox chain. The high potential 1e⁻-carrier III is required as an electron sink for reoxidation of the transformase radical HFl, which by itself is inactive in the dehydrogenation reaction.

In these systems, binary carrier contacts, whether of measurable lifetime or of shorter duration, implicate, therefore, two classes—"interflavin" and flavin–heme contacts and flavin–iron (sulfur) contacts. Figure 3 makes it clear that, even irrespective of whether one or two redox equivalents are being transferred, the geometry of these contacts (roughly spoken: π or σ) must be asked for, and if it is σ, the site of contact must be discussed, whether inner sphere or outer sphere.

To begin with the FeS cluster and the heme system, the terms inner and outer sphere are nonequivocal while the "inner sphere of the flavin molecule" with respect to metal contact is taken to be the chelating profile O(4α), N(5) (Figure 4). The outer sphere of a flavin in this context follows from the known structures of flavodoxins (7–12) which show that only the nitrogen-free part of the flavin nucleus reaches out into the solvent. Since the theory of tunnelling as essential e⁻-transfer pathway is getting less and less support these days, it appears already from the x-ray structures that C(8) of the flavocoenzyme might be an efficient outer-sphere contact site in electron transfer.

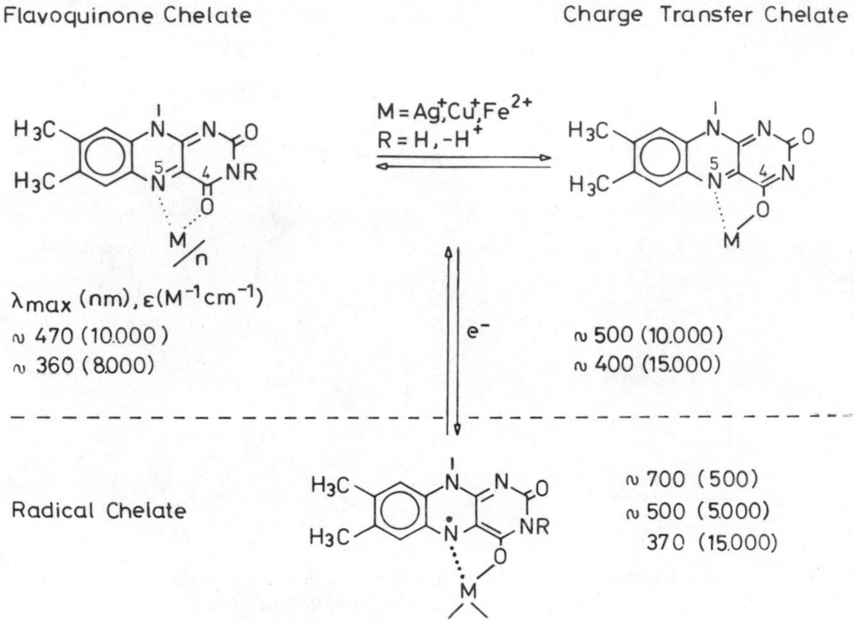

Flavoquinone Chelate   Charge Transfer Chelate

$$M = Ag^+, Cu^+, Fe^{2+}$$
$$R = H, -H^+$$

λ_{max} (nm), ε($M^{-1}cm^{-1}$)

∿ 470 (10.000)   ∿ 500 (10.000)
∿ 360 (8.000) e^- ∿ 400 (15.000)

- -

                                        ∿ 700 (500)
Radical Chelate   ∿ 500 (5.000)
                                        370 (15.000)

Figure 4.   Known classes of flavin–metal complexes (26)

Interflavin and Flavin–Heme Contacts

Independent support for interflavin σ-contacts comes from recent chemical studies by Favaudon and Lhoste (13, 14). The french authors describe, as already anticipated, a nearby diffusion-controlled dimer formation in aprotic polar medium as the first step in the interflavin contact between oxidized and reduced states, which would finally yield two flavin radicals. This dimer was shown to be not identical in any respect with the well known quinhydrone which can only be obtained in aqueous systems at high flavin concentrations. The long wave band in the absorption spectrum of the new dimer appears to be of charge transfer type, but with a highly reduced half width and better resolved shape than the flavoquinhydrone spectrum.

I propose, therefore, the structure of a σ-bonded, but still labile, 8,8′-dimer for this intermediate (Figure 5). When the rotation about the 8,8′-bond is free—as in the chemical system—the preferred conformation of the dimer will be the "hairpin," which permits an additional π-charge transfer between the two flavin halves even at an angle of 60°. The stability maximum will be reached in the monoanion, where the N(1)-protonated flavin half acts as the acceptor and the deprotonated one as the donor, in agreement with the pH-data (13). Hence, the proton at

N(1) is stabilizing the dimer, while blocking the N(5)-lone pair, preferably by metal, will support 8,8'-cleavage. A similar addition at C(8) was detected for the first time, when an intermediate was seen in the flavin-dependent photodecarboxylation of phenylacetate (*17, 18*), which was not identical with the known products of reductive alkylation, viz. 4a-benzyl-4a,5-dihydroflavin, and 5-benzyl-1,5-dihydroflavin. NMR evidence in favor of the structure 8-benzyl-1,8-dihydroflavin was obtained later on (*19, 20*). The 1,8-dihydroflavin chromophore, as present in the dimer (Figure 5) has a very characteristic absorption spectrum (*21*), showing a very intense and well resolved band in the 330-nm range, which might be verified in the 8,8'-dimer. Furthermore, 1,8-dihydroflavin is a N(1)-acid of a $pK \leq 6$, lower than that of 1,5-dihydroflavin ($pK_a \geq 6.5$). The benzyl 8-adduct is, therefore, stabilized at pH > 7 by the negative charge at N(1).

The numerous known flavin dimer species are reviewed in Table II. Apart from the normal very unstable quinhydrone π-complex (*15, 16*), which can only be observed in protic media, and apart from the stable "biflavins" obtained irreversibly by a Michael-type alkali-catalyzed self-condensation of flavoquinone (*22*), there are two further types of σ-dimers known besides the above one, namely the 7α,4a-product of phosphate-catalyzed photodimerization (*23*) and the 5,5'-deazaflavin dimer (*24*), which is discussed below.

In summary, it appears that "interflavin σ-contact" is a quite common phenomenon which may occur both in artificial (mostly irreversible) as

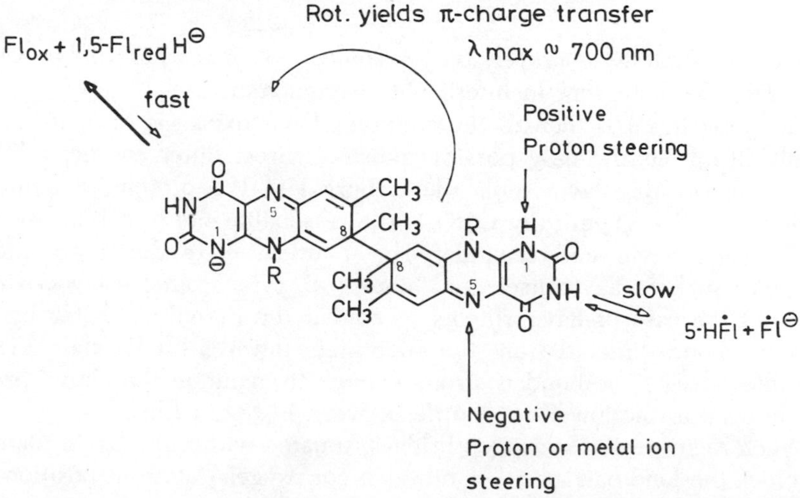

Figure 5. "*Interflavin contact*" *and* σ-*electron transfer (based upon the data of Ref. 13. In the syn-conformation the flavin halves are at an unstrained angle of* $\sim 50°$, *sufficient for efficient charge transfer.*

Table II. Structures of

*Position of
Interflavin
Linkage*   *Formation*[a]

π
$$\left\{ \begin{array}{l} \overset{\bullet}{Fl}_{ox} + \overset{\bullet}{Fl}_{red}H^- \\ HFl + Fl^- \end{array} \right\} \rightleftarrows Fl^-_{ox}\,Fl_{red}H_2$$

$(8\alpha,8'\alpha)\,\sigma$
$$2HFl_{ox} - 2H^+ \overset{\text{anhydr. base}}{\rightleftarrows} (\overset{\bullet}{Fl}^-)_2 \overset{-2e^-}{\rightleftarrows} (Fl_{ox})_2$$

$(8,8')\,\sigma$?
$$\left\{ \begin{array}{l} Fl_{red}H^- + Fl_{ox} \\ HFl + Fl^- \end{array} \right\} \underset{H^+,\,Me^{2+}}{\rightleftarrows} HFl - Fl^-$$

$(5,5')\,\sigma$
$$5\text{-deaza} \left\{ \begin{array}{l} Fl_{ox} + Fl_{red}H_2 \overset{h\nu}{\underset{2H^+}{\rightarrow}} \\ 2Fl_{ox} + 2e^-_{aq.} \rightarrow \end{array} \right\} (FlH)_2$$

$(7\alpha,4a)\,\sigma$
$$2HFl_{ox} \xrightarrow{h\nu\ HPO_4^{2-}\ cat.} HFl_{red} - Fl_{ox}H$$

$$HFe_{ox} \equiv$$

[a] Those cases involving deprotonation of flavoquinone are in italics. The defini-

well as in biocatalytic (reversible) systems. σ-Contact has to be considered as the first event in interflavin electron transfer.

Keeping in mind that all three x-rayed flavodoxins (7, 8, 9, 10, 11) exhibit C(7,8) as the only possible point of direct outer contact while all the rest of the flavin molecule is buried in the protein, we must finally admit that chemically C(8) is the reasonable site of orbital overlap between flavin and secondary heteroaromatic 1e⁻-donor–acceptor molecules such as flavin itself and, above all, cytochrome. Flavodoxins represent the case of flavoproteins scheduled for 1e⁻-only transfer by a hydrogen bond directed from the apoprotein towards N(5), stabilizing the radical HFl. The bond is strong enough to maintain the active protein reduced in the lower 1e⁻-shuttle between Fl_red and Fl.

Such regiospecific hydrogen bridges, situated within the flavin plane, can block the lone pair at either nitrogen (or oxygen) atom in position 5 or 1/2α and thus provide the link between active site structure and the choice of alternative 1e⁻- or 2e⁻-activities. The assignment of blockade site to the activity type requires knowledge of acidobasic properties of

Known Flavin Dimers

References

Decay

heat   *15*
dilution

irrevers.   *22*

H$^+$, weak   *13*
Me^{2+}

O$_2$, hν *24*

H$^+$, strong   *23*

tion is then $HFl \equiv Fl$.

the flavin nucleus (*4*), as discussed below. Hence, blockade does not mean that the access to the corresponding nitrogen atom is obstructed but that the atom is kept in the trigonal state. However, the deazaflavin case confirms by the formation of a stable dimer 5,5′-HdFl (for structure *see* below, Figure 7) that not only C(8), but also N(5) is a site, which is potentially active in 1e$^-$-transfer. In the unmodified flavin the 5,5′-contact must be sterically feasible in the same way, although it cannot lead to a stable (catalytically dead) 5,5′-dimer, but to labile flavin (radical) metal 4α,5-chelates, as discussed in the following chapter.

Flavin–Metal Contact

The σ-character of flavin–metal contacts is more easily accepted a priori, since kinetic stability of metal-N,O-ligation need not be apprehended. Metal chelates (bidentate or polydentate structures) may, however, exhibit enough thermodynamic advantages to preclude dissociation and thereby, electron transfer. Flavin, however, is established being a relatively weak metal chelator, for the following independent reasons (20).

1. $O(4\alpha)$ is always forming part of a carbonyl rather than a hydroxy group. Any $4\alpha,5$-chelation of metal to flavin must, therefore, occur at the expense of $3,4\alpha$-prototropic energy.

2. $N(5)$ is a very weak basic site in Fl_{ox} ($pK < 0$), a moderately strong one in Fl^- ($pK \sim 8.5$) and again very weak in Fl_{red}. In both HFl^-_{red} ($pK \sim 6.5$) and Fl_{ox} ($pK \sim 0.5$), $N(1)$ is the preferred protonation site. This explains why enforced blockade of the $N(5)$-lone pair favors the radical, i.e., $1e^-$-transfer, while enforced $N(1)$-blockade favors Fl_{ox} and Fl_{red} and, therefore, $2e^-$-transfer. If, however, the blockade is achieved by metal instead of protons, the attachment site at Fl_{ox} is switched from $N(1)$ towards $N(5)$ because of the $4\alpha,5$-bidentate mode of metal fixation. Hence, metal complexation as such is sufficient to shift the $2e^-$-transfer system into a $1e^-$-transfer system. For redox-active metal, i.e., $Fe^{II/III}$, the biologically essential shuttle (Figure 2) between fully reduced flavin and radical via the chemically reasonable intermediate complexes (Figure 4) is drawn in Figure 6, as compared with the inessential upper shuttle between radical and flavoquinone. In the complexes of the essential shuttle, metal–ligand spin delocalization is high while in Fl_{ox}-complexes, it is low.

3. $C(6)H$ causes steric hindrance of metal coordination in all flavin redox states to further prevent unfavorably high chelate stability.

Three types of flavin metal chelates have been distinguished according to Figure 4 (25).

1. Fl_{ox} as neutral ligand yields light red "flavoquinone chelates," whose stability is lower than that of the corresponding metal aquocomplexes in water. These chelates can be isolated, and their submolecular structure has been elucidated by NMR (26, 27) in aprotic polar solvents. The delocalization of metal d-electrons towards the ligand is very small, and any spin found in the ligand is caused rather by polarization than by delocalization mechanisms. This suggests strongly that these chelates are biologically irrelevant since they would not support flavin–metal e^--transfer, as reflected in Figure 6.

2. Upon addition of base, these flavoquinone chelates will rather dissociate by direct metal coordination of base (even in the case of hindered base molecules such as tertiary amines) than lose the proton at $N(3)$. This is true for all metal ions except very soft ones, such as Ag^+, Cu^+, and Fe^{2+}. The new metal chelate species thus obtained have been termed "charge-transfer chelates" because of their long wave absorption. They cannot be studied by NMR, owing to very small, but rapidly exchangeable equilibrium admixtures of ligand radicals nor by EPR (25) because of their diamagnetism.

3. Upon adding an electron, the oxidized chelate system changes towards the "radical chelate state." Its evaluation by EPR (28) shows intense electron delocalization from the radical ligand towards the metal and vice versa, independent of the metal nature (to a first approximation). The new d,d-absorption bands encountered in the near-infrared (29), however, are characteristic for the metal center. No such absorp-

tion has been verified before in biological systems, although this transition is not too strongly forbidden. Further search for such proof of flavin–iron contact remains tempting.

Flavohydroquinone, finally, has no affinity whatsoever for non-acceptor metal ions. With Fe^{3+} it yields the ferrous radical chelate by full electron transfer. Favaudon and Lhoste (*14*) have recently observed

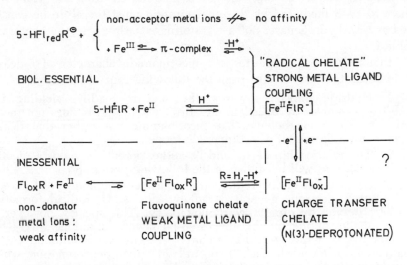

Figure 6. Biologically essential and inessential parts of e^--transfer between flavin and iron (proposal connecting Figures 2 and 4).

an intermediate in this process, which cannot be a chelate but may be a π-complex. With NADH the formation of a similar Fe^{III}-complex has been observed (*30, 31*). This is almost certainly a π-complex, since the reduced nicotinamide does not offer any bidentate or even monodentate profile for σ-fixation of metal ions. The NADH–Fe^{III}-π-complex is more stable, as compared with hypothetical dihydroflavin analogs, since NADH is not a $1e^-$donor, and its Fe^{III}-complex cannot rearrange with release of ferrous iron unless a second metal ion supports the process.

N(5)-Modified Flavocoenzymes

The essential difference between nicotinamide and flavin is in the nature of the acceptor site, i.e., CH for nicotinamide as compared with N for flavoquinone. This explains why 5-deazaflavins appear to be interesting flavin (anti)metabolites since they are expected to be functional analogs of nicotinamide, exhibiting at the same time the steric shape of a flavin. Indeed it turns out that incorporation of 5-deazaflavin into apo-flavoproteins blocks the electron transfer and oxygen activation properties (*32, 33, 34, 35, 36*). Retaining at least partially the (de)hydrogenase function. Hence, flavin-dependent transformases (*see* Table I) and oxidases turn into transhydrogenases with 5-deazaflavin as redox site. Care must be taken not to misinterpret the nature of the acceptor. Phenazonium methosulfate (PMS) (*36*), for example, is an ambiguous species quite like flavin; unlike nicotinamide, PMS can swallow radical electrons as well as $2e^-$-equivalents. A possible electron flow from succinate to PMS mediated by deazaflavoprotein must therefore be considered as transhydrogenase, not a transformase pathway in the context of Table I.

In its chemical properties, the nicotinamide character of 5-deazaflavin becomes most obvious from the following four observations.

1. Hydride transfer towards dFl_{ox} occurs readily, yielding 1,5-H_2dFl_{red}, as shown by Brüstlein and Bruice (*37*). This does not mean, however, that flavin behaves like nicotinamide but rather that deazaflavin is not a flavin but a nicotinamide model.

2. The relationship of $1e^-$ and $2e^-$-redox potentials yields a nicotinamide pattern, as will be shown in the following paragraph.

3. Upon enforced $1e^-$-reduction of deazaflavoquinone (dFl_{ox}), e.g., by hydrated electrons generated radiolytically (*24*), radical formation can be observed by rapid spectrophotometry. This species, however, is not an analog of the biologically essential flavosemiquinone 5-HFl but exhibits the structure 1-HdFl (Figure 7). This is ascertained by its relatively low pK and the lack of spectral shift upon deprotonation, which is analogous to the behavior of 1-HFl and by its decay, which, in contrast to the flavin radical decay, is irreversible and implies a dimerization instead of dismutation.

4. The same dimer can be obtained photochemically (*24*), starting either from dFl_{ox} and oxalate (reaction steps A,B) or from 1,5-dihydro-5-deazaflavin-5-carboxylate (step B alone, started with authentic substrate).

$$\text{A:} \quad dFl_{ox} + (COO)_2^{2-} \xrightarrow[\text{slow}]{h\nu,\, -CO_2} dFl^-_{red}\text{-5-}COO^-$$

$$\text{B:} \quad dFl^-_{red}\text{-5-}COO^- + dFl_{ox} \xrightarrow[\text{fast}]{h\nu,\, -CO_2} \text{1-HdFl-5,5'-dFlH-1'}$$

This dimer is colorless, in agreement with the structure shown in Figure 6. It cannot be reduced further and is oxidized only slowly in the dark.

Hence, the deazaflavin system can be forced to form radicals, but they are kinetically and thermodynamically unstable. Their structure is analogous to the biologically inessential N(1)-blocked red flavin radicals observed only in nonelectron transfer flavoproteins, in agreement with the fact that N(1)-blockade shifts the flavin system ready for $2e^-$-transfer only. The deazaflavoprotein radical, which has been observed by Hersh

(35) after artificial reduction, is definitely not an intermediate of biological relevance quite like all known flavin oxidase red radicals (4).

Consequently, photoreduction of dFl$_{ox}$ leads to adduct formation according to Reaction A even in cases where normal flavins would not yield stable adducts. These adducts frequently undergo secondary photoreactions with starting dFl$_{ox}$, as shown in B, yielding "radical dimers" which might obscure the picture at first sight.

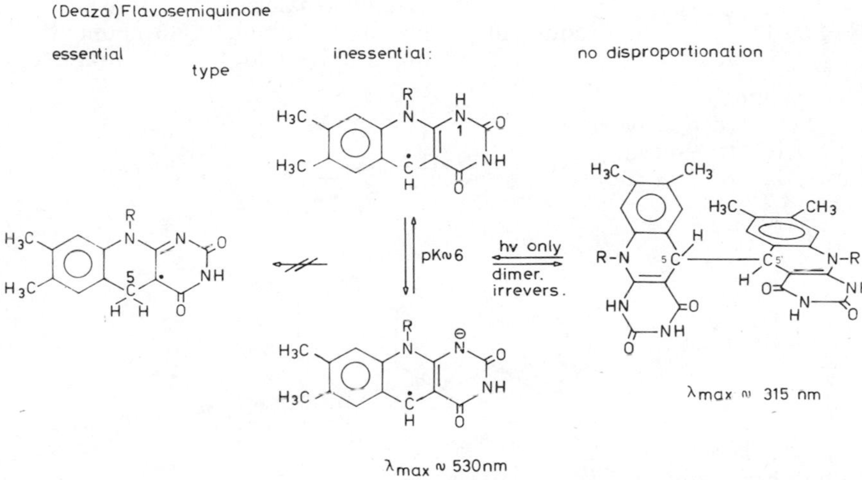

Figure 7. Structure and decay of 5-deazaflavin radicals. The 5-protonated tautomer, which would be analogous to the biologically essential blue flavosemiquinone, is not formed, and disproportionation, which is characteristic for the flavin system, does not happen.

Hence, replacement of flavin-N(5) by CH causes a deadlock of 1e$^-$-transfer. The reverse case of deadlocked 2e$^-$-transfer is to be expected from replacement of N(5) by sulfur. Accordingly, we have synthesized 5-thia-1,5-dihydroflavins (38), which are presently under biochemical evaluation. Thiaflavins exhibit stable radicals in rapid dismutation equilibria. In the neutral oxidized state, however, they rearrange slowly to yield sulfoxides (39), which are no longer flavin analogs. Hence, in analogy with flavin enzymes, thiaflavin-dependent 1e$^-$-transfer favors the lower 1e$^-$-shuttle between the radical and fully reduced states. Alteration of the flavin system by N(5)-modification is summarized in Table III, showing the tight correlation of structure and activity.

The Dynamics of 1e$^-$- vs. 2e$^-$-Oxidoreduction

In Table IV (40–47), the available data on redox potentials of free nicotinamide, flavin, deazaflavin, representative flavoproteins, and reac-

Table III. Correlation of Structure and Redox Activity for Flavin and Its N(5)-modified Derivatives

	Flavins (X = NH)	Deazaflavins (X = CH₂)	Thiaflavins (X = S)
Radical			
chemically	revers. disprop.	irrevers. dimer	stable
protein bound	stable	unstable	stable
Oxidized	quinoid	quinoid (flavin-analog)	non-quinoid (sulfoxide)
Reduced	hydroquinoid	non-hydroquinoid (NADH-analog)	hydroquinoid (flavin-analog)
Redox-shuttles			
2e⁻ (dehydrogenase)	+	+	−
Upper 1e⁻ (e⁻-storage)	+	−	−
Lower 1e⁻ (e⁻-transferase)	+	−	+

ᵃ Only the fully reduced (= dihydro) state is common (= isoelectronic) for all three systems.

Table IV. Bio-redoxpotentials

	Type of Transfer	E_1
		$Fl_{ox} \leftrightarrows Fl'$
Free flavin	1e⁻/2e⁻	−231
Flavodoxins	1e⁻	−190 to +130
NADP-reductase	1e⁻/2e⁻	not measured
D-Amino acid oxidase	2e⁻	artificial
Nicotinamide	2e⁻	NAD⁺ ⇌ NAD•
		−740
5-Deazaflavin	2e⁻	−650
5-Thiaflavin	1e⁻	irreversible
Dioxygen		$O_2 \rightleftarrows O_2^-$
		−330
Dithionite	1e⁻/2e⁻	$SO_2^- \rightleftarrows SO_2$
		−440

ᵃ For prevailing 1e⁻-transfer $E_1 > E_2 > E_1'$; for prevailing 2e⁻-transfer $E_1 < E_2$ prototropy; diffusion-controlled in the chemical system, conformation-controlled in

tants have been summarized. An upper $1e^-$-shuttle potential lower than the corresponding lower shuttle potential points towards radical instability and, thus, $2e^-$-only catalysts. Dithionite as widely used artificial donor is also ambiguous, quite like flavin while molecular oxygen as acceptor strongly prefers $2e^-$-transfer and, therefore, will yield only superoxide when coupled with a $1e^-$-only catalyst (5). Free dihydroflavin fulfills this requirement only under alkaline conditions (48) while oxidase dihydroflavin does not fulfill it at all.

The data of Table IV are in large part taken from a recent study by Blankenhorn (44). This author, furthermore, succeeds in showing that the mode of oxidoreduction, viz. $1e^-$- or $2e^-$-transfer, is not only reflected by thermodynamics but can be verified kinetically as shown in Figure 8.

When a series of dihydronicotinamides with variable substituents R' is oxidized either by flavin or by the $1e^-$-only nitroxide radical, the Nernst equation can be verified using the forward rate as well as the equilibrium constants, since the back reaction rate is independent of the nature of R'. In the case of the flavin acceptor, the Nernst number n turns out to be 2, as required for $2e^-$-transfer. Hence, the transformase cofactor flavin will react with nicotinamide in the $2e^-$-mode, showing that the latter is—for the biological acceptor—a $2e^-$-only agent and requires a very potent $1e^-$-only oxidant such as nitroxide for artificial $1e^-$-behavior.

One might, of course, argue that this statement is only valid for the chemical system while everything might be quite different in enzymes. On the other hand, what an apoprotein does to a cofactor is not magic, but chemistry. Apparently, nature overcomes the $2e^-$-only-limitation of its unique hydride activator, nicotinamide, by a mandatory transformer,

E_0' (pH 7) (mV)

E_2	E_1'	References
$Fl_{ox} \leftrightarrows Fl_{red}$	$Fl' \leftrightarrows Fl_{red}$	
-199	-167	40
-280 to -150	-420 to -377	41
-360	-400	42
-004	artificial	43
$NAD^+ \rightleftarrows NADH$	$NAD^{\bullet} \rightleftarrows NADH$	44
-320	$+100$	
-320	$+10$	44
irreversible	$+430$	45
$O_2 \rightleftarrows H_2O_2$	$O_2 \rightleftarrows 2H_2O$	46
$+290$	$+820$	
	$(SO_2)_2^{2-} \rightleftarrows 2\,SO_2$	46, 47
	-537	

$< E_1' \cdot 1e^-/2e^-$-mediating flavin systems show $E_1 \sim E_2 \sim E_1'$, because of rapid 1,5-the "transferase" protein.

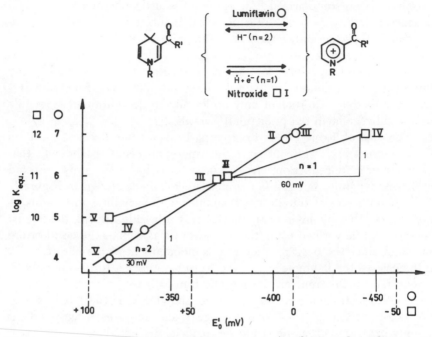

Figure 8. Nernst plots of equilibrium constants (left) and oxidation rate (2e⁻) and nitroxide (1e⁻), respectively, as counterparts (data from Ref. 44. CONH₂;(III), R = CH₃, R' = COOC₂H₅; (IV), R = CH₃,

flavin, which can uniquely be switched to and fro between 1e⁻-only and 2e⁻-only by selective blockage of electron pairs either at N(5) or at N(1) [or O(2α), respectively, which is indistinguishable from N(1)].

The question remains open whether or not there is a direct contact between flavin and metal in a 1e⁻-accepting binuclear iron cluster as, for example, in plant ferredoxin. Outer sphere 1e⁻-transfer through flavin position 8 remains the alternative. On the other hand, there can no longer be a question about direct and reversible electron transfer between dihydronicotinamide and metal proteins. Any such claims (49) must be taken with greatest reservation and require careful independent confirmation.

Meanwhile we can show (50) that the deazaflavin radical generated by photodissociation of the dimer as shown in Figure 7 can be utilized for selective reduction of flavodoxin radical. With the aid of this 1e⁻-transfer method Scherings, Haaker, and Veeger (51) were able to show that flavodoxin is the actual electron donor for the Azotobacter nitrogenase system and that N₂ reduction can be maintained in the light by an artificial chain system of the reactant sequence EDTA–deazaflavin–flavodoxin–nitrogenase. Hence, Azotobacter flavodoxin is actually a "carrier

$$E = \frac{RT}{n\,F}\left(\log k_1 - \log k_{-1}\right)$$

$$k_{-1} = \text{const. II } -\text{V}$$

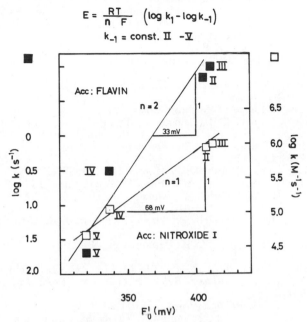

constants (right) of nicotinamides against redox potentials with lumiflavin Compound structures: (I), spirocyclohexylporphyrexide; (II), $R = CH_3$, $R' = R' = COCH_3$; (V), $R = ADP$-Rib, $R' = CONH_2$ (NADH).

II" of the "transformase" link (cf. Figure 2) between dinitrogen and reduced nicotinamide. Furthermore, another "1e⁻-only"-flavoprotein, the well-known "ETF," has been recognized now by Ruzicka and Beinert (52) as "carrier II" constituent of a new analogous "transformase" complex located between acyl-CoA and CoQ in mitochondrial respiration.

In summary, flavin-dependent biological 1e⁻-transfer still awaits full recognition of its ubiquitous importance in plant and animal, as well as microbial, metabolism.

Literature Cited

1. Gray, H. B., Clendening, P. J., Coyle, C. L., Hare, J. W., Holwerda, R. A., McArdle, J. V., McMillin, D. R., Rawlings, J., Rosenberg, R. C., Sailsutà, N., Solomon, E. I., Stephens, P. J., Wherland, S., ADV. CHEM. SER. (1977) **162**, 145.
2. Sutin, ADV. CHEM. SER. (1977) **162**, 156.
3. Hall, D. O., ADV. CHEM. SER. (1977) **162**, 227.
4. Hemmerich, P., *Prog. Chem. Org. Nat. Prod.* (1976) **33**, 451.
5. Massey, V., Hemmerich, P., in "The Enzymes," 3rd ed., P. D. Boyer, Ed., Vol. XII B, p. 191, Academic, New York, 1975.
6. Hemmerich, P., Schuman-Jorns, M., in "Enzymes: Structure and Function," J. Drenth, R. A. Oosterbaan, C. Veeger, Eds., p. 95, North-Holland, Amsterdam, 1972.

7. Eaton, W., Ludwig, M., *Biochemistry* (1975) **14**, 10.
8. Ludwig, M. L., Burnett, R. M., Darling, G. D., Jordan, S. R., Kendall, D. S., Smith, W. W., in "Flavins and Flavoproteins," T. P. Singer, Ed., p. 393, Elsevier, Amsterdam, 1976.
9. Andersen, R. D., Apgar, P. A., Burnett, R. M., Darling, G. D., Lequesne, M. E., Mayhew, S. G., Ludwig, M. L., *Proc. Natl. Acad. Sci. USA* (1972) **69**, 3189.
10. Watenpaugh, K. D., Sieker, L. C., Jensen, L. H., *Proc. Natl. Acad Sci. USA* (1973) **70**, 3857.
11. Lauterwein, J., Lhoste, J.-M., unpublished data.
12. Favaudon, V., Le Gall, J., Lhoste, J.-M., in "Flavins and Flavoproteins," T. P. Singer, Ed., p. 434, Elsevier, Amsterdam, 1976.
13. Favaudon, V., Lhoste, J.-M., *Biochemistry* (1975) **14**, 4731.
14. *Ibid.* (1975) **14**, 4739.
15. Kuhn, R., Ströbele, R., *Ber. Dtsch. Chem. Ges.* (1937) **70**, 753.
16. Gibson, Q. H., Massey, V., Atherton, N. M., *Biochem. J.* (1962) **85**, 369.
17. Walker, W. H., Hemmerich, P., Massey, V., *Helv. Chim. Acta* (1967) **50**, 2269.
18. Walker, W. H., Hemmerich, P., Massey, V., *Eur. J. Biochem.* (1970) **13**, 258.
19. Brüstlein, M., Knappe, W. R., Hemmerich, P., *Angew. Chem. Int. Ed. Eng.* (1971) **10**, 804.
20. Hemmerich, P., Bhaduri, A. P., Blankenhorn, G., Brüstlein, Haas, W., Knappe, W.-R., *Oxidases Relat. Redox Syst. Proc. Int. Symp. 2nd* (1973) **1**, 1.
21. Hemmerich, P., *Vitam. and Horm.* (1970) **28**, 467.
22. Hemmerich, P., Prijs, B., Erlenmeyer, H., *Helv. Chim. Acta* (1959) **42**, 2164.
23. Schöllnhammer, G., Hemmerich, P., unpublished data.
24. Duchstein, H. J., Fenner, H., Grauert, R., Blankenhorn, G., Hemmerich, P., Knappe, W.-R., Massey, V., Goldberg, M., Pecht, I., unpublished data.
25. Hemmerich, P., Lauterwein, J., in "Inorganic Biochemistry," G. L. Eichhorn, Ed., Vol. 2, p. 1168, Elsevier, Amsterdam, 1973.
26. Lauterwein, J., Hemmerich, P., Lhoste, J.-M., *Inorg. Chem.* (1975) **14**, 2152.
27. *Ibid.* (1975) **14**, 2161.
28. Müller, F., Eriksson, L. E. G., Ehrenberg, A., *Eur. J. Biochem.* (1970) **12**, 93.
29. Müller, F., Hemmerich, P., Ehrenberg, A., *Eur. J. Biochem.* (1968) **5**, 158.
30. Gutman, M., Eisenbach, M., *Biochemistry* (1973) **12**, 2314.
31. Gutman, M., Margalit, R., Scheiter, A., *Biochemistry* (1968) **7**, 2785.
32. Edmondson, D., Barman, B., Tollin, G., *Biochemistry* (1972) **11**, 1133.
33. Spencer, R., Fisher, J., Walsh, C., *Biochemistry* (1976) **15**, 1043.
34. Fisher, J., Spencer, R., Walsh, C., *Biochemistry* (1976) **15**, 1054.
34. Jorns, M., Hersh, L., *J. Biol. Chem.* (1975) **250**, 3620.
35. Hersh, L., Jorns, M., Peterson, J., Currie, M., *J. Am. Chem. Soc.* (1976) **98**, 865.
36. Grossman, S., Goldenberg, J., Kearney, E. B., Oestreicher, G., Singer, T. P., in "Flavins and Flavoproteins," T. P. Singer, Ed., p. 302, Elsevier, Amsterdam, 1976.
37. Brüstlein, M., Bruice, T., *J. Am. Chem. Soc.* (1972) **94**, 6548.
38. Janda, M., Hemmerich, P., *Angew. Chem. Int. Ed. Eng.* (1976) **15**, 443.
39. Rössler, H., Grauert, R., Fenner, H., Hemmerich, P., unpublished data.
40. Draper, R. D., Ingraham, L. L., *Arch. Biochem. Biophys.* (1968) **125**, 802.
41. Barman, B. G., Tollin, G., *Biochemistry* (1972) **11**, 4755.
42. Keirns, J. J., Wang, J. H., *J. Biol. Chem.* (1972) **247**, 7374.

43. Brunori, M., Rotilio, G. C., Antonini, E., Curti, B., Branzoli, U., Massey, V.,
    J. Biol. Chem. (1971) **246**, 3140.
44. Blankenhorn, G., *Eur. J. Biochem.* (1976) **67**, 67.
45. Massey, V., private communication.
46. Loach, P., in "Handbook of Biochemistry," 2nd ed., H. A. Sober, R. A.
    Harte, and E. K. Sober, Eds., p. J33, Chemical Rubber, Cleveland,
    1970.
47. Mayhew, S. G., Ludwig, M. L., in "The Enzymes," 3rd ed., P. D. Boyer,
    Ed., Vol. XII B, p. 118, Academic, New York, 1975.
48. Massey, V., Palmer, G., Ballou, D., *Oxidases Relat. Redox Syst. Proc. Int.
    Symp. 2nd* (1973) 25.
49. Kumar, S. A., Appaji Rao, N., Felton, S. P., Huennekens, F. M., Mackler,
    B., *Arch. Biochem. Biophys.* (1968) **125**, 436.
50. Massey, V., Hemmerich, P., unpublished data.
51. Scherings, G., Haaker, H., Vceger, C., unpublished data.
52. Ruzicka, F. J., Beinert, H., *Biochem. Biophys. Res. Commun.* (1975) **2**,
    622–631.

RECEIVED July 26, 1976.

19

Electrochemical and Spectroscopic Studies of Manganese(II, III, IV) Complexes as Models for the Photosynthetic Oxygen-Evolution Reaction

DONALD T. SAWYER, MARIO E. BODINI, LAURIE A. WILLIS, THOMAS L. RIECHEL, and KEITH D. MAGERS

Department of Chemistry, University of California, Riverside, CA 92502

The manganese complexes formed by gluconate ion, glucarate ion, sorbitol, mannitol, tartrate, glycerate, diethanolamine, triethanolamine, catechol, 4,5-dihydroxynaphthalene-2,7-disulfonate, salicylate, and 2,3-dihydroxy benzoate ion in basic media undergo oxidation–reduction which may parallel the behavior of the manganese group in photosystem-II of green plant photosynthesis. The redox chemistry of the complexes has been studied by polarography and controlled potential electrolysis. UV-visible spectrophotometry, ESR, and magnetic susceptibility measurements have been used to characterize the solution chemistry, formulas, and structures of the complexes. The kinetics for the reaction of the manganese gluconate complexes with molecular oxygen and with hydrogen peroxide have been determined. Mechanisms are postulated that are consistent with the electrochemical, spectroscopic, and kinetic data.

Manganese is an essential component of photosystem-II in green plant photosynthesis (*1, 2*), where it functions as an oxidation–reduction catalyst for the oxidation of water to molecular oxygen. In spite of the importance and long-term interest in biological oxygen production, there is no agreement as to the electron-transfer mechanism or to the manganese oxidation states and number of manganese atoms in the active group. One recent discussion proposed a two-quantum process in which the manganese is oxidized first from the 2+ to the 3+ state and then from

3+ to the 4+ state (*3*). Studies of the photoactivation process (*4*) and of the functional sites of manganese within photosystem-II (*5*) support this mechanism. Studies by Cheniae (*1, 6, 7*) thoroughly review the role of manganese and its photoactivation in photosystem-II. Clearly, manganese is the essential oxidant and cannot be replaced by any other metal.

In 1970 Kok and co-workers (*8*) put forth a four-step mechanism for the photosystem-II oxygen-evolution reaction. Experimental support for such a mechanism has been obtained from flash photolysis experiments (*9*). This in turn has led to the proposal of several alternative mechanistic models that fit the experimental data (*10*). The most recent investigations of photosystem-II indicate that about six manganese atoms are present per water oxidizing unit and that these are distributed into two distinct pools. The smaller, tightly bound pool has about one third of the total manganese and is not thought to be involved in electron transfer. Extraction of the more loosely bound manganese in the larger pool causes loss of chloroplast oxygen evolution (*5, 7, 11*). Based on this evidence, Cheniae (*7*) suggests that the four atoms in the larger manganese pool are necessary for oxygen evolution. However, other work (*12*) indicates that removing approximately one-half of the large-pool manganese causes a negligible decrease in the rate of oxygen evolution. This implies that only two manganese atoms are actually necessary to catalyze the oxidation of water. Whether two or four manganese atoms are necessary for oxygen evolution, higher valent forms of manganese appear to be necessary for the redox process. Water proton relaxation studies indicate that the manganese in dark-adapted chloroplasts exists in several oxidation states (*13*) and that the manganese changes its average oxidation state during the photosystem-II charge accumulation process (*14*).

In a recent discussion Earley (*15*) proposed a model system based on a three-manganese group with each metal undergoing the reaction sequence given in Reactions 1–4. These mechanistic steps are justified

$$MO + Q + h\nu \rightarrow MO^+ + Q^- \tag{1}$$

$$MO^+ + OH^- \rightarrow MOOH \tag{2}$$

$$2MOOH \rightarrow MO + MOO + H_2O \tag{3}$$

$$2MOO \rightarrow 2MO + O_2 \tag{4}$$

on the basis of previous flash photolysis studies of photosystem-II. Another manganese model compound:

$$[(Bipyr)_2Mn(III) \overset{O}{\underset{O}{<\!\!\!>}} Mn(IV(Bipyr)_2]^{3+}$$

has been put forth as a model for photosystem-II and its oxygen-evolution reaction (16). However, subsequent work failed to confirm that the compound yields molecular oxygen from water oxidation (17).

Most of the coordination chemistry for the higher oxidation states of manganese involves the 3+ oxidation state; a concise summary is provided by Cotton and Wilkinson (18). They include a discussion of the unusual structure for the manganese (III)–acetate complex:

$$[Mn(III)_3O(OAc)_6]^+OAc^- \cdot HOAc$$

In addition there has been a recent review of manganese porphyrin chemistry (19) as well as of the aqueous and chloro complexes of the porphyrin species (20). These discussions are complemented by a discussion of the electrochemistry of manganese porphyrin compounds (21). Related studies include the phthalocyanine complexes (22), the manganese(III)–hemoglobin structure (23), and the characterization of macrocyclic complexes of manganese (24, 25). Other relevant chemistry involves the Schiff-base complexes of Mn(III) (26, 27, 28).

The aqueous solution chemistry of Mn(III) has been reviewed thoroughly by Davis (29). A more recent review summarizes the coordination chemistry of the higher oxidation states of manganese and provides some insight to the electronic states and structure of such complexes (30). Most of our understanding of the electronic states of Mn(III) complexes depends on the review in 1966 by Dingle (31). This has been augmented by the useful set of spectroscopic data contained in the Ph.D. dissertation of Summers (32). The correlation between magnetism and valency of manganese compounds has been summarized in an early study by Goldenberg in 1940 (33).

Although the reaction chemistry of the higher oxidation states of manganese is complicated, its relevance to electrochemistry and potentially to biological systems has prompted numerous investigations. One of the more intriguing is the auto-oxidation of manganese(II) to a manganese(III) oxo-hydroxide precipitate (34). On the basis of product analysis and kinetic data, the rate-controlling step is concluded to be

$$Mn(II)(OH)_2 + O_2 \rightarrow Mn(III)(O)OH + HO_2 \qquad (5)$$

Reaction 5. Another aspect of the reaction chemistry of Mn(III) is demonstrated by the classic kinetic studies of the decomposition chemistry of manganese(III) oxalate complexes (35, 36). These studies derive from an earlier discussion of the synthesis of manganese(III)–oxalate compounds (37). A similar set of kinetic and synthetic investigations, have been summarized for the manganese(III)–malonate complexes (38, 39). Both sets of studies confirm that Mn(III) has a particularly strong affinity for oxygen-containing ligands and that such manganese–oxygen bonds apparently promote electron transfer to the manganese atoms.

Another interesting Mn(III) complex involves triethanolamine as the ligand system. The most recent discussion summarizes the polarography of the complex (40). It follows from a much earlier study which summarized an analytical procedure based on polarography after the air oxidation of manganous ion in the presence of triethanolamine (41). Some evidence that such a procedure might produce other than a simple complex is given in a spectrophotometric study of this system which indicates that peroxide complexes are initially formed upon air oxidation of manganese(II)–triethanolamine (42). Another study (40) indicates that a transiently stable Mn(IV) complex is formed by adding $K_3Fe(CN)_6$ to the Mn(II) complex.

An important complex of the Mn(III) ion is that formed by pyrophosphate in weakly acidic media. On the basis of early polarographic studies (43) the formula of the complex is concluded to be $Mn(III)(H_2P_2O_7)_3^{3-}$. Because the pyrophosphate molecule has structural similarities to the phosphate groups of ADP and ATP, such complexes are especially interesting in terms of structure and electron transfer mechanisms.

Another class of Mn(III) complexes involves the polyaminocarboxylic acid ligands. The earliest study appeared in 1962 (44) and was followed shortly by studies of the cyclohexane analog of EDTA as well as other derivatives of EDTA (45). A recent paper discusses the reactivity of the manganese(III)–diaminocyclohexanetetraacetate complex with hydrogen peroxide (46). A mechanism is proposed which involves complexation by the peroxide anion followed by subsequent electron transfer to produce the Mn(II) complex and the $HO_2\cdot$ radical. The results are interesting and indicate the potential for selective catalysis by the higher oxidation state manganese complexes.

The recent suggestion that the binuclear mixed valence manganese-(III, IV)–di-μ-oxo bis-dipyridyl complex photooxidizes water to oxygen (16) has revitalized interest in its catalytic properties. The complex was first synthesized in 1960 by Nyholm (47). A recent x-ray structure determination (48) indicates that in the solid state the two manganese atoms possess different oxidation states, bond lengths, and bond angles. Assuming the differences persist in solution, this is an especially interesting complex in terms of electrochemical studies of its electron transfer properties.

In contrast to the extensive array of aqueous studies of Mn(III) complexes, studies in aprotic media are much more limited. The most substantial structural and electronic characterizations have been for the β-diketones (49, 50, 51). The second class of complexes involves picolinic acid and 8-quinolinol as ligands (52). The latter study was limited to magnetic measurements and electronic spectral studies. The third class of conjugated ligand systems involves the dithiocarbamate (DTC) complexes of manganese (53).

Because of the current interest in the oxidation–reduction chemistry of manganese in biological systems and the limited number of complexes of Mn(III) and Mn(IV) that are stable in aqueous solution (18), a systematic search for complexing agents to stabilize the higher oxidation states of this element has been made. Previous experience has established that sodium gluconate, the salt of the carboxylic acid derivative of D-glucose, is especially effective for stabilizing strongly acidic ions in alkaline media (e.g., Fe(III), Bi(III), Sb(III), Ce(IV), and Os(VI) (54)) and resists oxidation more than most other sugar acids.

The goal of the present research has been to prepare and to characterize manganese complexes whose oxidation–reduction chemistry mimics that of photosystem-II. This should provide insight to the oxidation states and environment of the manganese group in the grana of the chloroplasts. A specific goal of the research has been to discover ligand and solvent systems to stabilize and to solubilize the 3+ and 4+ oxidation states of manganese.

The present paper summarizes the oxidation–reduction chemistry of the gluconate complexes of Mn(II), Mn(III), and Mn(IV) in alkaline media. Electrochemical, spectrophotometric, and magnetic susceptibility measurements have been used to establish the formulas and chemical characteristics of the complexes. The oxidation–reduction chemistry for the manganese complexes formed by other ligands with polyhydroxyl functions also has been determined.

The reactions of the several manganese gluconate complexes with molecular oxygen and hydrogen peroxide have been studied in terms of stoichiometry and reaction kinetics. Reaction mechanisms are proposed on the basis of the kinetic data. In addition, the thermodynamic and mechanistic characteristics of an ideal model system for photosystem-II are analyzed and evaluated.

Experimental

Polarographic measurements were made with either a three-electrode potentiostat, based on the use of solid state operational amplifiers (55), or a Sargent model XV polarograph. Controlled potential electrolyses were accomplished by use of a Wenking model 61-RH potentiostat.

The spectrophotometric measurements were made with either a Cary model 14 or a Perkin Elmer model 450 spectrophotometer. Quartz cells were used, and the reference cell contained equimolar concentrations of sodium hydroxide and sodium gluconate to that in the sample cell.

ESR spectra were obtained with a Varian model V-4500 spectrometer by using a standard flat-faced quartz cell. The magnetic susceptibilities of the manganese gluconate complexes were determined by the NMR method developed by Evans (56) and modified by Rettig (57). The inner tube was made from 3-mm o.d. Pyrex tubing, sealed with a nearly saturated solution of TMS* (sodium-2,2-dimethyl-2-silapentane-5-sulfonate) inside. The outer tube was a standard NMR tube and was filled with sample in a nitrogen-atmosphere glove box. The shift between the two TMS* peaks was measured with a Varian A-60D spectrometer at ambient temperature. The probe temperature was determined with an ethylene glycol standard. The magnetic susceptibilities were calculated by the method of Rettig (57), with values for diamagnetic corrections from Figgis and Lewis (58).

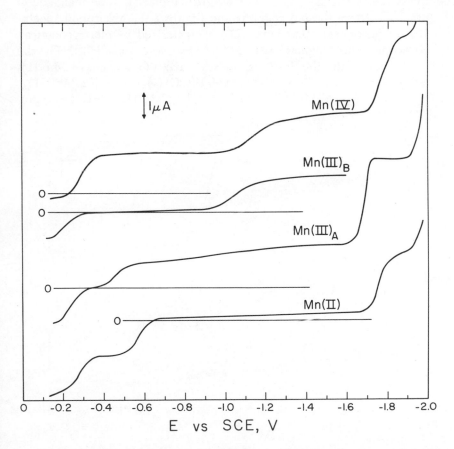

Figure 1. Polarograms for 1mF manganese gluconate complexes in 0.3F NaOH.

Roman numerals represent the oxidation states of the manganese ions, and A and B represent the bis and tris complexes of Mn(III), respectively. Sodium gluconate concentrations: Mn(II) and Mn(IV), 0.1F; Mn(III)$_A$, 0.002F; Mn(III)$_B$, 1F.

Reagents. The source of manganese ion for the gluconate studies was manganese(II) gluconate from Chas. Pfizer and Co.; the stock solutions were standardized by titration with EDTA (59). The source of gluconate ion was D-glucono-δ-lactone (Chas. Pfizer and Co.), which was recrystallized from ethylene glycol monomethyl ether (60). Manganese(III) acetate was synthesized by the procedure of Christensen (61); in some experiments it was used as the source of Mn(III). All of the other chemicals were reagent grade.

Results

Manganese–Gluconate Complexes. In alkaline media gluconate ion (GH_4^-; $C_6H_{10}O_7^-$, carbohydrate derivative of D-glucose) forms stable complexes with the 2+, 3+, and 4+ oxidation states of manganese. Figure 1 illustrates the polarograms and Figure 2 the absorption spectra for the four complexes that have been characterized by electrochemical, spectroscopic, and magnetic susceptibility measurements (62). On the basis of these data the four complexes have the formulas: Mn(II), $[Mn_2^{II}(GH_3)_4(H_2O)_2]^{4-}$; Mn(III)$_A$, $[Mn_2^{III}(GH_3)_4(OH)_2]^{4-}$; Mn(III)$_B$, $[Mn^{III}(GH_3)_3]^{3-}$; and Mn(IV), $[Mn_2^{IV}(GH_3)_4O_2(OH)_2]^{6-}$; GH_3^{2-} repre-

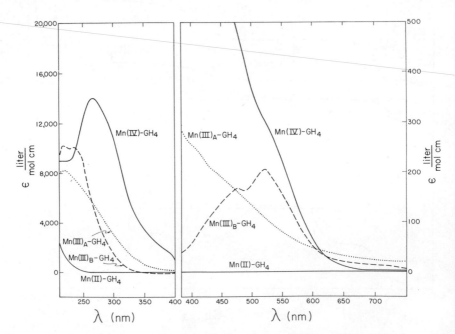

Figure 2.   Absorption spectra for 5mF manganese gluconate complexes in 0.3F NaOH.

Roman numerals represent the oxidation states of the manganese ions, and A and B represent the bis and tris complexes of Mn(III), respectively. Sodium gluconate concentrations: Mn(II) and Mn(IV), 0.1F; Mn(III)$_A$, 0.010F; Mn(III)$_B$, 1F.

sents the dianion of gluconic acid that results from removing the carboxylate proton and one of the secondary alcoholic protons.

The oxidation–reduction chemistry of the manganese–gluconate complexes can be expressed by Reactions 6–8.

$$[Mn_2^{II}(GH_3)_4(H_2O)_2]^{4-} + 2H_2O + 4e^-$$

$$\xrightarrow[\text{vs. SCE}]{-1.73V} 2Mn + 4GH_4^- + 4OH^- \tag{6}$$

$$[Mn_2^{II}(GH_3)_4(H_2O)_2]^{4-} + 2OH^-$$

$$\xrightarrow{-0.54\ V} [Mn_2^{III}(GH_3)_4(OH)_2]^{4-} + 2H_2O + 2e^- \tag{7}$$

$$[Mn_2^{III}(GH_3)_4(OH)_2]^{4-} + 4OH^-$$

$$\xrightarrow{-0.28\ V} [Mn_2^{IV}(GH_3)_4O_2(OH)_2]^{6-} + 2H_2O + 2e^- \tag{8}$$

At higher gluconate concentrations the $Mn(III)_A$ complex of Reaction 8 rearranges to a tris complex $[Mn(III)_B]$ (Reaction 9) (*see* Figures

$$[Mn_2^{III}(GH_3)_4(OH)_2]^{4-} + 2GH_4^- \overset{\text{slow}}{\rightleftharpoons} 2[Mn(GH_3)_3]^{3-} + 2H_2O \tag{9}$$

1 and 2) with an apparent equilibrium constant of 0.13. The latter species is reduced at much more negative potentials by the irreversible process in Reaction 10.

$$2[Mn^{III}(GH_3)_3]^{3-} + 4H_2O + 2e^-$$

$$\xrightarrow{-1.07\ V} [Mn_2^{II}(GH_3)_4(H_2O)_2]^{4-} + 2GH_4^- + 2OH^- \tag{10}$$

Figure 3 illustrates the ESR spectra for manganese(II) gluconate as a function of added base. In the absence of base the spectrum is analogous to that for $MnSO_4$ and represents the solvated $Mn(II)$ ion. Addition of base promotes the formation of the gluconate complex and the loss of hyperfine splittings in the ESR spectra. When two or more moles of base per $Mn(II)$ ion have been added, the spectrum becomes a single broad resonance.

The magnetic moments for the $Mn^{II}(GH_4)_2$ salt and for the $Mn(II)$, $Mn(III)$, and $Mn(IV)$ gluconate complexes are summarized in Table I. The latter species are formed conveniently by the stoichiometric addition of $K_3Fe(CN)_6$ to the $Mn(II)$ complex. Solutions (f) and (g) of Table I indicate that oxygenation of a solution of the manganese(II) gluconate complex initially forms the Mn(III) complex, and then, at a much slower rate, yields the Mn(IV) complex. When considered in total, the data provide convincing evidence that stable complexes of the 2+, 3+, and 4+ oxidation states of manganese are formed in alkaline gluconate solutions.

The difference between the magnetic moments for the Mn(II) salt and the Mn(II) complex (solutions (a) and (b) of Table I) indicates some degree of electron pairing for the complex. This in turn implies that the complex probably is binuclear.

The data in Table I for the oxidation of the Mn(II) complex by hydrogen peroxide (solutions (h) and (i)) indicate a one-to-one reaction stoichiometry. This is confirmed by electrochemical and spectrophotometric measurements and is rationalized in a later section.

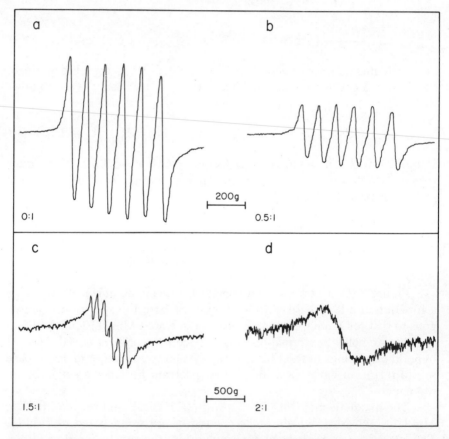

Figure 3.   ESR spectra for solutions of 8mF $Mn^{II}(GH_4)_2$ and 0.12F $LiGH_4$, with increasing mole ratios of base. (a) no base; (b) 4mF LiOH; (c) 12mF LiOH; (d) 16mF LiOH.

Table I. Magnetic Moments of Manganese Gluconate
Complexes from the NMR Method of Evans (56)

Species	NMR Shift (Hz)	μ (B.M.)		
		Uncorrected	Corrected	Theor. for (n) Electrons
(a) Mn(II) salt; 0.04F Mn(GH$_4$)$_2$	64	5.81	5.85	5.92 (5)
(b) Mn(II) complex; 0.04F Mn(GH$_4$)$_2$ 0.1F NaGH$_4$ 0.3F NaOH	54.5	5.24	5.38	5.92 (5)
(c) Mn(III) complex; solution (b) plus 0.04F K$_3$Fe(CN)$_6$	45	4.79	4.98	4.90 (4)
(d) Mn(III) complex; 0.04F Mn(OAc)$_4$ 0.1F NaGH$_4$ 0.3F NaOH	35	4.51	4.64	4.90 (4)
(e) Mn(IV) complex; solution (b) plus 0.08F K$_3$Fe(CN)$_6$	28	3.78	4.06	3.87 (3)
(f) Solution (b) after 1 min O$_2$ (brown)	44	4.71	4.87	4.90 (4)
(g) Solution (b) after 75 min O$_2$ (cherry red)	30	3.89	4.08	3.87 (3)
(h) Solution (b) plus 0.02F H$_2$O$_2$	49	4.96	5.11	4.90 (4)
(i) Solution (b) plus 0.04F H$_2$O$_2$	44	4.70	4.87	4.90 (4)

Manganese Complexes with Polyhydroxyl Ligands. Combinations of Mn(II) (from Mn^{II}(ClO$_4$)$_2$), Mn(III) (from Mn^{II}(ClO$_4$)$_2$ plus K$_3$Fe(CN)$_6$), and Mn(IV) (from Mn^{II}(ClO$_4$)$_2$ plus K$_3$Fe(CN)$_6$) with several polyhydroxyl ligands in alkaline media have been investigated by polarography. The results, summarized in Table II, indicate that aliphatic polyhydroxyl ligands stabilize the 4+ and 3+ oxidation states of manganese. In the case of the gluconate, glucarate, and tartrate ligands, the presence of a carboxylate group results in a soluble Mn(II) complex. Sorbitol and mannitol appear to stabilize and to solubilize Mn(III) and Mn(IV) as effectively as the ligands that contain carboxylate groups but do not form soluble Mn(II) complexes. The glucose configuration of the hydroxyl groups of sorbitol appears to form somewhat stronger complexes with Mn(III) and Mn(IV) than does the mannose configuration of the mannitol hydroxyls. The manganese complexes that are formed by glucarate ion, tartrate ion, sorbitol, and mannitol appear to have oxidation–

Table II. Oxidation–Reduction Chemistry of Manganese(II) Complexes

Half-Wave Potentials (V vs. SCE)

Ligand	$Mn(III) \rightleftarrows Mn(IV)$	$Mn(II) \rightarrow Mn(III)$
Gluconate	−0.28	−0.54
Glucarate	−0.28	−0.51
Tartrate	−0.22 [a]	−0.45
meso-Tartrate	−0.22 [a]	−0.46
Glycerate [b]	∼ 0.0	−0.49
Sorbitol	−0.32	—
Mannitol	−0.30	—
Triethanolamine	∼ 0.0	
Diethanolamine	∼ 0.0	
Catechol	—	
4,5-Dihydroxynaphthalene-2,7-disulfonate [c]	—	
Salicylate [d]		
2,3-Dihydroxybenzoate	—	

[a] Irreversible.
[b] 0.3F ligand.

reduction chemistry that is analogous to that for the manganese gluconate complexes, which is represented by Reactions 6–10.

The electrochemical data indicate that both triethanolamine and diethanolamine form stable bis complexes with Mn(II) and Mn(III) and transiently stable bis complexes with Mn(IV). The latter have sufficiently positive reduction potentials to oxidize mercury and to attack the ligand. These complexes appear to be mononuclear and not to be involved in equilibria as represented by Reaction 9.

The aromatic dihydroxylated ligands (represented by catechol, 2,3-dihydroxybenzoate, and 4,5-dihydroxynapthalene 2,7-disulfonate) stabilize the 3+ and 2+ oxidation states of manganese in alkaline media. The data for salicylate indicate that it forms a less stable Mn(III) complex than the dihydroxy ligands. Formation of the 4+ complex is precluded because the ligands are more easily oxidized than are the Mn(III) complexes.

Figure 4 illustrates the intriguing oxidation–reduction chemistry that is associated with the manganese-4,5-dihydroxynaphthalene 2,7-disulfonate system. The bottom curve represents the free ligand and the next curve the reduction of oxidized ligand. Curve C represents the reduction of the Mn(III) complex. When a two-fold excess of oxidizing agent $(K_3Fe(CN)_6)$ is added to a solution of the Mn(III) complex, curve D of Figure 4 results. Superficially the comparison of curves C and D leads to the conclusion that additional complex is formed. However, that is impossible because all of the available manganese is in the form of the

in Solutions that Contain 0.1F Ligand and 0.3F NaOH

Half-Wave Potentials (V vs. SCE)

$Mn(II) \rightleftarrows Mn(III)$	$Mn(III) \rightarrow Mn(II)$	$Mn(II) \rightarrow Mn(0)$
	−1.07	−1.73
	−1.19	−1.72
	−0.95	−1.70
	−0.91	−1.73
	−0.87	−1.68
	−1.09	−1.68
	−0.99	−1.67
−0.48		−1.72
−0.48		−1.72
−0.54		—
−0.54		−1.67
−0.35		−1.71
−0.47		−1.74

[c] 0.5F NaOH.
[d] Precipitate present in solution.

Mn(III) complex for the solution represented by curve C. Apparently, the excess $K_3Fe(CN)_6$ oxidizes the free ligand in the solution, and this in turn is catalytically reduced by the reduced form of the Mn(III) complex. Because of the presence of quinone-like materials in photosystem II (2), this represents a model that warrants additional study.

Reactions of Oxygen and Hydrogen Peroxide with Manganese Gluconate Complexes. The manganese(II) gluconate complex $[Mn_2^{II}-(GH_3)_4(H_2O)_2]^{4-}$, is susceptible to rapid oxidation by air. Figure 5 illustrates the extent of the reaction between the complex and molecular oxygen at 1 atm, as well as the concentrations of the product species. The data for this figure have been obtained by quenching the reaction with argon and then recording a polarogram to determine the concentrations of reactants and products. The reactivity of the Mn(II) complex with oxygen and the apparent reversibility of the reaction have been discussed in an earlier communication (63).

Figure 6 indicates the extent of the reaction of manganese(II) gluconate with molecular oxygen as a function of solution pH. Below pH 6 oxygen does not oxidize Mn(II), and from pH 6 to pH 12 the stable reaction products are the manganese(III) gluconate complexes. The latter reaction does not go to completion below pH 11.5. Above pH 12 the manganese(IV) gluconate complex begins to be formed, and at pH 14 and above the oxidation of the Mn(II) complex by the molecular oxygen to the Mn(IV) complex is stoichiometric.

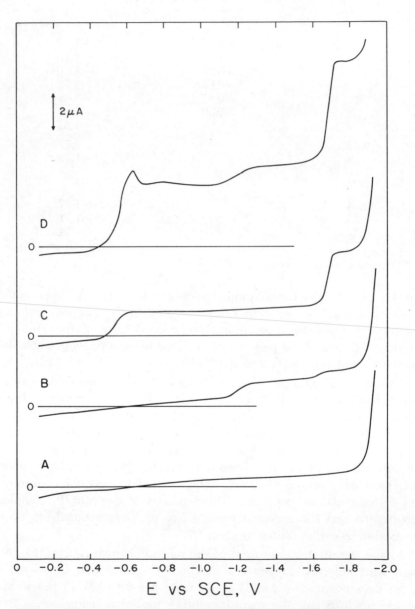

Figure 4. *Polarograms for 1mF manganese(II) perchlorate in a solu-*
tion that contains 0.1F 4,5-dihydroxynaphthalene 2,7-disulfonate and
0.5F NaOH.

(A) *ligand alone;* (B) *ligand plus 1mF* $K_3Fe(CN)_6$ *(represents reduction of*
oxidized ligand); (C *and* D) *addition of* $K_3Fe(CN)_6$ *to make original solution*
1mF and 3mF in this oxidant, respectively.

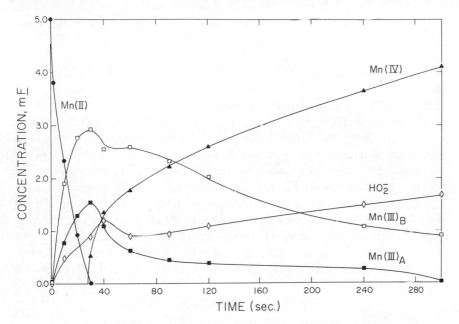

Figure 5. Concentrations of product species as a function of time for the reaction of 5mF Mn(II) gluconate in 0.1F NaGH₄ and 0.3F NaOH with O_2 at 1 atm. A and B represent the bis and tris gluconate complexes of Mn(III), respectively.

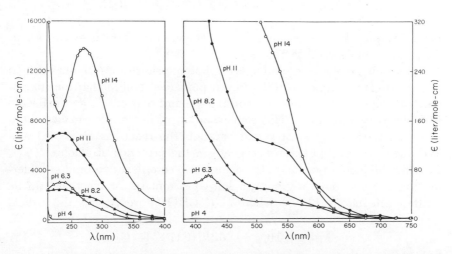

Figure 6. Spectra for solutions of 5mF Mn(II) gluconate in 0.1F $NaGH_4$ and O_2 at 1 atm as a function of solution pH

Reference to Figure 5 indicates that both forms of the manganese-(III) gluconate complexes, $[Mn_2^{III}(GH_3)_4(OH)_2]^{4-}$ and $[Mn^{III}(GH_3)_3]^{3-}$, are the initial products as well as HO_2^- ion. After all of the $Mn(II)$ complex is oxidized, the $Mn(III)$ complexes are oxidized by oxygen to the $Mn(IV)$ complex, $[Mn_2^{IV}(GH_3)_4O_2(OH)_2]^{6-}$, with the formation of additional HO_2^- ion. Polarographic and spectrophotometric studies of the combination of HO_2^- with the $Mn(II)$ complex have established that HO_2^- reacts with the $Mn(II)$ complex to yield the $Mn(III)$ complexes and oxidized ligand. The stoichiometries, kinetics, and mechanisms of the reactions of O_2 and HO_2^- with the manganese(II) and manganese(III) gluconate complexes are the subject of a separate detailed study (64).

On the basis of these prior electrochemical (62, 63) and kinetic (64) studies and the present results, the reaction stoichiometries for the manganese gluconate complexes with molecular oxygen and hydrogen peroxide can be expressed by a series of reactions. The primary reaction between the $Mn(II)$ complex and molecular oxygen is rapid (Reaction

$$Mn_2^{II}(GH_3)_4(H_2O)_2^{4-} + O_2 + OH^- \rightarrow$$
$$Mn_2^{III}(GH_3)_4(OH)_2^{4-} + HO_2^- + H_2O \tag{11}$$

11) with a second-order rate constant, k_1, of 2.8×10^4 l-mol^{-1} sec^{-1} at 25°. It is followed by a much slower oxidation of the complex by peroxide

$$Mn_2^{II}(GH_3)_4(H_2O)_2^{4-} + 2HO_2^- + 2GH_4^- \rightarrow$$
$$Mn_2^{III}(GH_3)_4(OH)_2^{4-} + (GH_4\overset{|}{C}HOH)_2^{2-} + 2OH^- + 2H_2 \tag{12}$$

ion (Reaction 12). The latter has a second-order rate constant, k_2, of 2.6×10^{-1} l-mol^{-1} sec^{-1} at 25°.

Both electrochemical and spectrophotometric studies of the reaction of the $Mn(II)$ complex with hydrogen peroxide confirm that the stoichiometry is one peroxide ion per $Mn(II)$ ion and that one ligand molecule is oxidized concurrently. This somewhat surprising result can be rationalized in terms of chemistry that parallels the reactions of $Fe(II)$ and hydrogen peroxide (Fenton's reagent) in the presence of alcohols (65). Leaving the ligand portion of the manganese complexes out and considering them as monomeric groups, a self-consistent set of reactions can be written (Reactions 13–15) where $[GH_4\overset{.}{C}HOH]^-$ represents the radical

$$Mn^{II} + HO_2^- \rightarrow Mn^{III}(OH^-) + \cdot O^- \tag{13}$$

$$\cdot O^- + GH_4^- \rightarrow [GH_4\overset{.}{C}HOH]^- + OH^- \tag{14}$$

$$2[GH_4CHOH]^- \rightarrow GH_4\text{---}\underset{\overset{|}{OH}}{CH}\text{---}\underset{\overset{|}{OH}}{CH}GH_4{}^{2-} \qquad (15)$$

that results from hydrogen atom abstraction from the terminal —CH$_2$OH group of the gluconate molecule by the hydroxyl anion radical. Although two parallel reactions to Reaction 15 occur with Fenton's reagent, (namely, oxidation of the radical to an aldehyde by either the oxidized metal ion or by a second hydroxyl radical) there is no evidence for such reactions in the case of the alkaline manganese(II) gluconate–hydrogen peroxide system.

The manganese(III) gluconate complexes are oxidized slowly to the Mn(IV) complex by molecular oxygen, which is reduced to peroxide ion (Reaction 16). A separate set of experiments has established that

$$Mn_2{}^{III}(GH_3)_4(OH)_2{}^{4-} + O_2 + 3OH^- \rightleftharpoons$$
$$Mn_2{}^{IV}(GH_3)_4O_2(OH)_2{}^{6-} + HO_2{}^- + H_2O \qquad (16)$$

this is a reversible process and that the reverse reaction is first-order in Mn(IV) complex and first-order in HO$_2^-$ ion and rapid. Although the data of Figure 7 initially were thought to indicate that the manganese-

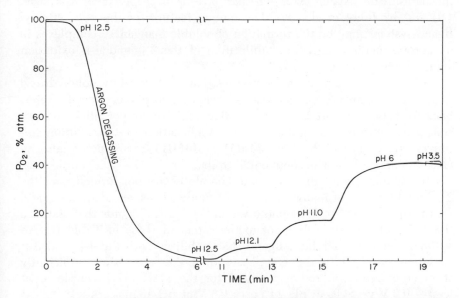

Figure 7. Partial pressure of O$_2$ for a solution that originally contained 5mF Mn(II), 0.1F NaGH$_4$, and 0.3F NaOH and was oxygenated for 20 min with O$_2$ at 1 atm before degassing with argon. pH lowered by addition of concentrated HClO$_4$ to a sealed cell. Partial pressures of O$_2$ measured with a Beckman membrane electrode system (model 1008).

(IV) gluconate complex oxidizes water to molecular oxygen when acidified, the results actually confirm the reversibility of Reaction 16.

Discussion and Conclusions

To understand the chemistry of the manganese that is associated with photosystem-II, the relevant coordination chemistry of the higher oxidation states of manganese must be characterized under media conditions that are consistent with the biological matrix. The polyhydroxyl ligand systems of the present study may be consistent with the kinds of groups that are likely to be bound to the manganese in the grana of the chloroplasts (2). Likewise, studies under conditions of low proton activity probably are consistent with the biological matrix of photosystem-II.

The manganese gluconate complexes, and probably several of the other polyhydroxyl complexes of Table II, represent a system that undergoes oxidation–reduction chemistry that parallels much that is observed for the manganese group in photosystem-II (2, 3, 4, 5). The apparent formation of a tetranuclear manganese complex at approximately pH 12 (62) is an interesting proposition, because it could provide the four oxidation equivalents that are required for the oxidation of water to an oxygen molecule. Such an oxidation center fits well in the schematic mechanisms for oxygen evolution that have been proposed by Kok et al. (8) and by Joliot et al. (9). The unique features of the polyhydroxyl ligand systems may be the formation of soluble manganese complexes in an aprotic medium and the stabilization of the 3+ and 4+ oxidation states of manganese.

Although the manganese complexes that are formed by polyhydroxyl ligands have many properties that are relevant to photosystem-II, reference to Tables II and III indicates that their redox potentials are not sufficiently positive to oxidize water. Acidification of the solution improves the situation, but the Mn(IV)–Mn(III) redox potentials, by calculation, would not become equal to that for the O_2–H_2O couple until pH 5. At this acidity the gluconate complexes are not formed, and the 3+ and 4+ states of manganese are not stable. Thus, an adequate model for the manganese group of photosystem-II requires ligands that stabilize the 4+ oxidation state, but not to the extent of those in Table II. An optimal system would be a ligand that stabilizes both the 4+ and 3+ oxidation states, with the complexation of the latter state sufficiently stronger to realize a redox potential for the (IV)–(III) couple of at least +0.2 V vs. SCE at pH 14 (or +0.5 V at pH 10).

In addition to thermodynamic considerations, the mechanistic challenge of the formation of the (O–O) group from two water molecules is a vital factor in the design of a realistic model. Systems that include di-μ-oxo bridging between two Mn(III) or —(IV) ions, such as the manganese(III,IV) bipyridyl complex (47, 48), the Schiff-base complexes (26), and possibly the manganese(IV) gluconate complex, may provide a sterically and kinetically convenient path for water oxidation. In the case of oxo-bridged binuclear Mn(IV) complexes, extensive electronic delocalization may be the basis for a low-energy, reversible pathway to the pairing of oxygen atoms. The essential role of manganese in the photosynthetic oxygen evolution reaction probably results from the unique oxidation–reduction chemistry of its 3+and 4+ states and from the potential for electronic delocalization in binuclear Mn(IV) complexes.

Table III. Oxidation–Reduction Potentials for Manganese(IV) Gluconate and for H_2O–O_2 Half-Reactions (66, 67)

Reaction	E' (V vs. SCE)
pH 14	
$Mn_2^{IV}(GH_3)_4O_2(OH_2)^{6-} + 2H_2O + 2e^- \rightleftharpoons Mn_2^{III}(GH_3)_4(OH)_2^{4-} + 4OH^-$	−0.40
$O_2 + 2H_2O + 4e^- \rightleftharpoons 4OH^-$	+0.15
$O_2 + H_2O + 2e^- \rightleftharpoons HO_2^- + OH^-$	−0.33
$HO_2^- + H_2O + 2e^- \rightleftharpoons 3OH^-$	+0.63
$O_2 + e^- \rightleftharpoons O_2^-$	−0.81
$O_2^- + H_2O + e^- \rightleftharpoons HO_2^- + OH^-$	+0.15
$Fe(CN)_6^{3-} + e^- \rightleftharpoons Fe(CN)_6^{4-}$	+0.11
pH 10	
$Mn_2^{IV}(GH_3)_4O_2(OH)_2^{6-} + 2H_2O + 2e^- \rightleftharpoons Mn_2^{III}(GH_3)_4(OH)_2^{4-} + 4OH^-$	+0.08
$O_2 + 2H_2O + 4e^- \rightleftharpoons 4OH^-$	+0.39
$O_2 + 2H_2O + 2e^- \rightleftharpoons H_2O_2 + 2OH^-$	−0.16
$H_2O_2 + 2H_2O + 2e^- \rightleftharpoons 4OH^-$	+0.93

An alternative approach to the achievement of an effective model system is the use of nonaqueous solvents. Such media may be representative of the biological matrix of photosystem-II because of extensive hydrogen bonding by the lipid and protein fractions in the grana. Also, if the manganese group of photosystem-II is membrane-bound, an aprotic medium probably is more analogous to its environment than an aqueous system. Not only will the redox potentials of the (IV)–(III) and (III)–(II) couples of complexed manganese be significantly shifted by solvent media, but more important, the redox chemistry for the water–oxygen couples will be substantially altered. These considerations are the basis of a current investigation to find a more realistic model for photosystem-II.

Literature Cited

1. Cheniae, G. M., *Ann. Rev. Plant Physiol.* (1970) **21**, 467.
2. Heath, R. L., *Int. Rev. Cytol.* (1973) **34**, 49–101.

3. Radmer, R., Cheniae, G. M., *Biochim. Biophys. Acta* (1971) **253**, 182.
4. Cheniae, G. M., Martin, I. F., *Biochim. Biophys. Acta* (1971) **253**, 167.
5. *Ibid.* (1970) **197**, 219.
6. Cheniae, G. M., Martin, I. F., *Plant Physiol.* (1969) **44**, 351.
7. Radmer, R., Cheniae, G. M., in "Primary Processes," in press.
8. Kok, B., Forbush, B., McGloin, M., *Photochem. Photobiol.* (1970) **11**, 457.
9. Joliot, P., Joliot, A., Bouges, B., Barbieri, G., *Photochem. Photobiol.* (1971) **14**, 287.
10. Mar, T., Govindjee, *J. Theor. Biol.* (1972) **36**, 427.
11. Blankenship, R. E., Ph.D. Dissertation, University of California, Berkeley, 1974.
12. Blankenship, R. E., Babcock, G. T., Sauer, K., *Biochim. Biophys. Acta* (1975) **387**, 165.
13. Wydrzynski, T., Zumbulyadis, N., Schmidt, P. G., Govindjee, *Biochim. Biophys. Acta* (1975) **408**, 349.
14. Wydrzynski, T., Zumbulyadis, N., Schmidt, P. G., Gutowsky, H. S., Govindjee, *Proc. Natl. Acad. Sci. U.S.A.* (1976) **73**, 1196.
15. Earley, J. E., *Inorg. Nucl. Chem. Lett.* (1973) **9**, 487.
16. Calvin, M., *Science* (1974) **184**, 375.
17. Cooper, S. R., Calvin, M., *Science* (1974) **185**, 376.
18. Cotton, F. A., Wilkinson, G., "Advanced Inorganic Chemistry," 3rd ed., p. 849–853, Interscience, New York, 1972.
19. Boucher, L. J., *Coord. Chem. Rev.* (1972) **7**, 289.
20. Boucher, L. J., *J. Am. Chem. Soc.* (1970) **92**, 2725.
21. Boucher, L. J., Garber, H. K., *Inorg. Chem.* (1970) **9**, 2644.
22. Lever, A. B. P., *Adv. Inorg. Chem. Radiochem.* (1965) 27.
23. Moffat, K., Loe, R. S., Hoffman, B. M., *J. Am. Chem. Soc.* (1974) **96**, 5259.
24. Bryan, P. S., Dabrowiak, J. C., *Inorg. Chem.* (1975) **14**, 296.
25. *Ibid.* (1975) **14**, 299.
26. Boucher, L. J., Herrington, D. R., *Inorg. Chem.* (1975) **13**, 1105.
27. Prabhakaran, C. P., Patel, C. C., *J. Inorg. Nucl. Chem.* (1969) **31**, 3316.
28. Yarino, T., Matsushita, T., Masuda, I., Shenia, K., *J. Chem. Soc. Chem. Commun.* (1970) 1317.
29. Davies, G., *Coord. Chem. Rev.* (1969) **4**, 199.
30. Levason, W., McAuliffe, C. A., *Coord. Chem. Rev.* (1972) **7**, 353.
31. Dingle, R., *Acta Chem. Scand.* (1966) **20**, 33.
32. Summers, J. C., Ph.D. Dissertation, The University of Florida, 1968.
33. Goldenberg, N., *Trans. Faraday Soc.* (1940) **37**, 847.
34. Kessick, M. A., Morgan, J. J., *Environ. Sci. Technol.* (1975) **9**, 157.
35. Taube, H., *J. Am. Chem. Soc.* (1947) **69**, 1418.
36. *Ibid.* (1948) **70**, 1216.
37. Cartledge, G. H., Ericks, W. P., *J. Am. Chem. Soc.* (1936) **58**, 2061.
38. Bullock, J. I., Patel, M. M., Salmon, J. E., *J. Inorg. Nucl. Chem.* (1969) **31**, 415.
39. Cartledge, G. H., Nichols, P. M., *J. Am. Chem. Soc.* (1940) **62**, 3057.
40. Kitagawa, T., *J. Chem. Soc. Japan; Pure Chem. Sec. J.* (1960) **81**, 572.
41. Issa, I. M., Issa, R. M., Hewaidy, I. F., Omar, E. E., *Anal. Chim. Acta* (1957) **17**, 434.
42. Nightingale, E. R., Jr., *Anal. Chem.* (1959) **31**, 146.
43. Kolthoff, I. M., Watters, J. I., *Ind. Eng. Chem. Anal. Ed.* (1943) **15**, 8.
44. Toshino, Y., Ouchi, A., Tsunoda, T., Kokima, M., *Can. J. Chem.* (1962) **40**, 775.
45. Hamm, R. E., Suwyn, M. A., *Inorg. Chem.* (1967) **6**, 139.
46. Jones, T. E., Hamm, R. E., *Inorg. Chem.* (1974) **13**, 1940.
47. Nyholm, R. S., Turco, A., *Chem. Ind.* (1960) 74.
48. Plaksin, P. M., Stoufer, R. C., Mathew, M., Palenik, G. J., *J. Am. Chem.*

Soc. (1972) **94**, 2121.

49. Avdeef, A., Costamagna, J. A., Fackler, J. P., Jr., *Inorg. Chem.* (1974) **13**, 1854.
50. Cartledge, G. H., *J. Am. Chem. Soc.* (1952) **74**, 6015.
51. Fackler, J. P., Jr., Avdeef, A., *Inorg. Chem.* (1974) **13**, 1864.
52. Ray, M. M., Adhya, J. N., Biswas, D., Poddar, S. N., *Aust. J. Chem.* (1966) **19**, 1737.
53. Cambi, L., Cagnasso, A., *Atti. Accad. Naz. Lincei. Cl. Sci. Fis. Mat. Nat. Rend.* (1931) **13**, 404.
54. Sawyer, D. T., *Chem. Rev.* (1964) **64**, 633.
55. Goolsby, A. D., Sawyer, D. T., *Anal. Chem.* (1967) **39**, 411.
56. Evans, D. F., *J. Chem. Soc.* (1959) 2003.
57. Rettig, M. F., University of California, Riverside, private communication.
58. Figgis, B. N., Lewis, J., "The Magneto Chemistry of Complex Compounds," *Mod. Coord. Chem.* (1960).
59. Welcher, R. J., "The Analytical Uses of Ethylenediamine Tetraacetic Acid," p. 217, D. Van Nostrand Co., New York, 1958.
60. Sawyer, D. T., Bagger, J. B., *J. Am. Chem. Soc.* (1959) **81**, 5302.
61. Brauer, C., "Handbook of Preparative Inorganic Chemistry," Vol. 2, 2nd ed., p. 1469, Academic, New York, 1965.
62. Bodini, M. E., Willis, L. A., Riechel, T. L., Sawyer, D. T., *Inorg. Chem.* (1976) **15**, 1538.
63. Sawyer, D. T., Bodini, M. E., *J. Am. Chem. Soc.* (1975) **97**, 6588.
64. Bodini, M. E., Riechel, T. L., Willis, L. A., Sawyer, D. T., *J. Am. Chem. Soc.* (1976) **98**, 8366.
65. Walling, C., *Acc. Chem. Res.* (1975) **8**, 125.
66. Latimer, W. M., "Oxidation Potentials," 2nd ed., p. 39, Prentice-Hall, New York, 1952.
67. Parsons, R., "Handbook of Electrochemical Constants," Butterworths, London, 1959.

RECEIVED July 26, 1976. This work was supported by the U. S. Public Health Service–NIH under Grant No. GM22761. We are grateful for a Predoctoral Fellowship to M.E.B. from the Catholic University of Chile.

Molybdenum-Containing Biomolecules

Molybdoenzymes: The Role of Electrons, Protons, and Dihydrogen

EDWARD I. STIEFEL, WILLIAM E. NEWTON, GERALD D. WATT, K. LAMONT HADFIELD, and WILLIAM A. BULEN[1]

Charles F. Kettering Research Laboratory, Yellow Springs, Ohio 45387

The biochemistry of molybdenum enzymes and the coordination chemistry of molybdenum are each discussed as background for understanding the role of molybdenum in enzymes. Electron transfer pathways and spectroscopic data implicate the molybdenum site in substrate reactions. For xanthine oxidase there is evidence for involvement of proton transfer in substrate oxidation. For nitrogenase, data on dihydrogen inhibition of nitrogen fixation and HD formation (under dideuterium and dinitrogen) can be interpreted in terms of a bound diimide-level intermediate. Several mechanistic schemes are possible for ATP utilization, dihydrogen evolution, and substrate reduction by nitrogenase. For other molybdoenzymes, oxo transfer and coupled proton-electron transfer processes are alternative mechanistic possibilities. Molybdenum may be uniquely suited for its biochemical role.

Molybdenum is the only element of the second transition row known to have a natural biological function. It is also considerably less abundant in the earth's crust than the first transition-row elements which play key biological roles—iron, cobalt, copper, and manganese. The relative scarcity of molybdenum, coupled with the great importance of the biological processes for which it is essential, has led to consideration of the potential insight which this may give concerning the origin of life on earth. In particular, Crick and Orgel (*1*) have suggested that the use of molybdenum by terrestrial microorganisms may (weakly) support a

[1] Deceased

directed panspermia hypothesis for the origin of life on earth. This hypothesis claims that life did not spontaneously originate on earth but rather was sent to earth from somewhere else in the universe. Crick and Orgel reasoned that if life originated on earth, it is unlikely that an element as rare as molybdenum would have been chosen for such an important task as nitrogen fixation. On the other hand, if life originated elsewhere, where molybdenum was abundant, then the use of this element would not be at all unusual.

These interesting arguments can be faulted on two grounds. First, while molybdenum is indeed relatively rare in the earth's crust or in the earth as a whole, this is not the case in seawater (2, 3, 4). In fact, according to recent estimates, the concentration of molybdenum is either comparable with (2, 3) or perhaps slightly exceeds (4) that of manganese, iron, and copper. While this situation may result from the presence of life and/or from an oxidizing atmosphere of more recent origin (5), at the very least it opens the possibility that molybdenum may have been reasonably abundant in the ancient waters where life supposedly arose.

A second argument against the extraterrestrial origin of life would be valid if molybdenum were the only available metal which, when incorporated into a protein system, could catalyze certain reactions. If this were the case, then even if molybdenum were relatively rare, it would be worth the effort for the microorganisms to extract it from the environment. The organisms which learned how to use molybdenum (for example, to fix nitrogen) would then have an evolutionary advantage over organisms which did not. The selective survival of the molybdenum-using (nitrogen-fixing) species would ensure the continued use of molybdenum by future generations.

The question then arises as to what chemical features of molybdenum make it uniquely suitable for the biological reactions in which it participates. In this chapter we first discuss some of the information gained from biological studies of molybdenum enzymes paying particular attention to nitrogenase and to xanthine oxidase. For nitrogenase, we focus on the relationship between dihydrogen, dinitrogen, and the enzyme where there is evidence for sequential two-electron–two-proton processes in the production of ammonia from dinitrogen. For xanthine oxidase, we summarize the data which implicate proton transfer (in addition to electron transfer) as a feature of the molybdenum site. Some of our recent results and some trends in molybdenum coordination chemistry help to determine reasonable possibilities for the action of molybdenum in these systems. Finally, some of the mechanistic proposals are considered and evaluated in terms of the most recent results from biological and inorganic systems.

Molybdoenzymes: Occurrence and Biological Importance

Nitrogenase. Nitrogenase is the enzyme which catalyzes the reduction of dinitrogen to ammonia. The process of biological nitrogen fixation is responsible for most of the fixed nitrogen input into the biosphere, outweighing the enormous amount of dinitrogen that is fixed (converted to ammonia) by the Haber process (6). The contrast between the Haber process and the biological process is startling. The former requires high temperature, high pressure, and dihydrogen as a feedstock for both energy and reducing power (7). The latter process works (often in air in vitro) under ambient conditions (i.e., room temperature and at 0.8 atm of dinitrogen) and uses (ultimately) solar energy as input either directly, or indirectly through the carbohydrates which are produced by photosynthesis. Thus, the overall biological process makes use of readily available air, water, and sunlight to effect the desired transformation.

The ability to fix nitrogen is exclusively a property of prokaryotic organisms—bacteria and blue-green algae (8, 9). The nitrogen-fixing bacteria include both free-living species and those which live symbiotically with higher plants. Among the free living species there are strict aerobes (such as *Azotobacter vinelandii*), facultative anaerobes (such as *Klebsiella pneumonae*), and strict anaerobes (such as *Clostridium pasteurianum*). Additionally, some photosynthetic bacteria (e.g., *Rhodospirullum rubrum* and *Chromatium*) are known to fix nitrogen. The most prominent of the symbiotic fixing species are members of the genus *Rhizobium* which fix nitrogen when living as bacteroids in the root nodules of leguminous plants (soybeans, peas, alfalfa). Recently, some species of rhizobia have been induced to fix nitrogen in a free-living state (10–15), showing conclusively that the genes for nitrogen fixation are in the microorganisms and not in the plant. The rather stringent conditions required to observe the nitrogen fixation by free-living rhizobia in culture suggests that additional organisms may be found to fix nitrogen under more precisely controlled conditions. Among the blue-green algae, nitrogen fixation occurs in both unicellular (such as *Gleocapsa*) and filamentous (such as *Anabaena*) species (9). The blue-green algae are both photosynthetic and nitrogen-fixing and are thus remarkably self-sufficient organisms.

Nitrogenase has been successfully isolated from several bacterial species (8) and seems to be remarkably similar in all cases. To date, no pure preparations of nitrogenase have been obtained from blue-green algae (9).

Nitrate Reductase. Nitrate reductase is found widely distributed among plants and microorganisms and catalyzes the reduction of NO_3^- to NO_2^- (16, 17, 18, 19). The physiological role of this enzyme depends on the organism. Often the enzyme nitrite reductase, which catalyzes the reduction of NO_2^- to NH_3, is found in addition to the nitrate reductase. In this case, nitrate reductase plays an assimilatory role, being responsible for the first step in the conversion of NO_3^- to NH_3. In other organisms, the nitrate serves as the terminal electron acceptor in an electron transport system, similar to O_2 or SO_4^{2-}, and the nitrate reductase plays a respiratory or dissimilatory role.

Unlike the nitrogenases, which do not vary much in composition and physical properties with the source, nitrate reductases vary considerably from one organism to the next. The only feature which they all seem to have in common is an absolute requirement for the presence of molybdenum.

Xanthine Dehydrogenase. This enzyme catalyzes the oxidation of xanthine to uric acid and is found in various microorganisms (including bacteria and fungi) and animals (including insects, fish, birds, and mammals) (19, 20, 21). In some bacteria and fungi, xanthine can serve as the sole nitrogen source. Coupling this fact with the presence of molybdenum in nitrogenase and nitrate reductase, we find that in each case a molybdenum enzyme plays a role in nitrogen assimilation and is, in fact, responsible for the first step in this process. Molybdenum appears to play a role in the metabolism of nitrogen similar to that played by first transition-row elements (iron, manganese, and copper) in the metabolism of oxygen.

Xanthine Oxidase. This enzyme is very closely related to the xanthine dehydrogenase systems (19, 20, 21), and in some cases, the oxidase and dehydrogenase are interconvertible forms of the same enzyme (22). Xanthine oxidase is found in a variety of mammalian systems including man. In most organisms, the oxidation of xanthine to uric acid is followed by further degradation of the uric acid. However, in man and some other primates, uric acid is the terminal species in purine catabolism and is excreted through the kidneys. Excess accumulation of uric acid leads to the syndrome called gout. Nowadays, gout is often treated with inhibitors of xanthine oxidase, and the nature of these inhibitors and their reaction with xanthine oxidase has given insight into the functioning of the enzyme (23, 24).

Aldehyde Oxidase. This enzyme is usually found in similar locations to xanthine oxidase or dehydrogenase and has been isolated from insects, birds, and mammals (20, 21). Aldehyde oxidase seems to be a poor choice of name for this enzyme because, while it oxidizes aldehydes to carboxylic acids, it also accepts a variety of purines and pyrimidines as oxidizable substrates. For example, aldehyde oxidase catalyzes the conversion of 2-hydroxypyrimidine to uracil and of adenine to 8-hydroxyadenine (25). It appears that xanthine oxidase and aldehyde oxidase are a set of purine and pyrimidine hydroxylases with a rather broad range of substrate specificity.

Sulfite Oxidase. This enzyme, isolated from bovine (26, 27) and chicken liver (28), catalyzes the oxidation of sulfite to sulfate. This is possibly a crucial function in animals as SO_3^{2-} (or SO_2, its gaseous precursor) is toxic while SO_4^{2-} is relatively innocuous. For example, one of the first signs of molybdenum deficiency in rats is a greatly increased susceptibility to SO_2 poisoning (28). In addition, a human child born without sulfite oxidase activity did not survive for very long (29).

Other Enzymes. Molybdenum has been suggested as a component of an enzyme possessing CO_2 reductase or formate dehydrogenase activity (30, 31, 32). In the latter case, the unique observation has been made that tungsten can substitute for molybdenum while maintaining activity (33). This enzyme is also postulated to contain selenium (32). A NADPH dehydrogenase from a mitochondrial fraction may contain molybdenum based upon the observation of a Mo(V) EPR signal (34). Authentication of these findings may lengthen the list of molybdenum enzymes.

The Molybdenum Cofactor. Based on genetic evidence, Cove and Pateman (35) suggested that xanthine dehydrogenase and nitrate reduc-

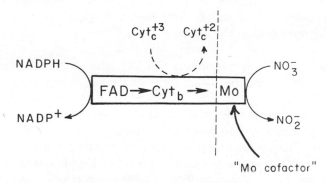

Figure 1. Nitrate reductase from Neurospora crassa *—composition and presumed electron transfer sequence* (16, 18)

tase of the fungus *Aspergillus nidulans* have a common molybdenum-containing unit. The work of Nason and co-workers with the *nit-1* mutant of *Neurospora crassa* also points to a "molybdenum cofactor" common to all molybdenum-containing enzymes (36, 37, 38). The *Neurospora* nitrate reductase, as shown schematically in Figure 1, catalyzes the reduction of NO_3^- by NADPH and contains FAD and a b-type cytochrome in addition to molybdenum. The full enzyme also has NADPH:cytochrome c reductase activity. The *nit-1* mutant produces an enzyme that cannot reduce NO_3^- but still possesses the NADPH:cytochrome c reductase activity. Significantly, the full enzyme activity towards nitrate (as well as other properties of the enzyme) can be restored by treatment of *nit-1* extracts with the neutralized, acid-hydrolysis product of any of the abovementioned molybdenum enzymes. These enzymes donate a molybdenum-containing group (38) which leads to the in vitro assembly of the intact and active nitrate reductase. Simple molybdenum complexes are unable to activate the *nit-1* extracts nor are neutralized acid-hydrolysis products of non-molybdenum enzymes. The donated molybdenum-containing factor can arise from enzymes isolated from mammals or bacteria. In studies using bacterial extracts, the factor was dialyzable (39) and is thus presumed to be a low molecular weight molybdenum compound.

Russian workers claim to have isolated low molecular weight molybdenum-containing peptides from nitrogenase and xanthine oxidase which are active in the reconstitution of the *nit-1* mutant (40, 41). Zumft (42) also claims to have separated two low molecular weight molybdenum-containing fractions from nitrogenase which also show reconstituting activity. Unfortunately, at present, the detailed nature of the molybdenum-containing fractions is virtually unknown, although the Russian workers claim that it is (at least predominantly) a small peptide. Nonetheless, the mere existence of a common cofactor indicates a common structural feature in all molybdoenzymes and opens the possibility that the molybdenum sites in the various enzymes operate in a mechanistically similar manner.

Recently, Brill and co-workers (43, 44) have isolated mutant strains of *Azotobacter vinelandii* which produce an inactive nitrogenase component. This component can be reactivated by treatment with the neutralized acid-hydrolysis products of other nitrogenases (which themselves become inactive on such a treatment) but not apparently with any other molybdenum enzymes. This may either reflect a difference between the cofactor in nitrogenase and other molybdenum enzymes or may be caused by the reconstitution conditions used which may not have been sufficiently varied to allow for different molybdenum oxidation states to be attained. In any event, the chemical characterization and authentication of the molybdenum cofactor should reveal some of the intimate details of the molybdenum site.

Biochemistry of Nitrogenase

General Considerations. The nitrogenase enzyme consists of two separately isolable proteins—the molybdenum–iron protein (Component I, Fraction I, molybdoferredoxin) and the iron protein (Component II, Fraction II, azoferredoxin). The most recent work on nitrogenase components from a variety of organisms indicates great similarity in molecular weight, number of subunits, and molybdenum, iron and inorganic sulfide contents. For *Azotobacter vinelandii*, the organism for which we present our experimental results, the molybdenum–iron protein has a molecular weight of 226,000, four subunits, two molybdenum, roughly 24 iron, and 22 labile sulfide ions per molecule (45). The iron protein has a molecular weight of around 65,000 with four iron and four labile sulfide ions per mole. With *Azotobacter vinelandii* (and to date only for this organism), it is possible to isolate a 1:1 complex of the iron and molybdenum–iron proteins by avoiding the use of DEAE cellulose during the preparation. This nitrogenase complex has been exclusively used for the reaction studies discussed below. Its preparation has been described in detail elsewhere (46, 47).

An input–output scheme for nitrogenase is shown in Figure 2. The material in the box represents the catalytic entities—the iron protein, the molybdenum–iron protein, and Mg^{2+} ions. Input consists of a reduc-

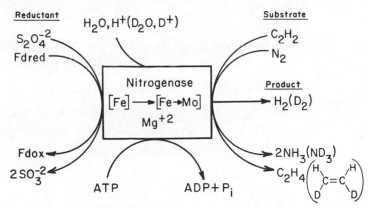

Figure 2.   Nitrogenase—input–output diagram (7, 8, 9, 52)

ing agent, ATP, and a source of protons (H_2O). The reducing agent can be ferredoxin or flavodoxin in vivo, but in assay systems in vitro, dithionite ($S_2O_4^{2-}$) invariably serves this function, being oxidized in the process by two electrons to SO_3^{2-}. The ATP is hydrolyzed during nitrogenase turnover to ADP and P_i. Under optimal (in vitro) conditions, 4–5 moles of ATP are hydrolyzed per mole of electron pairs passing through the enzyme (48). An unusual feature of nitrogenase which contributes to the difficulty in its study is the fact that it does not require a reducible substrate. In the absence of reducible substrate (vide infra), the enzyme system turns over and evolves dihydrogen via the so-called "ATP-dependent hydrogen evolution" reaction which requires the same inputs as nitrogen fixation (49). Thus, nitrogenase is not easily studied in a fully reduced state because this state will give rise to dihydrogen evolution. Perhaps in the future, rapid detection techniques will allow some glimpse of this key state.

The output for nitrogenase consists of dihydrogen and (as appropriate) reduced substrate. The presence of reducible substrates curtails the dihydrogen evolution reaction (although often not completely) (50, 51, 52, 53) and diverts electrons from dihydrogen production to substrate reduction. The natural substrate is dinitrogen which undergoes a six-electron reduction to ammonia. Other substrates have also been found, and their reactions are listed in Table I. The reducible substrates (52, 53) include molecules with triple bonds (acetylenes, nitriles, isonitriles, and cyanide) or reactive double (or potentially triple) bonds (azide, nitrous oxide, allene). Significantly, the acetylenes are reduced by a two-electron process to ethylenes (with no trace of ethane). If D_2O replaces H_2O as the proton source, then dideuterium is evolved, ND_3 is formed from dinitrogen, and acetylene is reduced exclusively to cis-dideuteroethylene (54).

Table I. Substrate Half-Reactions for Nitrogenase

$$N_2 \quad + 6H^+ + 6e^- \rightarrow 2NH_3$$

$$N_2H_4 + 2H^+ + 2e^- \rightarrow 2NH_3$$

$$HCN + 6H^+ + 6e^- \rightarrow CH_4 + NH_3$$

$$N_2O \quad + 2H^+ + 2e^- \rightarrow N_2 + H_2O$$

$$HN_3 \quad + 2H^+ + 2e^- \rightarrow N_2 + NH_3$$

$$RNC + 6H^+ + 6e^- \rightarrow RNH_2 + CH_4$$

$$RCN + 6H^+ + 6e^- \rightarrow RCH_3 + NH_3$$

$$C_2H_2 + 2H^+ + 2e^- \rightarrow C_2H_4$$

$$2H^+ + 2e^- \rightarrow H_2$$

$$[ATP + H_2O \xrightarrow{Mg^{2+}} ADP + P_i]$$

2nd International Conference on Chemistry and Uses of Molybdenum

Dihydrogen Reactions of Nitrogenase. One of the puzzles which nitrogenase has presented lies in its reactions involving dihydrogen. In early studies, dihydrogen was considered a reducing agent for dinitrogen, and the nitrogenase enzyme was thought to catalyze the Haber process reaction (Reaction 1). For example, crude cell-free extracts of Clostri-

$$3H_2 + N_2 \rightarrow 2NH_3 \qquad (1)$$

dium pasteurianum could indeed use dihydrogen as the reductant for dinitrogen (55). However, it is now clear that this process depends upon the presence of the enzyme hydrogenase (56) which catalyzes Reaction

$$H_2 + 2Fd_{ox} \rightarrow 2H^+ + 2Fd_{red} \tag{2}$$

2. Here, dihydrogen reduces oxidized ferredoxin (Fd_{ox}) to Fd_{red} which can then serve as the electron donor for the nitrogenase-catalyzed reduction of dinitrogen to ammonia (Figure 2). Thus, Reaction 1 is catalyzed only when hydrogenase, nitrogenase, and ferredoxin are present and requires ATP hydrolysis as a co-reaction.

When purified nitrogenase is incubated under dinitrogen and dihydrogen, the dihydrogen surprisingly inhibits nitrogen fixation (57, 58, 59, 60). Kinetic studies seem to display a competitive inhibition pattern, although more detailed studies in progress in our laboratory indicate that true competitive inhibition is not present here (57, 58, 59, 60).

A related observation concerns the appearance of HD in the gas phase when nitrogenase turns over in the presence of dinitrogen and dideuterium in H_2O (58, 59, 60). Likewise, nitrogenase turnover in D_2O with dihydrogen and dinitrogen in the gas phase causes HD formation. This process has uniformly been called "HD exchange" although there is no evidence for the formation of HDO in the aqueous phase (58, 59, 61). The formation of HD absolutely depends on the presence of dinitrogen, with low levels of dinitrogen sufficing to give moderate amounts of HD. It would appear from the kinetic data (58, 59, 60) that dinitrogen is, in a sense, a catalyst for the HD formation reaction. No other substrate leads to HD formation nor does HD formation occur in the absence of reducible substrates.

Electron Balance Studies on the Dihydrogen Reactions of Nitrogenase. A continuing project at the Kettering Laboratory is the detailed analysis of the inputs and the outputs for nitrogenase. A remarkable observation about nitrogenase is that its turnover rate is independent of the detailed nature of the output of the enzyme (50). As discussed above, electrons moving through nitrogenase can cause dihydrogen evolution, acetylene reduction, nitrogen fixation, or, depending upon the conditions, various combinations of these activities. However, the utilization rate of $S_2O_4{}^{2-}$ (reductant) and the hydrolysis rate of ATP are each totally independent of the distribution of electrons in these products. Furthermore, even in a dihydrogen-inhibited nitrogen-fixing system, the turnover rate (as measured by $S_2O_4{}^{2-}$ or ATP utilization) is unaffected. These data strongly suggest that the rate-determining step for nitrogenase turnover occurs prior to substrate reduction. Dihydrogen inhibition therefore affects the distribution of products but not the turnover rate of the enzyme. In order to probe the nature of the dihydrogen inhibition reaction, careful electron balance studies have been performed on the nitrogenase complex from *A. vinelandii*. The preliminary results of such studies have been briefly discussed ($46, 47, 58, 59$), and the experimental details and bulk of the data will appear elsewhere (60).

The experimental design is simple. A given sample of nitrogenase equipped with an ATP-generating system, Mg^{2+}, and reductant is allowed to turn over without substrate (case I), and the dihydrogen production is monitored. The amount of dihydrogen produced is found to be equal (within experimental error) to the dithionite oxidized (50). Therefore, the electron balance equation is:

$$2[H_2] = 2[S_2O_4^{2-}] \qquad (3)$$

where the brackets enclose the varying number of moles of indicated product formed or reductant oxidized per unit time.

Using the same nitrogenase preparation, dinitrogen is added to the reaction flask, and dihydrogen evolution and ammonia production are measured in the same reaction vessel. Under these circumstances (case 2), the electron balance Equation 4 obtains:

$$3[NH_3] + 2[H_2] = 2[S_2O_4^{2-}] \qquad (4)$$

This is in full agreement with the need for six electrons to reduce each dinitrogen, i.e., three per ammonia formed.

In case 3, again with the same preparation, a given amount of dihydrogen is introduced into the reaction flask in addition to the dinitrogen. When dihydrogen is present, the equation for electron balance remains the same as above [Equation 4]. However, at the same dinitrogen level, less ammonia and more dihydrogen are produced per unit time compared with case 2 above. Case 3 is, of course, the dihydrogen inhibition reaction, and as expected, it is found to shift electrons from ammonia to the formation of dihydrogen.

The clue to what is happening comes in case 4, when the reaction is analyzed at a level of dideuterium equal to that of dihydrogen used in case 3. Here, the HD, dihydrogen, and ammonia produced are measured in the flask. It is found that the ammonia level in case 4 is the same as in case 3. Thus, as expected, dideuterium and dihydrogen are equivalent in their ability to inhibit reduction (ammonia formation). However, the dihydrogen produced is found to be the same as in case 2, when no dihydrogen or dideuterium is present. Thus, the presence of dideuterium and by inference dihydrogen does not effect the ATP-dependent dihydrogen evolution reaction, and only dinitrogen reduction is effected. However, for case 4, we find that Equation 4 does not balance. However, Equation 5 does balance, meaning that one electron is required for the

$$3[NH_3] + 2[H_2] + 1[HD] = 2[S_2O_4{}^{2-}] \tag{5}$$

formation of each molecule of HD. Over a wide range of dideuterium and dinitrogen pressures, electron balance can be achieved only by adding this term in HD.

In summary, the key experimental findings are:

1. Nitrogenase turnover rate (electron flow) is independent of reducible substrate, electron distribution among a mixture of substrates, or the presence of dihydrogen.

2. Only dinitrogen reduction is inhibited by dihydrogen or dideuterium.

3. In the presence of dideuterium and dinitrogen, HD is produced in a reaction which uses one electron to form each HD.

4. Deuterium does not affect dihydrogen evolution in either the presence or absence of dinitrogen.

These results taken together suggest strongly that dihydrogen and dideuterium divert electrons from dinitrogen reduction, which in the former case leads to dihydrogen production, but in the latter case leads to HD formation. Thus, it appears that the dihydrogen inhibition of dinitrogen reduction and the dinitrogen-dependent HD formation reactions of nitrogenase are manifestations of the same phenomenon. This finding is interpreted on a molecular level in the following section.

Intermediates in the Fixation of Dinitrogen. There are still no spectroscopically detected intermediates in the reduction of dinitrogen to ammonia. We believe, however, that the electron balance studies with nitrogenase under dinitrogen/dihydrogen and dinitrogen/dideuterium atmospheres point to the existence of such intermediates. The stoichiometry for the HD formation reaction as determined experimentally is shown in Reaction 6. At first glance, this resembles the substrate reactions

$$2H^+ + 2e^- + D_2 \rightarrow 2HD \tag{6}$$

(Table I), and at one time it was thought (62) that dideuterium was in fact a nitrogenase substrate. However, the key fact remains that dinitrogen, at least in catalytic amounts, is required for HD formation and that in the process, the reduction of dinitrogen to ammonia is inhibited (57–61, 63). To explain these features, we suggest that a two-electron reduction product of dinitrogen is reactive towards dihydrogen or dideuterium. In analogy to the reduction of acetylene in D_2O to *cis*-$C_2H_2D_2$, this product is postulated to be a bound *cis*-diimide species (45).

As shown in Figure 3, the first step in dinitrogen reduction could involve binding of dinitrogen to the enzyme, with the mode of binding left unspecified. In analogy to the reduction of acetylene to ethylene, reduction by two electrons and two protons is postulated to produce a bound *cis*-diimide species with somewhat exposed N–H bonds. The bound diimide could then react with dihydrogen in a six-center reaction to reform dinitrogen (bound or unbound) and generate an additional dihydrogen molecule. The reaction, as expressed in Reaction 7, is effec-

$$E\text{-}N_2H_2 + H_2 \rightarrow E + N_2 + 2H_2 \tag{7}$$

tively the decomposition of bound diimide to its elements. This process is exothermic for free diimide by roughly 35 kcal/mole. The reaction of dihydrogen with *cis*-N_2H_2 is allowed by orbital symmetry considerations. The regeneration of dinitrogen is significant in two respects. First, it shows that dinitrogen is not reduced to ammonia, and so ammonia production is inhibited. Second, it means that dinitrogen is effectively a catalyst in the production of dihydrogen by a route which is not identical to the simple ATP-dependent dihydrogen evolution reaction.

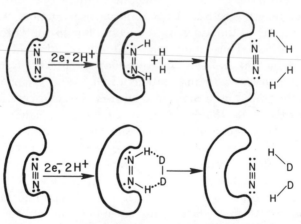

Figure 3. Nitrogenase—scheme of H_2 inhibition and HD production reactions

The reaction with dideuterium is shown in the lower part of Figure 3. Here, the inhibition of nitrogen fixation leads to the production of HD according to Reactions 8 and 9. Clearly, the overall process agrees pre-

$$E\text{-}N_2 + 2e^- + 2H^+ \rightarrow E\text{-}N_2H_2 \tag{8}$$

$$E\text{-}N_2H_2 + D_2 \rightarrow E + N_2 + 2HD \tag{9}$$

cisely with the electron balance studies which implicate two electrons per pair of HD molecules formed. Thus, the mechanism involves the diversion of electrons from N_2H_2 to dihydrogen or HD in agreement with experiment.

The postulation of the N_2H_2-level intermediate clearly implies that nitrogenase functions in a stepwise manner, with hydrazine therefore implicated as the second enzyme-bound intermediate. The question then arises as to whether hydrazine is a reducible substrate for nitrogenase.

Reduction of Hydrazine by Nitrogenase. At pH 7.2–7.4, the normal assay conditions, nitrogenase can catalyze the reduction of hydrazine at only a very slow rate. However, at these pH values, hydrazine is present largely as the hydrazinium ion, $N_2H_5^+$, and it is possible that this cationic species cannot serve as a substrate. To test this idea, the pH of the solution was raised in steps to pH 8 where the enzyme remains active, and substantial amounts of neutral N_2H_4 are present. The production of ammonia closely paralleled the increase in pH. At pH 8, hydrazine is reduced at $\sim 20\%$ of the rate of dihydrogen evolution. In all respects, the reduction of hydrazine behaves like that of other substrates, with ATP and $S_2O_4^{2-}$ being required and dihydrogen evolution being decreased by an amount commensurate with the amount of hydrazine reduced (45). Thus, hydrazine is reducible by nitrogenase, and although there is still no direct evidence, this result establishes the potential of a bound hydrazine intermediate in the overall process of dinitrogen reduction.

Finally, the hydrazine reduction reaction is unaffected by either dihydrogen or dideuterium, i.e., under either, there is no inhibition of ammonia formation, and under dideuterium, there is no HD production. Assuming that bound hydrazine reacts in the same manner as added hydrazine, then the dihydrogen and dideuterium effects must occur prior to the formation of bound hydrazine, and again a diimide-level species is implicated.

In short, the electron balance and hydrazine reduction studies strongly implicate bound N_2H_2 and bound N_2H_4 in the catalytic reduction of dinitrogen by nitrogenase. In this respect, the key site of nitrogenase is shown to be a two-electron two-proton reagent. As elaborated below, this finding points to a possibly greater similarity between nitrogenase and other molybdenum enzymes than might be otherwise thought.

Biochemistry of Xanthine Oxidase and Other Molybdenum Oxidases

General Considerations. Much experimental information is available concerning the role of molybdenum in xanthine oxidase (19, 20). In early work (prior to 1970), there was much confusion in the literature because of the presence of various inactive forms of the enzyme. It is now known that both demolybdo and desulfo forms of xanthine oxidase were present in most early preparations and remain present in many current preparations as well (20, 64).

A major advance in the elimination of catalytic site inhomogeneity in new preparations came with the development of an affinity chromatographic method (65) for purifying the enzyme. This method made use of the known high affinity of xanthine oxidase for alloxanthine when the enzyme is in a fully reduced state. By attaching alloxanthine to a polymeric matrix, a selective absorption of active enzyme was achieved.

Evidence Concerning the Molybdenum Site. Despite considerable debate, there is at present a good deal of agreement as to the overall mode of functioning of xanthine oxidase (20, 65–70). Furthermore, the EPR spectroscopic properties indicate that the molybdenum sites in aldehyde oxidase and (to a somewhat lesser extent) sulfite oxidase are very similar in nature to that in xanthine oxidase.

Several lines of evidence indicate that the molybdenum oxidases use molybdenum in the oxidation states VI, V, and IV. In the oxidized (or resting in oxygen) enzyme, there is generally no EPR signal. Upon reduction with less than stoichiometric amounts of substrate (or reductant), a Mo(V) EPR signal appears which disappears when further reductant is added (20). A reasonable interpretation invokes Mo(VI) as being present in the resting oxidase and Mo(V) as an intermediate state in the reduction process. A second line of evidence for Mo(VI) concerns the well known antagonism which tungsten displays for molybdenum in a variety of systems. WO_4^{2-}, when used in place of MoO_4^{2-} in culture media, plant food, or animal feed, causes the formation of either a demolybdoenzyme or a tungsten-substituted protein (71, 72). In sulfite oxidase, the tungsten-substituted protein has been characterized (72). It totally lacks enzymatic activity and, in view of the greater difficulty in reducing W(VI) compared with Mo(VI), undoubtedly contains W(VI). The tungsten protein is immunologically identical to its molybdenum analog. Substrate (SO_3^{2-}) cannot significantly reduce tungsten to an EPR-active state but $S_2O_4^{2-}$ (a more powerful reductant) can produce a fully EPR-active W(V) state. Significantly, the W(V) EPR signal will not disappear when excess reductant is present. The inability to achieve the (IV) state may be responsible for the inability of the tungsten-protein to turn over catalytically, which in turn implicates a Mo(IV) state in the catalytic cycle.

There is, however, more direct evidence for the presence of Mo(IV) in the cycle of xanthine oxidase. This evidence comes from the experiments of Massey and co-workers (24) who used alloxanthine (1) to trap the enzyme in its reduced state. A strong complex is formed between the reduced enzyme and alloxanthine, and excess alloxanthine and reductant can be removed. The enzyme is then reoxidized with $Fe(CN)_6^{3-}$, and two electrons per molybdenum center are found after the electrons required for the reoxidation of the iron–sulfur and flavin groupings are accounted for. If the oxidized molybdenum state is Mo(VI), then Mo (IV) is implicated as the reduced state.

An additional crucial piece of information emerges from the alloxanthine study (*24*). Thus, it was shown that one alloxanthine binds to the enzyme per active molybdenum site. This result clearly implies that the molybdenum site is mononuclear. If a dinuclear site were involved, then it would be unlikely to require two alloxanthine molecules for inhibition and would be expected to be at least partially inhibited with one alloxanthine/two molybdenum. Also, a difference in binding constant would be expected for the second compared with the first bound alloxanthine, but none is found. This result, coupled with the lack of evidence for Mo(V)–Mo(V) spin–spin interactions in the EPR spectra, clearly implicates a mononuclear site, and it would seem that xanthine oxidase possesses two full catalytic units, each containing one molybdenum, one flavin, and two Fe_2S_2 units (*20*). Other molybdenum oxidases also contain paired prosthetic groups and subunits, and it is likely that they each have two catalytic units per molecule.

(1)   (2)

Electron spin resonance spectroscopy has given tremendous insight into the nature of the overall xanthine oxidase reactions as well as into the nature and function of the molybdenum site (*19, 20*). During turnover, the EPR signal from a single Mo(V) group appears in the spectra of all molybdenum oxidases. The g and A values implicate at least one sulfur atom in the molybdenum coordination sphere, (*73, 74*) but until more definitive models become available, the detailed nature of the site must remain obscure. One very important feature of the Mo(V) EPR signal from the oxidases is the near-isotropic proton superhyperfine splitting of 10–14 gauss. The proton responsible for this splitting is exchangeable as evidenced by the rapid collapse of the splitting when D_2O replaces H_2O as the solvent. The proton has an apparent pK_a of ~8 in xanthine oxidase. For sulfite oxidase, where the EPR spectra are relatively simple, a clear titration curve is seen with a pK_a of 8.2. Finally, Bray and Knowles (*75*) were able to demonstrate using 8-deuteroxanthine (2) that the proton responsible for the hyperfine splitting originates in the 8-position of the substrate xanthine.

This proton seems to be transferred to the enzyme in conjunction with the two-electron transfer process. In view of the prominence of the proton, its possible location on the enzyme is of considerable importance. The fact that its splitting of the molybdenum signal is nearly isotropic suggests that it is unlikely to be a molybdenum hydride. Nevertheless, this possibility has not been totally ruled out because, as noted by Edmondson et al. (76), it is possible that this isotropy is apparent and not real. If the components of the hyperfine coupling tensor differ in sign and if the isotropic splitting is exactly half the anisotropic splitting, the apparent isotropy would be explained. On the other hand, protons on protein atoms directly bound to molybdenum may be responsible for the observed splitting, and in view of mechanistic considerations and recent work on inorganic systems considered below, this seems most reasonable.

Electron Transfer and Substrate Half-Reactions

Each of the molybdenum enzymes is a complex entity containing molybdenum and other redox-active prosthetic groups. These enzymes are designed to catalyze redox reactions by providing a low energy pathway for electrons to transfer from reductant to oxidant. In most cases, and certainly in the physiological reaction, the electrons enter and leave the enzyme at different sites. In the simpler cases like sulfite oxidase, a definite sequence of electron transfer within the protein can be formulated. However, more sophisticated treatments for the other proteins reveal that the electron carriers within the protein achieve a distribution of electron occupancy depending upon their inherent potentials and the total electronic charge of the enzyme (69). The inherent redox potential of a given group may be changed by the presence of substrate or presumably by a conformational change in the protein. The question which concerns us here is the interaction of the molybdenum site with the external medium. Although the question has not been answered to total satisfaction in all cases, it seems clear that for the molybdenum oxidases, the molybdenum site of the enzyme is the one which interacts with the oxidizable substrate. In contrast, in the molybdenum reductases (although here the evidence is not strong), the molybdenum site interacts with the reducible substrate. In either event, the molybdenum site interacts with the named substrate for the reaction and either accepts or donates electrons. In a sense, the single site of the enzyme is like the electrode of an electrochemical cell which (with respect to the medium) carries out a chemical half-reaction. Thus, to comprehend the role which molybdenum plays in enzymes, scrutiny of the substrate half-reactions is appropriate.

The substrate half-reactions are displayed in Tables I and II. In each case, a two-electron process seems to be involved. Only in nitrogenase are greater numbers of electrons transferred, and the discussion earlier in this paper summarizes the evidence that these processes occur in two-electron steps. The two-electron reaction of the molybdenum site never appears to be simply an electron transfer reaction. In the case of nitrogenase, each substrate takes up an equal (or greater) number of protons to form the product. In the other molybdenum enzymes, proton transfer and addition or removal of H_2O are also required. In each case, however, there is at least one proton transferred in the same direction as the pair of electrons. These data, taken in conjunction with the EPR evidence for proton transfer from the substrate to the active site in xanthine oxidase, suggest that the molybdenum site in all the enzymes

Table II.   Substrate Half-Reactions for Molybdoenzymes (74)

Nitrate reductase
$$NO_3^- + 2H^+ + 2e^- \rightarrow NO_2^- + H_2O$$

Xanthine oxidase
$$Xanthine + H_2O \rightarrow uric\ acid + 2H^+ + 2e^-$$

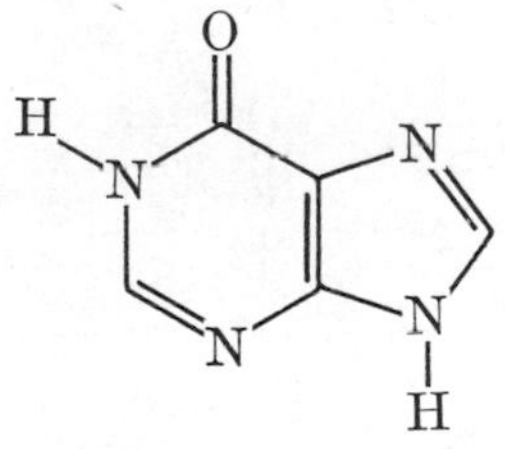

$$+ H_2O \rightarrow xanthine + 2H^+ + 2e^-$$

Hypoxanthine

Aldehyde oxidase

$$RCHO + H_2O \rightarrow RCOOH + 2H^+ + 2e^-$$

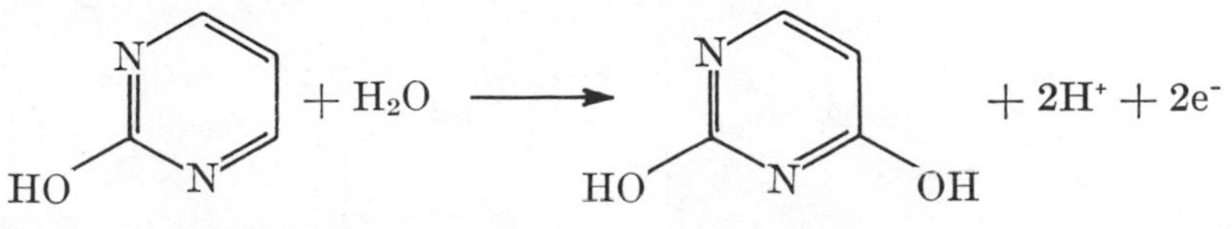

$$+ H_2O \longrightarrow \qquad + 2H^+ + 2e^-$$

Sulfite oxidase

$$SO_3^{2-} + H_2O \rightarrow SO_4^{2-} + 2H^+ + 2e^-$$

is in some way responsible for both proton and electron transfer processes (*66*).

Molybdenum Coordination Chemistry

Compared with other transition metals in biological redox systems, the oxidation states likely to be used by molybdenum are very high (*74*). As discussed previously, the IV, V, and VI states are a likely set of participants in molybdenum oxidases, and while the II and III states remain viable for molybdenum reductases, it nevertheless seems likely that higher oxidation states will be found in these enzymes as well. Indeed, the substitution of tungsten for molybdenum in both nitrate reductase and nitrogenase indicates this likelihood as it is much more difficult to obtain the lower oxidation states of tungsten.

When we examine the oxidation states of molybdenum, there are some key trends which become apparent (*74*). First, the higher oxidation states are always found to be coordinated by deprotonated ligands. In the most common case, these ligands are waters, which when fully deprotonated, are designated oxo groups. The compounds of Mo(IV), Mo(V), and Mo(VI) with dithiocarbamates (*74, 77, 78, 79*) nicely illustrate the structural variety as well as the presence of oxo groups. Thus the complexes (**3, 4, 5, 6**) show the presence of a single oxo group in the

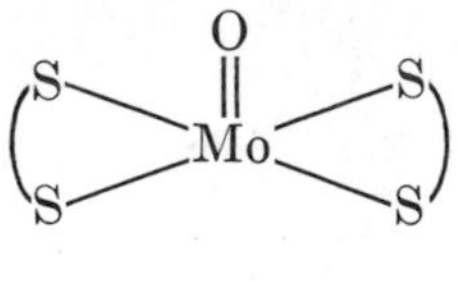

(3)

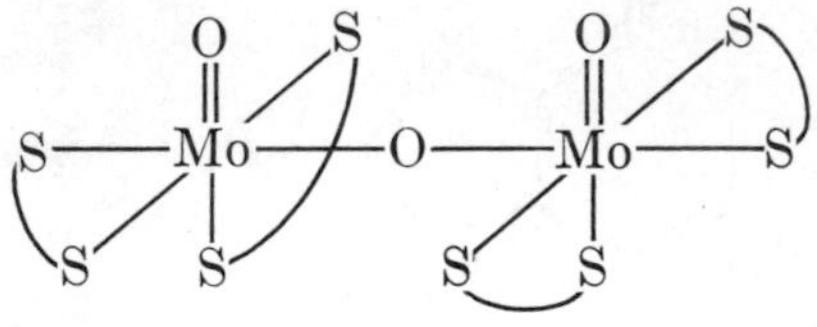

(4)

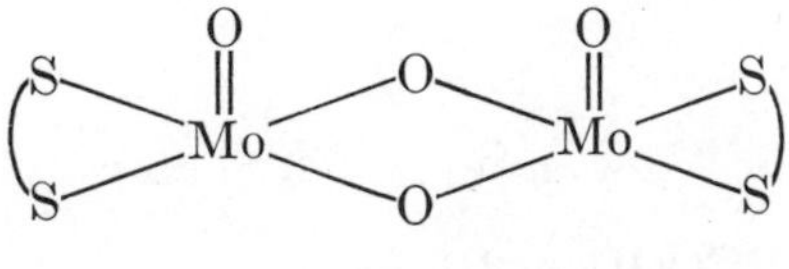

(5)

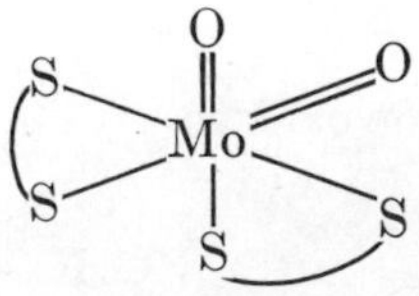

(6)

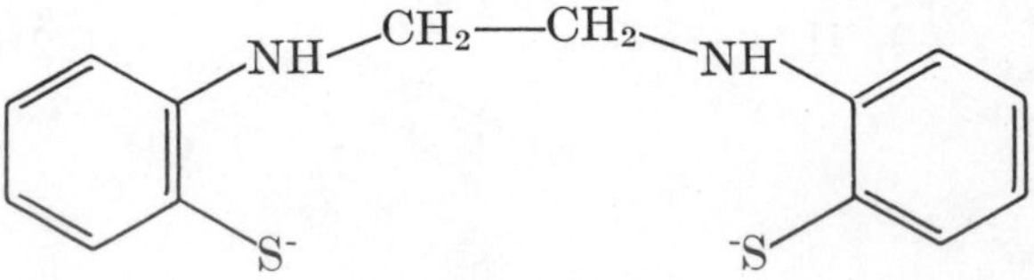

(7) = L

Table III. ΔH Values for Reactions of Dithiocarbamate Complexes of Molybdenum[a]

Reaction	ΔH $(kcal/mol)$
$MoO_2(dtc)_2 + SO_3^{2-} \rightarrow MoO(dtc)_2 + SO_4^{2-}$	-28.5 ± 5.5
$MoO_2(dtc)_2 + CH_3CHO \rightarrow MoO(dtc)_2 + CH_3COOH$	-32.0 ± 4.0
$MoO(dtc)_2 + H_2O \rightarrow MoO_2(dtc)_2 + H_2$	$+30.1 \pm 4.2$
$3MoO(dtc)_2 + N_2 + 3H_2O \rightarrow 3MoO_2(dtc)_2 + 2NH_3$	$+51.7 \pm 12.4$
$MoO(dtc)_2 + NO_3^- \rightarrow MoO_2(dtc)_2 + NO_2^-$	-4.4 ± 4.2

[a] Data from Ref. *87*, derived from experimental values in 1,2-dichloroethane at 25°C.

Mo(IV) species (3) and two in the Mo(VI) species (6). This result suggests the possibility of oxygen atom transfer reactions (*80, 81*) with Mo(IV) extracting an oxo from a substrate (e.g., $NO_3^- \xrightarrow{- [O]} NO_2^-$) or Mo(VI) donating an oxo (e.g., $SO_3^{2-} \xrightarrow{[O]} SO_4^{2-}$). The nitrate reduction serves as a distinct model for nitrate reductase, but Mo(II) (*82*), Mo(III) (*83*), and Mo(V) (*84, 85, 86*) compounds can also reduce nitrate to nitrite. So at present, such model reactions offer no help in our deliberations about molybdenum enzymes.

Recently, thermodynamic studies have been carried out in our laboratory (*87*) to evaluate the possible participation of these complexes in model reactions. The ΔH values for relevant reactions are listed in Table III. The Mo(IV)–Mo(VI) couple with dithiocarbamate ligands would exothermically execute the SO_3^{2-}/SO_4^{2-} or CH_3CHO/CH_3COOH conversions. On the other hand, there is a highly endothermic reaction when the Mo(IV)/Mo(VI) couple is used to effect the production of dihydrogen from water or the production of ammonia from dinitrogen. In the case of the NO_3^-/NO_2^- conversion, there is a very small exothermicity associated with the reaction. These results show that the dithiocarbamate complexes could be used to model the sulfite oxidase or aldehyde oxidase reactions but not the nitrogenase reaction. However, the redox properties of the Mo(IV)/Mo(VI) couple vary substantially with ligand (*88*), and these results therefore do not vitiate the possibility of an Mo(IV)/Mo(VI) couple being present in nitrogenase with a different set of donor atoms.

Another aspect of the deprotonated (acidic) ligand effect manifests itself when the oxo groups are removed. For example, reaction of the tetradentate ligand (7) with $MoO_2(C_5H_7O_2)_2$ gives a Mo(VI) complex of the form MoO_2L (*89*), analogous to the dithiocarbamate complex of Mo(VI). In contrast, the reaction of MoO_4^{2-} with *o*-aminobenzenethiol

$$MoO_4^{-2} + 3 \; [\text{2-aminothiophenol}] + 2H^+ \xrightarrow[\text{EtOH-H}_2\text{O}]{H^+} \left[Mo \begin{array}{c} \text{H} \\ \text{N} \\ \text{S} \end{array} \right]_3 + 4H_2O \qquad (10)$$

proceeds (67, 68, 89) according to Reaction 10. The product is formally a Mo(VI) complex wherein the coordinated ligand is a deprotonated amine. The formation of this product is understood in the light of the presence of Mo(VI) and its acidity-enhancing properties (67, 68). Thus, when the more acidic aquo ligands (oxos) are removed from the coordination sphere, the acidity manifests itself in the ionization of a coordinated amine ligand which ordinarily would not be considered as a potentially ionizable grouping. There are numerous examples in coordination chemistry which show the effect of oxidation number on ligand acidity. Consideration of a large number of examples (90) reveals that the pK_a of a coordinated ligand atom decreases by about 6–10 units per unit change in the oxidation number of the metal atom. This effect is illustrated in molybdenum chemistry by the aquo ions (74). In Mo(VI) chemistry in strong acid solutions, the principal species is thought to be $[MoO_2-(H_2O)_4]^{2+}$ while for Mo(III) in acid solution, the ion $[Mo(H_2O)_6]^{3+}$ is present. Similarly, for Mo(III), the species $[Mo((NH_2)_2C_6H_4)_3]^{3+}$ is formed in contrast to the result for Mo(VI) discussed above where the comparable species $Mo(NHSC_6H_4)_3$ was found. The results from coordination chemistry illustrate that ligands coordinated to molybdenum can engage in proton transfer reactions which, through the effect of oxidation state on pK_a, can be coupled to electron transfer reactions.

$$Mo_2O_4(S_2CNEt_2)_2 + 4 \; [\text{2-aminothiophenol}] \longrightarrow 2 \, Mo(S_2CNEt_2) \left[\begin{array}{c} \text{NH} \\ \text{S} \end{array} \right]_2 + 4H_2O \qquad (11)$$

Recent electron spin resonance studies in our laboratory add weight to the notion that protons on coordinated nitrogen participate in catalytic steps. Reaction 11 was discovered (91), leading to the isolation of $Mo(S_2CNEt_2)(NHSC_6H_4)_2$. The monomeric Mo(V) complex formed displays superhyperfine splitting (92) from two equivalent nitrogen atoms as well as two equivalent hydrogen atoms as illustrated in Figure 4. The coupling constants which have been confirmed by preparation of the N–CH$_3$ and N–D complexes are $A_N = 2.4$ and $A_H = 7.4$ gauss. The facile preparation of the N-deutero complex from its N-proteo analog and CH_3OD attests to the exchangeability of the proton in question. This signal, with its relatively large value of A_H and low value for A_N, reaffirms the possibility of N–H groups being coordinated to molybdenum in enzymes. While these compounds do not represent models for the molybdenum site of enzymes, they nevertheless illustrate that the key proton(s) involved in the catalytic step may be associated with ligand atom(s) bound directly to molybdenum.

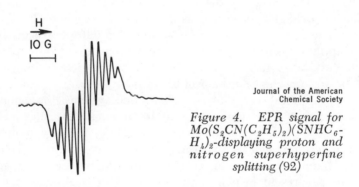

H
→
10 G
⊢―――⊣

Journal of the American
Chemical Society

Figure 4.   EPR signal for
$Mo(S_2CN(C_2H_5)_2)(SNHC_6-$
$H_4)_2$-*displaying proton and*
nitrogen superhyperfine
splitting (92)

Mechanistic Considerations

Mechanistic speculations about the molybdoenzymes must be considered to be in their infancy with the possible exception of those for xanthine oxidase. Although the detailed structural nature of the molybdenum site is unknown, there is sufficient information from biochemical and coordination chemistry studies to allow informed arguments to be drawn. Here we first discuss evidence for the nuclearity of the molybdenum site and then discuss both oxo-transfer and proton-electron transfer mechanisms for molybdenum enzymes. A final discussion considers the unique aspects of nitrogenase and the possible reasons for the use of molybdenum in enzymes.

Mononuclear vs. Dinuclear Sites. All molybdenum enzymes contain two molybdenum atoms. Dinuclear molybdenum complexes are well known in the chemistry of Mo(VI), Mo(IV), and Mo(III) and play a dominant role in the chemistry of Mo(V). The juxtaposition of the biochemical and inorganic chemical numerology has led to the suggestion that this may be more than mere coincidence and that the use of dinuclear molybdenum in the catalytic sequences requires that the molybdenum content of the enzymes be doubled. Furthermore, there are attractive catalytic schemes which make use of Mo(V) complexes. In particular, the oxo transfer reaction useful in the oxidation of tertiary phosphines (93) led to the discovery (78, 79, 93) of Reaction 12 in which a dinuclear

$$Mo_2O_3(R_2dtc)_4 \rightleftarrows MoO_2(R_2dtc)_2 + MoO(R_2dtc)_2 \qquad (12)$$

Mo(V) complex disproportionates as it dissociates to produce mononuclear Mo(IV) and Mo(VI). As Mo(IV) and Mo(VI) are directly interconvertible by an oxo transfer reaction, they are viable participants in catalytic cycles. A dinuclear Mo(V) species of this nature can thus supply either the oxidizing or reducing member of this couple and presents a mechanism by which molybdenum enzymes can channel reducing or oxidizing power. Several inorganic reactions have recently been explained using this scheme (80, 81). To date, however, Reaction 12 only applies when the ligand is a dithiocarbamate or dithiophosphate. Nevertheless, were there known dinuclear active sites in enzymes, this would be an important mechanism to consider.

It appears, however, that in well studied systems, the evidence for dinuclear sites is outweighed by that for mononuclear sites. The case for xanthine oxidase seems most explicit. Here, the two molybdenum atoms are accompanied by two FAD groups as well as two each of two different types of Fe_2S_2 cluster. All components, including subunits, appear to be present in pairs and most modern treatments invoke two separate catalytic units, each involving one molybdenum, one FAD, and one of each of the Fe_2S_2 systems. The experimental support for this is impressive. First, it is clear that each molybdenum atom can accept two electrons from substrate. This implies that if a dinuclear site were present, it would be required to accept four electrons. It is not clear why four electrons should be added to a site which catalyzes a two-electron substrate reaction. Second, as discussed previously, the inhibitor alloxanthine binds to a reduced form of the enzyme containing Mo(IV) with only one very tight binding constant and a stoichiometry of precisely one alloxanthine per one molybdenum. These data are very difficult to accommodate in a dinuclear model. Finally, despite an extraordinary amount of EPR work, no spin–spin broadening interaction between Mo(V) sites has been observed. Therefore, either the Mo(V) state is always accompanied by a diamagnetic molybdenum partner [Mo(IV) or Mo(VI)], or the individual molybdenum atoms are far apart. The latter interpretation seems far more likely.

The question now arises as to why xanthine oxidase has a molecular weight of 300,000 when the size of a catalytic subunit is only 150,000. Perhaps the answer lies in its particular integration into cellular physiology or is simply caused by the proclivity of the enzyme to maximize the number of active sites per unit surface area; i.e., when two active monomers join together, the number of active sites goes from one to two, but the exposed surface area increases by a factor less than two. This argument is a familiar one in heterogeneous catalysis where attempts are often made to maximize the number of active sites per unit surface area to produce more efficient catalysts.

The other molybdenum enzymes each contain duplicate prosthetic groups and paired subunits in addition to two molybdenum atoms. Many of the experiments performed for xanthine oxidase have also been carried out with aldehyde oxidase and sulfite oxidase, and there is no evidence for chemical Mo–Mo coupling in these enzymes. Thus, in oxidases, the evidence for mononuclear molybdenum sites appears strong, and in view of the duplicate subunits and composition found, it is reasonable to assume a similar situation in reductases as well. However, at present, insufficient information bars a full generalization.

Oxo Transfer Mechanisms. Except for nitrogenase, all substrate half-reactions involve the addition or removal of oxygen. The simplest manner of representing these reactions, involves the direct transfer of an oxygen atom to or from substrate, e.g., Reactions 13 and 14. Furthermore,

$$NO_3^- \rightarrow NO_2^- + [O] \tag{13}$$

or
$$SO_3^{2-} + [O] \rightarrow SO_4^{2-} \tag{14}$$

it is known that, at least with some reactants, various molybdenum complexes will undergo such an apparently simple oxo transfer (*81, 93*), e.g., Reaction 15. This observation suggests that oxygen atom transfer

$$MoO_2(R_2dtc) + P(C_6H_5)_3 \rightarrow MoO(R_2dtc)_2 + OP(C_6H_5)_3 \tag{15}$$

is a reaction worth considering for the molybdenum enzymes (*77*). Such a mechanism for nitrate reductase could involve Reaction 16 where

$$\tag{16}$$

cleavage of the N–O bond presumably occurs concertedly with the formation of the multiple Mo–O bond. One of the problems with this type of mechanism for nitrate reductase involves the removal of the oxo group on molybdenum to regenerate the open site. Heretofore, oxo removal reactions generally required strong acid (74, 78, 79) or an oxo removal agent such as a phosphine (74, 81, 93, 94). Recently, thiol ligands have, under certain conditions, also removed oxo groups (91, 95, 96, 97). In most cases unfortunately, a sulfur-donor ligand replaced the oxo group. However, in other instances (95, 96), removal of an oxo group with concomitant reduction of Mo(VI) to Mo(IV) by two electrons has been affected by thiols which are oxidized to the disulfide in the process

$$\mathrm{Mo(VI)O_2(R_2dtc)} + 2C_6H_5SH \rightarrow$$
$$\mathrm{Mo(IV)O(R_2dtc)_2} + C_6H_5SSC_6H_5 + H_2O \tag{17}$$

(Reaction 17). Oxo removal to leave an open site might be effected in this manner in enzymes particularly under hydrophobic conditions. The related problem in nitrogenase may be overcome by ATP, which may function in oxo removal from molybdenum.

For the molybdenum oxidases, the reverse oxo transfer reaction can be postulated wherein an oxomolybdenum(VI) species donates oxo to substrate. For example, the oxidation of aldehydes (Reaction 18) can

$$\tag{18}$$

be affected by a Mo(VI) species. Although the reaction is stoichiometrically acceptable, it is not clear how the aldehyde C–H bond is **activated** for cleavage. A similar problem occurs for xanthine oxidation. For this reason, and to make use of the experimental evidence for proton transfer, the schemes involving coupled electron–proton transfer were proposed (66, 67, 68) and are discussed below.

Coupled Proton–Electron Transfer Mechanisms. The suggestive evidence for proton transfer in xanthine oxidase has been discussed above. The key piece of experimental information is the proton superhyperfine splitting in the Mo(V) EPR signal of xanthine oxidase. Model studies (89, 92) have indicated that a coordinated nitrogen is the likely location of the proton (although coordinated oxygen is not eliminated). Coordination chemistry further shows that proton transfer can be coupled to electron transfer through the effect of oxidation state on the pK_a of coordinated ligands (66, 67, 68). The combined biochemical and inorganic information leads to a mechanistic suggestion for xanthine oxidase which is depicted in Figure 5.

The evidence for the presence of various molybdenum oxidation states has been presented previously. The resting enzyme (upper right of Figure 5) is assumed to contain Mo(VI). In this high oxidation state, at least some of the ligands on molybdenum must be deprotonated and a nitrogen atom of the protein is so depicted in the figure. Substrate xanthine can then coordinate to Mo(VI) through its 9-nitrogen. The C=N of the 8- and 9-purine positions is then polarized by the Mo(VI) causing the 8-carbon to become susceptible to nucleophilic attack. Although there is evidence for a protein bound persulfide being the nucleophilic agent (65), for simplicity in this scheme, OH⁻ assumes that role. (Were the persulfide involved, it would have to be subsequently hydrolyzed by OH⁻ or H_2O anyway.)

In conjunction with the nucleophilic attack at the 8-carbon, two electrons could flow from xanthine to produce Mo(IV) which requires a concomitant decrease in pK_a of the protein nitrogen, resulting in transfer

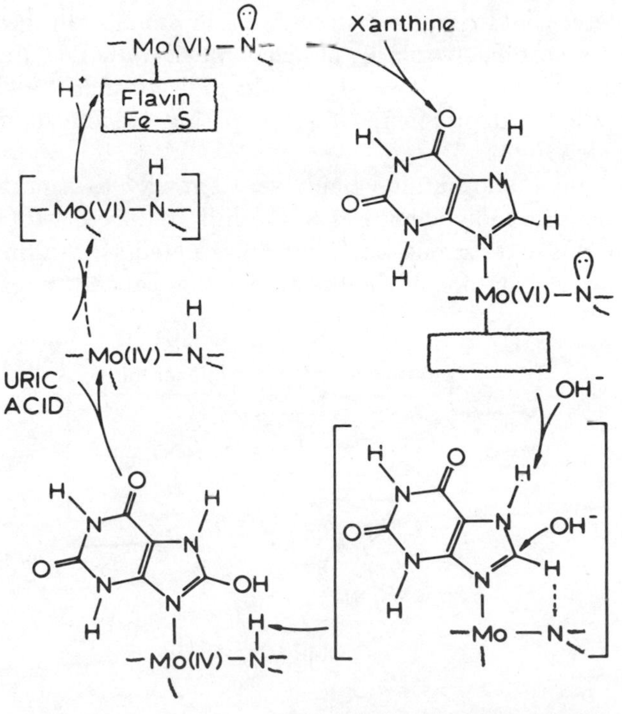

1st International Conference on Chemistry and Uses of Molybdenum

Figure 5. Proposed coupled proton–electron transfer scheme for xanthine oxidase activity (66, 67, 68)

of the proton to the molybdenum site. This process couples the transfer of a proton from substrate with the two-electron transfer. Uric acid (or its persulfido precursor) is now coordinated to the Mo(IV), and reactivation of the site must involve dissociation of product, two-electron oxidation of molybdenum, and loss of a proton from coordinated nitrogen. The order of these events is not clear and, in fact, may not be susceptible to temporal designation.

It is clear, however, that this two-electron reactivation process occurs in two sequential one-electron steps which cause the appearance of the Mo(V) EPR signal. Furthermore, the proton (originally from substrate) may remain in place on the protein nitrogen during oxidation from Mo(IV) to Mo(V) and may thus be responsible for the superhyperfine splitting. In support of this mechanism, the use of 8-deuteroxanthine causes this signal to appear initially in its deutero form. A key aspect of this signal is the apparent pK_a of 8 for this proton in the Mo(V) state. The effect of oxidation state on pK_a requires that in the Mo(IV) state, the pK_a of this proton would be very high (perhaps 14 or greater) whereas in the Mo(VI) state this pK_a would be quite low (perhaps 2 or lower). Thus, the (IV) state would contain a strongly basic protein nitrogen in agreement with its postulated role in cleaving the C–H bond. Furthermore, for the Mo(VI) state, the low pK_a would indicate that the coordinated nitrogen would be deprotonated thus preparing the site to re-enter the catalytic cycle (Figure 5).

The inhibitor alloxanthine binds very strongly to xanthine oxidase but only when the molybdenum is in the fully reduced [Mo(IV)] state. In Figure 6, this extra strong binding is interpreted as resulting from the possession by alloxanthine of the full recognition capability for the active

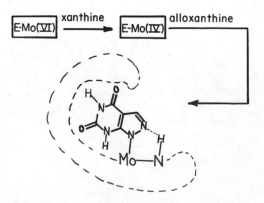

Figure 6. Suggested binding mode for alloxanthine to the Mo(IV) site in the xanthine oxidase

site coupled with the formation of an additional hydrogen bond with the enzyme through the key proton. Furthermore, this tight enzyme-inhibitor complex clearly resembles the proposed transition state (in the coupled proton–electron transfer mechanism) for the catalyzed reaction, as many such complexes do.

Many experimental observations on xanthine oxidase activity are correlated by this scheme, and at present, there appear to be no major inconsistencies. The coupled proton–electron transfer scheme (*66, 67, 68*) has been successfully incorporated into an overall mechanistic scheme (*69*) which explains, with great economy, a large amount of rather demanding data, from both kinetic and electron uptake experiments.

None of the other molybdenum enzymes have been studied as thoroughly as xanthine oxidase, but both aldehyde oxidase and sulfite oxidase display data consistent with their use of the Mo(IV)–Mo(V)–Mo(VI) triad of oxidation states during catalysis. Their Mo(V) EPR signals are similar to those from xanthine oxidase, including the large near-isotropic superhyperfine splitting for a single proton with a pK_a close to 8 (*20*). These observations require structural similarities of the molybdenum sites and suggest mechanistic similarities as well. Thus, coupled electron–proton transfer processes have been suggested for both aldehyde oxidase and sulfite oxidate (*66, 67, 68*). For the former, a mechanism closely akin to xanthine oxidase action is suggested (Reaction 19) in which C–H bond

$$\tag{19}$$

breaking is assisted by nucleophilic attack (again possibly by persulfide (*97*)) with the proton being transferred to the molybdenum site in conjunction with the electron transfer process. In fact, xanthine oxidase will oxidize aldehydes (*98*), and aldehyde oxidase will handle a variety of purines (*25*). The key feature of the molybdenum site is its ability to abstract in a concerted manner two electrons and a proton from substrate coupled with a nucleophilic attack on the carbon bearing the proton to be transferred.

The coupled proton–electron transfer mechanism can also be applied to the molybdenum reductases. For nitrate reductase, a scheme such as Reaction 20 is possible. A Mo(IV)–Mo(VI) couple is used to illustrate this, and while such a couple is viable for some nitrate reductases, the Mo(II)–Mo(IV) or the Mo(III)–Mo(V) couple could also be accommodated

$$
\begin{array}{ccc}
\underset{\underset{\displaystyle |}{\overset{\displaystyle \mathrm{N}}{\overset{\displaystyle \|}{}}}{\mathrm{O}}} & & \\
\mathrm{O}\diagdown\underset{\displaystyle |}{}\diagup\mathrm{O} & \longrightarrow & \underset{\underset{\displaystyle \mathrm{O}\diagup}{\overset{\displaystyle \mathrm{N}}{\overset{\displaystyle \|}{}}}{\overset{\displaystyle \mathrm{O}^{-}}{}}} + \mathrm{OH^{-}} \\
\end{array}
$$

within the proton–electron transfer scheme. This scheme is simply the reverse of that for the oxidases. The lower oxidation state would have a protonated lingand (here shown as nitrogen) and during two-electron reduction of substrate, this ligand would increase in acidity and transfer its proton to an oxygen atom on nitrate, making the oxygen a better leaving group and facilitating nitrite production.

This process can be contrasted directly with the oxo transfer scheme (Reaction 16) discussed above. In either case, the cleavage of the N–O bond is assisted by the binding of oxygen to an electrophile (to molybdenum itself in the oxo transfer mechanism or to proton(s) in the coupled proton–electron transfer scheme). Although the coupled proton–electron transfer mechanism would possibly have the advantage of leaving an open site on molybdenum to restart the cycle, there is no strong data to support either of these mechanisms at present.

For nitrogenase, the situation is less certain. First, the metal(s) present at the active site, the oxidation state(s), and the state(s) of aggregation are virtually unknown at present. Thus, any suggestion made for nitrogenase must be viewed as highly speculative and useful only to the extent to which it suggests further experiments on enzymes or model systems. With this in mind, coupled electron–proton transfer schemes can be suggested for nitrogenase. Again, while a (IV)–(VI) couple is used to illustrate the process, other two-electron couples are also possible. The key step in such a process, as visualized in Figure 7a for acetylene reduction, involves the coupled transfer of two protons and two electrons to substrate. The known cis stereochemistry of the addition is consistent with this proposal.

For dinitrogen reduction, a two-metal-site hypothesis (52, 53, 66–68, 96), as shown in Figure 7b, might be invoked with dinitrogen first binding end-on (perhaps to iron) and then additionally binding side-on to the same site at which acetylene reduction occurs. The first reduction product of dinitrogen would then be a bound cis-diimide species in agreement with the interpretation of the dihydrogen inhibition and HD production reactions of nitrogenase discussed earlier. The molybdenum site could then be reactivated twice more, with hydrazine and finally ammonia being sequentially formed. For nitrogenase there are clearly additional experimental observations which remain to be integrated with and must reflect upon the eventual mechanistic conclusions.

Nitrogenase—Additional Considerations. Nitrogenase differs from all other molybdenum enzymes in several important ways. It is the only known molybdenum enzyme to consist of two separately isolable proteins and to require ATP hydrolysis in its catalytic cycle. Its substrate half-reactions are the only ones which do not (in any obvious way) involve transfer of oxygen atoms. The nitrogenase system evolves dihydrogen when supplied with ATP and reductant in the absence of reducible substrate. All substrate reactions (but not ATP-dependent dihydrogen evolution) are inhibited by carbon monoxide. The presence of substrate appears to curtail the dihydrogen evolution reaction but, for some substrates at least (including dinitrogen), it seems impossible to eliminate dihydrogen evolution completely (*46, 58, 59*). In this section, we address dihydrogen evolution first and then the ATP utilization reaction of nitrogenase.

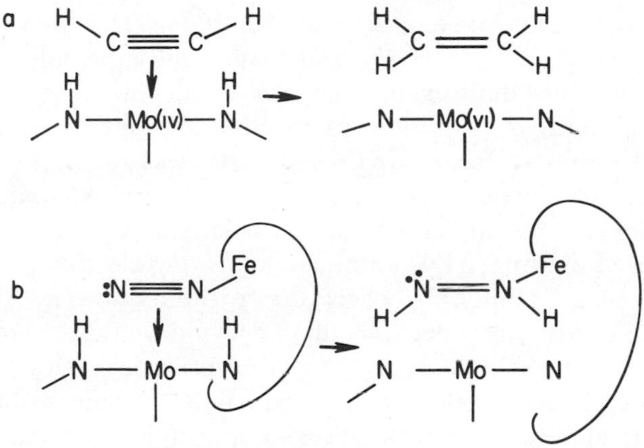

Figure 7. Proposed proton–electron transfer step for nitrogenase. (a) C_2H_2 reduction to C_2H_4; (b) N_2 reduction to bound N_2H_2 (66, 67, 68).

Dihydrogen Evolution Reaction. The dihydrogen evolution reaction of nitrogenase is certainly a thermodynamically reasonable one, i.e., the site which reduces dinitrogen should have sufficient potential to evolve dihydrogen. The molecular mechanism by which this arises is totally unknown although this will not stop us from speculating. In view of the lack of inhibition by carbon monoxide and its ATP dependence, it is assumed that dihydrogen evolution in nitrogenase occurs by a process different from that which occurs in hydrogenase (*56*). As hydrogenase contains only iron–sulfur clusters and no molybdenum (*56*), the molybdenum site (the presumed substrate reduction site) may be the location of the dihydrogen evolution reaction in nitrogenase. In the coupled proton–electron transfer scheme, the fully activated site having the potential to donate two protons and two electrons can react to evolve dihydrogen either with or without the aid of water. If water were involved, an Mo–O linkage might be formed with ATP helping to remove that oxo ligand from molybdenum. Alternatively, it is possible that a metal hydride or dihydride is involved which, upon reductive elimination or reaction with protons, produces dihydrogen. It is not possible to distinguish between these possibilities at present.

The question now arises as to why some substrates (like acetylene) completely curtail dihydrogen evolution while others (such as dinitrogen) do not. Most data indicate that the evoltuion of one mole of dihydrogen per mole of dinitrogen reduced is approached at high levels of dinitrogen (51, 58). A related problem involves the fact that acetylene is a noncompetitive inhibitor of nitrogen fixation while dinitrogen is a competitive inhibitor of acetylene reduction (51, 57). Several hypotheses have been advanced to explain these facts.

THE LEAKY SITE HYPOTHESIS (51, 99, 100). This hypothesis envisions nitrogenase as an electron sink which must be full (at least six electrons) to reduce dinitrogen. However, a full sink may not always be maintained because of limitations in electron flow, and the sink may then leak two electrons to form dihydrogen. If acetylene is present, it can accept these two electrons from the sink. The acetylene–dinitrogen inhibition studies can then be rationalized as follows. Acetylene can overcome the presence of dinitrogen by continually removing electron pairs from the sink and keeping it empty. However, dinitrogen requires six electrons for reduction, and therefore a partially filled enzyme would remain accessible to acetylene. Thus, high dinitrogen concentrations cannot effectively eliminate acetylene reduction. In this model, the inability of high dinitrogen totally to eliminate either dihydrogen or acetylene reduction is attributed to the inability of the enzyme to keep the sink full. Using the commonly accepted notion that the iron protein supplies electrons, this model would predict that as the component ratio ($[Fe]/[Mo–Fe]$) is increased and the sink is kept more nearly full, the dihydrogen evolution and acetylene reduction reactions could be more nearly quenched by dinitrogen. However, as most dihydrogen evolution and acetylene reduction experiments have been carried out using nitrogenase complex or a fixed $[Fe]/[Mo–Fe]$ ratio, this question still awaits an experimental answer.

THE FOUR-ELECTRON HYPOTHESIS. This hypothesis, apparently favored by Shilov (101, 102), postulates that nitrogenase works by a series of two four-electron processes. In the reduction of dinitrogen, the first step would be the production of N_2H_4 while the second would be the production of $2NH_3$ and H_2 which nicely explains the 1:1 stoichiometry for dinitrogen and dihydrogen discussed above. The residual reduction of acetylene in the presence of dinitrogen is explained by a sequential disassociation of ammonia and dihydrogen from the enzyme with the activated hydrogen (or protons and electrons) on the enzyme being able to reduce acetylene (but only to ethylene). This model does not fully explain why acetylene is only reduced to ethylene by nitrogenase in the absence of dinitrogen if the four-electron process is really the basic step.

THE EQUILIBRATING HYDROGEN/NITROGEN MODEL. This model (52, 103) postulates, in close analogy to inorganic systems, that a dihydride site on the enzyme is reactive towards dinitrogen binding in a manner exactly analogous to the reactions of known molybdenum and iron dihydrides (e.g., Reactions 21 and 22). The 1:1 $N_2:H_2$ stoichiometry is

$$FeH_4(PEtPh_2)_3 + N_2 \rightarrow FeH_2N_2(PEtPh_2)_3 + H_2 \tag{21}$$

$$MoH_4(dppe)_2 + 2N_2 \rightarrow Mo(N_2)_2(dppe)_2 + 2H_2 \tag{22}$$

directly explainable in this model. Although the apparent competitive inhibition of nitrogen fixation by dihydrogen is also explainable in such a model, our recent results (60) indicate the absence of true competitive inhibition in nitrogenase such that this latter point is moot. Further, if such an equilibrium were operative, dihydrogen inhibition of other substrate reactions would be probable. However, in practice, only dinitrogen reduction is inhibited, and the likely mechanism for that process is described earlier in this paper.

THE REDUCTIVE ELIMINATION HYPOTHESIS. The 1:1 stoichiometry of dihydrogen evolved and dinitrogen reduced can also be explained if formation and displacement of dihydrogen at the molybdenum site were an integral part of dinitrogen reduction but were unnecessary for acetylene reduction, i.e., two-electron and six-electron substrates are reduced via related but somewhat different catalytic cycles (96). Dihydrogen evolution might occur by reductive elimination from a molybdenum dihydride, thus making an electron pair on molybdenum available for substrate reduction. With dinitrogen, this dihydrogen elimination might be required to initiate each reduction cycle whereas for acetylene, the dihydrogen evolution is required only initially to prime the site for multiple acetylene reductions, i.e., the site remains primed after acetylene reduction but requires repriming after the six-electron dinitrogen reduction. The reason for the difference may involve the larger number of oxo groups produced on molybdenum during nitrogen fixation and the consequent requirement for their removal. ATP may also function (as discussed below) in this site clearing (oxo removal) which results in dihydrogen evolution. The nonreciprocal nature of the mutual inhibition of dinitrogen and acetylene could be explained by dinitrogen reduction requiring an initial activation at a second metal (iron) at the site, while acetylene does not. Further experimental elaboration is clearly needed to distinguish these possibilities.

The ATP Utilization Reaction. Under optimum conditions, nitrogenase requires the hydrolysis of 4–5 moles of ATP per electron pair transferred (48, 58, 59). As this represents the expenditure of 100 kcal/mole per dinitrogen reduced, a clearly pertinent question arises as to the reason for the ATP requirement. Among the many suggestions for the role of ATP are those involving ATP in a conformational change of either the iron and/or molybdenum–iron proteins. While this function of ATP may well be important, here we focus on those suggestions involving specific chemical interactions between ATP and the molybdenum site. Such suggestions (which are not necessarily preclusive of a concurrent conformational change) generally involve ATP in the generation of an open coordination site on molybdenum.

We have noted previously that Mo–O linkages pervade the chemistry of molybdenum in high oxidation states and that the removal of Mo–O is not an easy task. ATP, by its ability to act as either a phosphorylating agent or a proton source, could facilitate the oxo removal reaction. One possibility is that ATP couples phosphorylation of the Mo–O group with reduction of the site by two electrons. The rationale for the coupled electron–phosphoryl group transfer is similar to that for the coupled electron–proton transfer with the electrophilic phosphoryl group serving in place of the proton. Thus, as shown in Reaction 23, Mo–O could,

$$
\text{R—O—}\overset{\overset{\displaystyle O}{\|}}{\underset{\underset{\displaystyle O^-}{|}}{P}}\text{—O—}\overset{\overset{\displaystyle O}{\|}}{\underset{\underset{\displaystyle O^-}{|}}{P}}\text{—O—}\overset{\overset{\displaystyle O}{\|}}{\underset{\underset{\displaystyle O^-}{|}}{P}}\text{—O}^- + \text{O=Mo(VI)} \xrightarrow{2e^-}
$$

(ATP)

$$
\text{R—O—}\overset{\overset{\displaystyle O}{\|}}{\underset{\underset{\displaystyle O^-}{|}}{P}}\text{—O—}\overset{\overset{\displaystyle O}{\|}}{\underset{\underset{\displaystyle O^-}{|}}{P}}\text{—O}^- + {}^-\text{O—}\overset{\overset{\displaystyle O}{\|}}{\underset{\underset{\displaystyle O^-}{|}}{P}}\text{—O—Mo(IV)} \qquad (23)
$$

(ADP)

$$
\text{P}_i + \text{Mo(IV)} \longleftarrow
$$

upon accepting electrons, become more nucleophilic and be attacked by the terminal phosphoryl group of ATP. Transfer of this group to molybdenum would produce ADP and a molybdenum phosphate. The phosphate might then dissociate (perhaps assisted by acid formed from hydrolysis of additional ATP) to form the open active molybdenum site.

Alternatively, it is well known (74) that oxo group removal coupled with reduction is facilitated by the presence of acid. Protonation of oxo groups produces hydroxo or aquo ligands which are superior leaving groups. The hydrolysis of ATP near pH 7 produces protons, and this hydrolysis may be enzymically controlled to produce these protons near the oxo group which would facilitate the coupled reduction and oxo removal on the molybdenum site. In either case, the role of ATP would appear to be the production of an open site on molybdenum and in some model systems (104, 105), ATP may also be serving this function. Further experimental elaboration of the ATP reactions is desirable in both enzyme and model systems.

Why Use Molybdenum? At this point we return to the question posed at the outset of this paper, viz., what is unique in the chemistry of molybdenum which allows it to participate in the enzyme systems discussed in this paper? Some tentative answers may now be given. First, molybdenum can undergo two-electron transfer reactions at potentials compatible with biochemical reactions. This criterion eliminates the congeners, chromium and tungsten, as chromium is too oxidizing in the (VI) state while the lower states of tungsten may be too strongly reducing. Furthermore, molybdenum can couple these two-electron transfer reactions to either an oxide transfer or a proton transfer. The high oxidation states available to molybdenum make it particularly suitable for these tasks. As further details of the molybdenum sites in enzymes become available, refinements in our mechanistic proposals and model building studies should cast further light on the role of molybdenum in enzymes.

Literature Cited

1. Crick, F. H. C., Orgel, L. E., *Icarus* (1973) **19**, 341.
2. Chappell, W. R., Meglen, R. R., Runnells, D. D., *Icarus* (1974) **21**, 513.
3. Jukes, T. H., *Icarus* (1974) **21**, 516.
4. Egami, F., *J. Mol. Evol.* (1974) **4**, 113.
5. Orgel, L. E., *Icarus* (1974) **21**, 518.
6. Green, R. V., in "Handbook of Industrial Chemistry," J. A. Kent, Ed., Van Nostrand Reinhold, New York, 1974.
7. Newton, W. E., Nyman, C. J., Eds., "Proceedings of the First International Symposium on Nitrogen Fixation," Washington State University, Pullman, Washington, 1976.
8. Quispel, A., Ed., "The Biology of Nitrogen Fixation," North Holland, Amsterdam, 1974.
9. Stewart, W. D. P., Ed., "Nitrogen Fixation by Free-living Microorganisms," Cambridge University, Cambridge, 1975.
10. Pagan, J. D., Child, J. J., Scowcraft, W. R., Gibson, A. H., *Nature* (1975) **256**, 407.
11. Kurz, W. G. W., LaRue, T. A., *Nature* (1975) **256**, 407.
12. Dilworth, M. J., Elliot, J., McComb, J. A., *Nature* (1975) **256**, 409.
13. Keister, D. L., *J. Bacteriol.* (1975) **123**, 1265.

14. Reporter, M., Hermina, N., *Biochem. Biophys. Res. Commun.* (1975) **64**, 1126.
15. Tjepkema, J., Evans, H. J., *Biochem. Biophys. Res. Commun.* (1975) **65**, 625.
16. Hewitt, E. J., in "Plant Biochemistry," Northcote, D. H., Ed., University Park, Baltimore, 1974.
17. Beevers, L., Hageman, R. H., *Ann. Rev. Plant Phys.* (1969) **20**, 495.
18. Orme-Johnson, W. H., Jacob, G., Henzl, B., Averill, B. A., ADV CHEM. SER. (1977) **162**, 389.
19. Bray, R. C., Swann, J. C., *Struct. Bonding* (1972) **14**, 107.
20. Bray, R. C., in "The Enzymes," P. D. Boyer, Ed., 3rd ed., Academic, New York, 1975.
21. Wurzinger, K. H., Hartenstein, R., *Comp. Biochem. Physiol.* (1974) **49B**, 171.
22. Waud, W. R., Rajagopalan, K. V., *Arch. Biochem. Biophys.* (1976) **172**, 354.
23. Elion, G. B., *Ann. Rheum. Dis.* (1966) **25**, 608.
24. Massey, V., Komai, H., Palmer, G., Elion, G. B., *J. Biol. Chem.* (1970) **245**, 2837.
25. Krenitsky, T. A., Neil, S. M., Elion, G. B., Hitchings, G. H., *Arch. Biochem. Biophys.* (1972) **150**, 585.
26. Cohen, H. J., Fridovich, I., *J. Biol. Chem.* (1971) **246**, 367.
27. Cohen, H. J., Fridovich, I., Rajagopalan, K. V., *J. Biol. Chem.* (1971) **246**, 374.
28. Cohen, H. J., Drew, R. T., Johnson, J. L., Rajagopalan, K. V., *Proc. Natl. Acad. Sci. USA* (1973) **70**, 3655.
29. Mudd, S. M., Irreverre, F., Laster, L., *Science* (1967) **156**, 1599.
30. Andreesen, J. R., Ljungdahl, L. G., *J. Bacteriol.* (1973) **116**, 967.
31. Enoch, H. G., Lester, R. L., *J. Biol. Chem.* (1975) **250**, 6693.
32. Ljungdahl, L. G., Andreesen, J. R., *FEBS Lett.* (1975) **54**, 279.
33. Ljungdahl, L. G., *Trends Biochem. Sci.* (1976) **1**, 63.
34. Dervartanian, D. V., Bramlett, R., *Biochem. Biophys. Acta* (1970) **220**, 443.
35. Pateman, J. A., Cove, D. J., Rever, B. M., Roberts, D. B., *Nature* (1964) **201**, 58.
36. Nason, A., Antoine, A. D., Ketchum, P. A., Frazier, W. A., Lee, D. K., *Proc. Natl. Acad. Sci. USA* (1970) **65**, 137.
37. Nason, A., Lee, K.-Y., Pan, S.-S., Ketchum, P. A., Lamberti, A., DeVries, J., *Proc. Natl. Acad. Sci. USA* (1971) **68**, 3242.
38. Lee, K.-Y., Pan, S.-S., Erickson, R., Nason, A., *J. Biol. Chem.* (1974) **249**, 3941.
39. Ketchum, P. A., Swarin, R. S., *Biochem. Biophys. Res. Comm.* (1973) **52**, 1450.
40. Ganelin, V. L., L'vov, N. P., Sergeev, N. S., Shaposhnikov, G. L., Kretovich, V. L., *Dokl. Akad. Nauk., SSSR* (1972) **26**, 1236.
41. L'vov, N. P., Ganelin, V. L., Alikulov, Z., Kretovich, V. L., *Izv. Akad. Nauk., SSSR, Ser. Biol.* (1975) 371.
42. Zumft, W. G., *Ber. Dtsch. Bot. Ges.* (1974) **87**, 135.
43. Nagatani, H. H., Shah, V. K., Brill, W. J., *J. Bacteriol.* (1974) **120**, 697.
44. Brill, W. J., ADV. CHEM. SER. (1977) **162**, 402.
45. Bulen, W. A., in "Proceedings of the First International Symposium on Nitrogen Fixation," W. E. Newton and C. J. Nyman, Eds., Washington State University, Pullman, Washington, 1976.
46. Bulen, W. A., LeComte, J. R., Lough, S. L., Land, R., Hadfield, K. L., Watt, G. D., unpublished data.
47. Bulen, W. A., LeComte, J. R., in "Advances in Enzymology," A. San Pietro, Ed., Vol. XXIV, Part B, p. 456, Academic, New York, 1972.

48. Watt, G. D., Bulen, W. A., Burns, A., Hadfield, K. L., *Biochemistry* (1975) **14**, 4266.
49. Bulen, W. A., Burns, R. C., Lecomte, J. R., *Proc. Natl. Acad. Sci., USA* (1965) **53**, 532.
50. Watt, G. D., Burns, A., *Biochemistry* (1977) **16**, 264.
51. Rivera-Ortiz, J. M., Burris, R. H., *J. Bacteriol.* (1975) **123**, 537.
52. Hardy, R. W. F., Burns, R. C., Parshall, G. W., ADV. CHEM. SER. (1971) **100**, 219.
53. Zumft, W. G., Mortenson, L. E., *Biochim. Biophys. Acta* (1975) **416**, 1.
54. Dilworth, M. J., *Biochem. Biophys. Acta* (1966) **12**, 285.
55. Hardy, R. W., Burns, R. C., in "Iron Sulfur Proteins," W. Lovenberg, Ed., p. 66, Academic, Nek York, 1973.
56. Chen, J.-S., Mortenson, L. E., *Biophys. Biochim. Acta* (1974) **371**, 283.
57. Huang, J. C., Chen, C. H., Burris, R. H., *Biochim. Biophys. Acta* (1973) **292**, 256.
58. Hadfield, K. L., Ph.D. Thesis, Brigham Young University, 1970.
59. Hadfield, K. L., Bulen, W. A., *Biochemistry* (1969) **8**, 5103.
60. Newton, W. E., Bulen, W. A., Hadfield, K. L., Stiefel, E. I., Watt, G. D., in "Recent Developments in Nitrogen Fixation," W. E. Newton, J. R. Postgate, and C. Rodriguez-Barrueco, Eds., Academic, London, in press.
61. Jackson, E. K., Parshall, G. W., Hardy, R. W. F., *J. Biol. Chem.* (1968) **243**, 4952.
62. Kelly, M., *Biochem. J.* (1968) **109**, 322.
63. Turner, G. L., Bergersen, F. J., *Biochem. J.* (1969) **115**, 529.
64. Dalton, H., Lowe, D. J., Pawlik, R. T., Bray, R. C., *Biochem. J.* (1976) **153**, 287.
65. Edmondson, D., Massey, V., Palmer, G., Beacham, L. M., Elion, G. B., *J. Biol. Chem.* (1972) **247**, 1597.
66. Stiefel, E. I., *Proc. Nat. Acad. Sci., USA* (1973) **70**, 788.
67. Stiefel, E. I., Gardner, J. K., in "Proceedings of the First Internat. Conference on Chemistry and Uses of Molybdenum," P. C. H. Mitchell, Ed., Climax Molybdenum Co., London, 1974.
68. Stiefel, E. I., Gardner, J. K., *J. Less-Common Met.* (1974) **36**, 521.
69. Olson, J. S., Ballou, D. P., Palmer, G., Massey, V., *J. Biol. Chem.* (1974) **249**, 4363.
70. *Ibid.* (1974) **249**, 4350.
71. Johnson, J. L., Waud, W. R., Cohen, H. J., Rajagopalan, K. V., *J. Biol. Chem.* (1974) **249**, 5056.
72. Johnson, J. L., Cohen, H. J., Rajagopalan, K. V., *J. Biol. Chem.* (1974) **249**, 5046.
73. Bray, R. C., Meriwether, L. S., *Nature* (1966) **212**, 467.
74. Stiefel, E. I., *Prog. Inorg. Chem.* (1976) **22**, 1.
75. Bray, R. C., Knowles, P. F., *Proc. Roy. Soc.* (1968) **A302**, 351.
76. Edmondson, D., Ballou, D., Van Heuvelen, A., Palmer, G., Massey, V., *J. Biol. Chem.* (1973) **248**, 6135.
77. Schneider, P. W., Bravard, D. C., McDonald, J. W., Newton, W. E., *J. Am. Chem. Soc.* (1972) **94**, 8640.
78. Newton, W. E., Corbin, J. L., Bravard, D. C., Searles, J. E., McoDnald, J. W., *Inorg. Chem.* (1974) **13**, 1100.
79. Newton, W. E., Bravard, D. C., McDonald, J. W., *Inorg. Nucl. Chem. Lett.* (1975) **11**, 553.
80. Mitchell, P. C. H., Scarle, R. D., *J. Chem. Soc. Dalton* (1975) 2552.
81. Chen, G. J.-J., McDonald, J. W., Newton, W. E., *Inorg. Chem.* (1976) **15**, 2612.
82. McDonald, J. W., personal communication.
83. Ketchum, P. A., Taylor, R. C., Young, D. C., *Nature* (1976) **259**, 202.

84. Garner, C. D., Hyde, M. R., Mabbs, F. E., Routledge, V. I., *Nature* (1974) **252**, 579.
85. Garner, C. D., Hyde, M. R., Mabbs, F. E., Routledge, V. I., *J. Chem. Soc. Dalton* (1975) 1180.
86. Taylor, R. D., Spence, J. T., *Inorg. Chem.* (1975) **14**, 2815.
87. Watt, G. D., McDonald, J. W., Newton, W. E., in "Second International Conference on Chemistry and Uses of Molybdenum," P. C. H. Mitchell, Ed., Climax Molybdenum Co., London, in press.
88. Howie, J. K., Sawyer, D. T., *Inorg. Chem.* (1975) **15**, 1892.
89. Gardner, J. K., Pariyadath, N., Corbin, J. L., Stiefel, E. I., unpublished data.
90. Basolo, F., Pearson, R. G., "Mechanisms of Inorganic Reactions," 2nd ed., Wiley, New York, 1967.
91. Newton, W. E., Chen, G. J.-J., McDonald, J. W., *J. Am. Chem. Soc.* (1976) **98**, 5386.
92. Pariyadath, N., Newton, W. E., Stiefel, E. I., *J. Am. Chem. Soc.* (1976) **98**, 5388.
93. Barral, R., Bocard, C., Seree de Roch, I., Sajus, L., *Kinet. Catal.* (1973) **14**, 130.
94. Barral, R., Bocard, C., Seree de Roch, I., Sajus, L., *Tetrahedron Lett.* (1972) 1693.
95. Jowitt, R. N., Mitchell, P. C. H., *J. Chem. Soc.* (A) (1969) 2632.
96. Newton, W. E., Corbin, J. L., McDonald, J. W., in "Proceedings of the First International Symposium on Nitrogen Fixation," W. E. Newton and C. J. Nyman, Eds., Washington State University, Pullman, Washington, 1976.
97. Branzoli, U., Massey, V., *J. Biol. Chem.* (1976) **249**, 4346.
99. Burris, R. H., Orme-Johnson, W H., in "Proceedings of the First International Symposium on Nitrogen Fixation," W. E. Newton and C. J. Nyman, Eds., Washington State University, Pullman, Washington, 1976.
100. Davis, L. C., Shah, V. K., Brill, W. J., *Biochem. Biophys. Acta* (1975) **403**, 67.
101. Shilov, A. E., Likhtenstein, G. I., *Izv. Akad. Nauk SSSR Ser. Biol.* (1971) 518.
102. Shilov, A. E., *Russ. Chem. Rev.* (1974) **43**, 378.
103. Hardy, R. W. F., Burns, R. C., Parshall, G. W., "Inorganic Biochemistry," Vol. 2, G. L. Eichhorn, Ed., Elsevier, New York, 1973.
104. Schrauzer, G. N., *Angew. Chem. Int. Ed.* (1975) **14**, 514.
105. Krushch, A. P., Shilov, A. E., Vorontsova, T. A., *J. Am. Chem. Soc.* (1974) **96**, 989.

RECEIVED July 26, 1976. This is contribution No. 565 from the Charles F. Kettering Research Laboratory.

21

Molybdenum in Enzymes

W. H. ORME-JOHNSON, C. S. JACOB, M. T. HENZL, and B. A. AVERILL

Department of Biochemistry, College of Agricultural and Life Sciences, University of Wisconsin–Madison, Madison, WI 53706

Current concepts of the chemical nature and role of molybdenum-containing enzymes are reviewed. Methods for molybdenum in enzymes, spectroscopic manifestations of the metal, and the characteristics of molybdenum-deficient enzymes are discussed, with particular attention to xanthine oxidase, sulfite oxidase, and nitrate reductase, in which Mo^{5+} (and Mo^{3+} in some cases) species are readily demonstrated. Nitrogenase is presumed to use molybdenum in a catalytic step, but no direct evidence for its participation in catalysis is yet available.

O f the elements in the fifth period of the periodic table, only molybdenum, iodine, and tin now are known to be required by living organisms. One of the fascinating puzzles of metallobiochemistry addressed in this symposium is the explanation of those unique properties of molybdenum that have led to its incorporation in a series of oxidoreductases. These enzymes catalyze processes that, as chemists, we do not consider uniquely to require molybdenum. Thus the evolutionary development and preservation of a need for this metal, the satisfaction of which doubtless requires concerted physiological effort by many organisms, hints at a range of catalytic specialization which we do not understand now. The tacit recognition of this fact has led to the current surge in the study, exemplified by the contribution of Enemark et al. (1), of the chemistry of the element.

To come to grips with the general problem posed above, we have to ask and answer a series of questions which on reflection seem rather obvious but which are general to current studies on many metalloenzymes. These include:

(1) Establishing reliable estimates of the molybdenum content of each enzyme molecule and each active site

(2) Deciding whether given atoms function in structural, catalytic, or dual roles

(3) Determining the formal oxidation states enjoyed by the metal during various phases of catalysis

(4) Determining the relative thermodynamic properties of these various states.

We have to decide in each case whether an identifiable low molecular weight "molybdenum cofactor" is used, whether coordinating groups provided by the protein are also directly used, and whether coordination of substrates to the metal occurs during catalysis. In addition, we must understand the integration of the properties of the metal with the other catalytic resources of the protein. We seek nothing less than a complete account of the chemistry of molybdenum in these proteins; at this point a statement of these wishes must be immediately followed by an admission of our profound ignorance of these affairs. We will briefly examine the methodological avenues that currently seem open and then present a short discussion of the present position in regard to certain molybdenum-containing enzymes. It is hoped that the overall effect of these remarks will be stimulating to readers with various talents and ranges of expertise. Clearly the investigation of these catalytic systems will continue to require the participation of workers from several branches of chemistry, physics, and biology.

Methods Used to Study Molybdenum in Enzymes

Perhaps the most elementary question to be asked about molybdoproteins is, what is the molybdenum content? Most workers have used the method of Clark and Axley (2), in which complete digestion of the specimen in hot mineral acid is followed by colorimetric determination as the toluenedithiol complex. About 5 nmoles of molybdenum are required, which for many enzymes is the order of a mg of protein. In principle, neutron activation and x-ray fluorescence methods might be at least as sensitive, but they have not been widely applied to these systems (see, however, Ref. 3). A number of workers have not been able to obtain accurate results with atomic absorption methods using flames, but it appears that the newer graphite furnace technique may be more satisfactory (4). These problems are of continuing concern because even homogeneous molybdoprotein preparations have not always exhibited integral metal-to-protein ratios. The case of milk xanthine oxidase, doubtless the most intensively studied of all molybdoenzymes, is instructive (5). Crystalline preparations contain a minimum of three types of molecules: the native, active form with two molybdenums per molecule; partially active molecules depleted of molybdenum; and a form which has two molybdenums per molecule but is inactive, presumably because it lacks a critical persulfide group. Bray and Swann suggest that molybdenum-depleted forms of milk xanthine oxidase occur depending on the nutritional status of the cow. Johnson et al. (*6, 7*) have shown that rat liver xanthine oxidase and sulfite reductase are both depleted of molybdenum when the animals are maintained on a diet containing tungsten, which prevents efficient incorporation of molybdenum. These phenomena have their analogs in the nutrition of microorganisms, as discussed by Brill (*8*), and emphasize the caveat that molybdoprotein preparations may be homogeneous as to their protein moiety but heterogeneous with respect to their metal content. This may be inconvenient to rectify except by nutritional manipulation before isolation of the protein begins.

The antagonism of biological molybdenum incorporation by metals nearby in the periodic table deserves further comment. Notton and Hewitt (*9*) have given a useful summary of earlier work in this area. Attempts by several workers to produce vanadium-substituted nitrogenase have not been decisive, but the more logical antagonist tungsten has proven useful. Such an antagonist may block uptake of molybdenum by the cells or may prevent incorporation into the apoprotein. This may take place by blockage of cellular processing of molybdenum or by replacement of molybdenum by tungsten in the finished protein. Even in the same tissue both processes may be evident. Johnson et al. (*6*) found that the livers of tungsten-treated rats produce apoxanthine oxidase devoid of molybdenum but cross-react with antibodies to normal rat liver xanthine oxidase. The tungsten-induced apoprotein had the normal ratio of FAD:non-heme iron (1:4) and was devoid of tungsten and molybdenum. The apoenzyme oxidized NADH but was unable to oxidize xanthine. The same animals contained a hepatic sulfite oxidase apoprotein, as judged by the presence of antigen reacting with antibodies to normal sulfite oxidase (*7*). This apoprotein possessed no molybdenum and no sulfite oxidizing activity but did contain the b_5 cytochrome of normal sulfite oxidase. In addition, the protein contained tightly bound tungsten incorporated into about 35% of the molybdenum-free molecules. Thus, in the same tissue tungsten antagonism is manifest both in the prevention of the formation of the molybdenum–protein complex, and, in the case of sulfite oxidase, in the replacement of molybdenum by tungsten in the enzyme. The finding that upon reduction, the protein-bound tungsten yielded EPR near $g = 1.87$, similar to that of molybdenum in the native enzyme (near $g = 1.97$), suggests that the replacement by tungsten is at the active site of the enzyme.

As interesting as these results are by themselves, perhaps their major importance may lie in allowing one to prepare an apoprotein system that can be used to test for possible low molecular weight "molybdenum cofactors." Indeed, Brill and his co-workers (8) have shown that the presence of molybdenum in the medium is required by *Azotobacter vinelandii* for the depression of nitrogenase, i.e., for the synthesis of the enzyme when fixed nitrogen (NH_4^+) levels in the medium are low. They found that tungstate replaces molybdate in this function, since the polypeptide chains comprising the molybdenum–iron protein of nitrogenase are synthesized in the presence of tungstate, but tungsten is not incorporated into the enzyme. The apoprotein thus formed can be reactivated by acid extracts of the molybdenum–iron protein of nitrogenase from several bacterial species, forming the basis of a sensitive test system for the "molybdenum cofactor." Pateman et al. (10) found mutants of *Aspergillus nidulans* lacking both nitrate reductase and xanthine dehydrogenase and suggested that these mutants lack a cofactor common to the two enzymes. Nason and his co-workers initially proposed the existence of a "molybdenum cofactor." This was based on their discovery that nitrate reductase activity in extracts of a nitrate reductaseless mutant of *Neurospora crassa* (nit-1) was elicited by the addition of acid-treated extracts of various molybdoenzymes (11). These extracts could not be replaced by molybdate or other well characterized low molecular weight molybdenum compounds, although subsequent studies (12) indicated that radioactivity from $^{99}MoO_4^{2-}$, added to the mixture of acid extract and nit-1 extract, was incorporated into the nitrate reductase fraction as resolved during sucrose density gradient centrigugation. Furthermore, molybdate exerted a protective effect on the principle extracted by acid from the donor molybdoproteins. These studies suggest that the "molybdenum cofactor" in the extracts may be in equilibrium with molybdate, i.e., the molybdenum dissociates from another essential component, since molybdate alone is ineffective. This further suggests that, unlike the case of prosthetic groups such as flavins and hemes but as in the example of iron–sulfur centers (13, 14), one may have to use rather delicate means, perhaps including nonaqueous solvents, to suppress hydrolysis in order to recover "molybdenum cofactor(s?)" intact for further study.

It is obvious that x-ray cyrstallographic methods will be the final arbiter of the structural features of molybdoproteins, but until such structures are obtained, and even afterwards as far as dynamic features are concerned, spectroscopic methods must be used to gain insight into the nature of these catalysts. Electronic spectroscopy so far has been of little use here since molybdenum complexes in general appear to exhibit broad weak absorptions. In proteins these are always buried under absorptions from hemes, flavins, and iron–sulfur centers. Massey et al., (15) discovered that pyrazolo [3,4-*d*] pyrimidines will bind Mo(IV) in milk xanthine oxidase that had been reduced with xanthine in the absence of oxygen. When oxygen was readmitted, the iron–sulfur centers and FAD were reoxidized while the Mo(IV) remained as the pyrimidine complex. Difference spectra between such preparations and the native oxidized enzyme revealed the presence of a band presumed the from the Mo(IV) complex, with molar absorbancies of about 7×10^{-3} l·mole^{-1}·cm^{-1} and absorbance maxima between 375 and 500 depending on the substituents on the pyrazolo [3,4-d] pyrimidine. This in situ colorimetric determination is the only example so far of an absorbance change in an enzyme which reasonably could be ascribed to a change in the oxidation state of molybdenum.

Low temperature EPR has provided almost all of the spectroscopic information about molybdoproteins to date. The method is applicable to the Mo(III) and Mo(V) oxidation states, with most of the published information relating to the latter state. Mo(V) is generally easily recognizable, with a slightly anisotropic set of g-values in the region of $g = 1.96$–1.99 (*see* Ref. 5 and references therein). In cases of doubt enrichment of the specimen in molybdenum isotopes with $I = 5/2$ can verify the assignment, as elegantly shown in the experiments of Bray and Meriwether (*16*) with bovine milk xanthine oxidase. The EPR of Mo(III) complexes has not been studied nearly as extensively, but a series of high spin Mo(III) compounds examined in this laboratory (*17*) all showed broad axial signals, with $g_1 \sim 4$, $g_{11} \sim 2$, as is expected for a $4d^3$ ion (*18*). Additionally, the combination of EPR with freeze-quenched sample preparation (*19*) allows one to extend the time resolution of the technique into the msec range. The sensitivity of the method is such that $10^{-6}M$ solutions of paramagnetic species may be used. The EPR technique has several disadvantages. It does not detect Mo(II), Mo(IV), or Mo(VI), and it may not detect Mo(III) or Mo(V) if these species are strongly coupled to other paramagnets, as is often the case with dimeric low molecular weight complexes of Mo(V). When EPR techniques (including ENDOR and ELDOR) are applicable, the results may be impressive as in the finding (Ref. 5 and references therein) that a substrate-donated proton (as evidenced by a proton hyperfine splitting) appears near Mo(V) which is produced rapidly in the presence of purine substrates. Furthermore, the combination of EPR and absorbance rapid reaction techniques, as performed by Olson et al. (*20*) for milk xanthine oxidase, in favorable cases may allow the calculation of the relative probabilities of populating the various combinations caused by oxidation cofactors existing in multiple oxidation states. Thus, the most probable sequence of events during a catalytic cycle may be evaluated. It is clear, however, that additional spectroscopic techniques are required for less favorable cases, the extreme being nitrogenase where no EPR as yet ascribable to molybdenum has been seen. Spectroscopic methods other than the low energy forms of spectroscopy (magnetic resonance methods) are intrinsically of lower sensitivity unless suitable radiation sources are developed, so that methods such as ESCA and EXAFS (21) will require heroic experimentation. Nonetheless, we are seemingly forced to these measures in order to carry out our program.

The Role of Molybdenum in Xanthine Oxidase, Nitrogenase, and Nitrate Reductase

Table I (adapted from Ref. 5) lists the types of molybdoenzymes recognized so far. Representative sources and literature references are given, but many other sources of each protein are known in general. We will give a brief guide to the literature and discuss a few cases where either information about the molybdenum complex has been obtained or there is some hope of obtaining such information.

Xanthine Oxidase. This molybdoenzyme is readily available from cows' milk in gram quantities (28) and is relatively stable, which accounts for the fact that it is by far the most intensively studied molybdoenzyme. Bray and Swann (5) have reviewed comprehensively the earlier literature, and recent papers by Olson et al. (20) summarize combined kinetic and thermodynamic approaches to the states of the prosthetic groups during catalysis. Two molybdenum, four iron–sulfur centers, and two FAD groups are present in each molecule. An important point raised by Edmondson, et al. (29) is that the rates of internal electron transfer among the prosthetic groups appear to be much more rapid than turnover. Olson et al., (20) deduced that the reduction potentials of the two processes $Mo(VI) \longleftrightarrow Mo(V) \longleftrightarrow Mo(IV)$ were —60 and —31 mv, respectively, relative to the redox potential for one of the iron–sulfur centers (center II) in the molecule. Thus, at equilibrium one can never have more than a small fraction of molybdenum as

Table I.

Name	Source
Nitrogenase MoFe protein	*Clostridium pasteurianum*
NADH dehydrogenase	*Azotobacter vinelandii*
Respiratory nitrate reductase	*Escherichia coli*
Assimilatory nitrate reductase	*Neurospora crassa*
Sulfite oxidase	bovine liver
Aldehyde oxidase	rabbit liver
Xanthine oxidase	cow's milk
Xanthine dehydrogenase	chicken liver

a Adapted from Ref. 5.

Mo(V). The results of Edmondson et al. (29) imply that these equilibria are maintained during catalysis. The groups coordinated to molybdenum in xanthine oxidase are not known, but some hints are available. Four main types of Mo(V) differentiable by EPR spectroscopy are observed in preparations of xanthine oxidase (5).

1. Inhibited, which is formed in the presence of formaldehyde or methanol during turnover with a normal substrate,

2 and 3. Rapid and Very Rapid, whose appearances depend on the conditions used for their elicitation and which are caused by molybdenum in the active enzyme,

4. Slow, which is from inactivated enzyme formed, for example, by cyanolysis which yields thiocyanate (30).

Bray and Swann suggest, on the basis of the changes in the EPR of Mo(V) after inactivation of the enzyme, that molecules yielding Slow lack a persulfide group which may be a ligand of molybdenum in the native enzyme. They also discuss the fact that the Mo(V) species Rapid and Slow show proton hyperfine splittings in their EPR which disappear when the enzyme is in D_2O, indicating that a protonatable group is near the molybdenum. On the evidence of experiments with 8-deutero-xanthine (31), the proton abstracted from purines is transitorily deposited in the position occupied by the exchangable proton, suggesting that the binding site for the oxidizable substrate is also in the immediate vicinity of the molybdenum.

Nitrogenase. This molybdoenzyme is composed of two proteins, both of which are required for the MgATP-dependent six-electron reduction of dinitrogen to ammonia. In the absence of other substrates, the enzyme reduces protons to hydrogen. Both enzyme components are air sensitive, and strict exclusion of oxygen is required during isolation and subsequent experimentation. A recent review of the enzymology of nitrogenase is given by Orme-Johnson and Davis (32), and the pro-

Molybdoenzymes[a]

Typical Substrates				
Reducing	Oxidizing	MW	Other Constituents	References
ferredoxin	N_2	220,000	Fe	22
NADH	quinones	n.d.	Fe, FMN	23
artificial donors[b]	NO_3^-	750,000	Fe	3, 24
NADPH	NO_3^-	220,000	heme, FAD	this work
SO_3^{2-}	O_2	110,000	heme	25
aldehydes	O_2	300,000	Fe, FAD	26
purines, aldehydes	O_2	275,000	Fe, FAD	20
purines, aldehydes	NAD	300,000	Fe, FAD	27

[b] In vitro; in vivo probably coupled to formate oxidation.

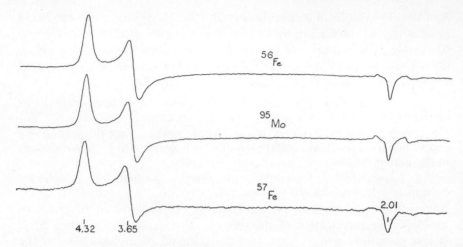

Figure 1.   EPR spectra of the molybdenum–iron protein of Azotobacter vinelandii, *enriched as indicated by isolation of the protein from bacteria grown on
isotopically enriched media.*

The magnetic field strength (abscissa) increases from left to right, and the ordinate is
an arbitrary function of the first derivative of the microwave power absorption. Some
values of $g = 0.71445 \, \nu/H$, where ν is the microwave frequency and H is the magnetic field strength, are marked along the abscissa. Microwave frequency, 9.2 GHz;
microwave power, 3 mw; modulation frequency, 100 KHz; modulation field, 10 gauss;
sweep rate, 500 gauss/min; time constant, 0.25 sec; temperature, 13°K.

ceedings of a symposium on nitrogen fixation held in 1974 have appeared
(33). The composition of the molybdenum–iron protein component has
not yet been settled. It appears to be composed of two kinds of subunits
and to contain about 24 non-heme iron atoms, 24 acid labile sulfur atoms,
and one or two molybdenum atoms. Except for the EXAFS measurements
reported by Cramer et al. (21), no spectroscopic manifestations of the
molybdenum in nitrogenase have yet been described. On the other hand,
the spectroscopy of the iron centers is a rich and expanding area. Combination of low temperature EPR and Mössbauer spectroscopy (34) suggests that the molybdenum–iron protein as isolated contains two centers
with four iron atoms per center. These centers are in a $S = 3/2$ spin
state and yield EPR as shown in Figure 1. Examination of [95]MoFe or
Mo[57]Fe protein from *Azotobacter vinelandii* (grown on isotopically
enriched media) showed that although the [57]Fe gives a small (ca. 7
gauss) hyperfine broadening on the feature near $g = 2$, the [95]Mo speci-
mens show no such broadening. In the steady state of catalysis, when
reductant, the iron protein, and MgATP are in excess, this signal
disappears. Carbon monoxide is a powerful noncompetitive inhibitor of
nitrogenase, and, when carbon monoxide is present in the steady state
mixture, two EPR signals arise depending on the [CO]/[MoFe] ratio
(35). These signals are evidently from iron–sulfur clusters, in oxidation

states corresponding to reduced ferredoxin or oxidized high potential iron–sulfur proteins (*see* Ref. 36) and references therein). As shown in Table II, these signals also exhibit hyperfine broadening when [57]Fe is present in molybdenum–iron protein; none is seen when [95]MoFe protein is used. These negative data cannot exclude the presence of molybdenum in the centers represented by these EPR signals, since the coupling constant for the [95]Mo nuclear interactions may be small. On the other hand, for many known Mo(V) complexes the [95]Mo coupling constants are comparable with the linewidths of the resonances dealt with here (37). It may be that the molybdenum atoms in nitrogenase are isolated, at least from the iron centers observed by EPR. The apparent intractible nature of the molybdenum centers in not yielding recognizable Mo(V) or Mo(III) EPR further suggests that if these states are utilized, either their existence is transitory and/or thermodynamically unfavored, or they are coupled to other paramagnets to form non-Kramers systems. The situation is presently extremely unsatisfactory, so that the outcome of the EXAFS experiments as well as possible ENDOR and static suscepti- bility studies on the system will be of great interest. Anticipating that molybdenum will be at the nitrogen-reducing site of nitrogenase, several workers have studied possible abiological model systems using molyb- denum complexes. Useful summaries of recent work are given by several authors in the volumes edited by Newton and Nyman (33).

Nitrate Reductases. These enzymes have two important func- tions (38). They serve as a terminal oxidation system in some micro- organisms, and in many microorganisms, molds, fungi, and higher plants they are utilized in the assimilation of nitrogen into the cell via production of ammonia, which is subsequently metabolized into cell components.

Table II. Nuclear Hyperfine Effects[a] on EPR Signals Elicited under Carbon Monoxide

Isotope Tested	High pCO		Low pCO		
	2.17	2.05	2.08	1.97	1.98
Av$_1$ natural	34[a]	32	17	14	14
Av$_1$ [95]Mo	34	33	18	14	14
Av$_1$ [57]Fe	40	44	30	30	29
Av$_2$ [57]Fe	34	32	17	14	14
—— [13]CO	33	32	17	14	14

[a] Line width at half height (gauss). Av$_1$, the molybdenum–iron protein; Av$_2$, the iron protein of nitrogenase from *Azotobacter vinelandii*; High pCO, [CO] >> [MoFe protein]; low pCO, [CO] = [MoFe protein]. The line widths are given for features at the g-values indicated in the column headings. Otherwise identical mix- tures of Av$_1$ and Av$_2$ containing isotopically enriched components, as indicated in the left column, were incubated at 30°C in the presence of the indicated relative amount of carbon monoxide and with a MgATP generating system for 45″ to insure that a steady state had been achieved. The samples were then frozen and examined under the EPR conditions of Figure 1. From Ref. 35.

The assimilatory enzyme from the mold *Neurospora crassa* has been intensively studied for over two decades, particularly by Nason and his collaborators. Thus, Nason and Evans (*39*) identified FAD as a prosthetic group in the enzyme; Nicholas, Nason, and McElroy (*40*) showed that molybdenum was required for the synthesis of nitrate reductase; Nicholas and Nason (*41*) suggested its presence in the enzyme; Garrett and Nason (*42*) showed that a *b*-type cytochrome (cytochrome b_{557}) co-purifies with this nitrate reductase; and Nason et al. (*11*) suggested, from in vitro complementation experiments with nitrate reductaseless mutants, that the enzyme consists of at least two components required for activity. These workers have suggested that the electron transfer pathway is:

$$\text{NADPH} \rightarrow \text{FAD} \rightarrow \text{cytochrome } b_{557} \rightarrow \text{Mo} \rightarrow \text{NO}_3^-$$

The purified enzyme (*43*) was subjected to gel electrophoretic examination and metal and prosthetic group analysis. Analytical results are shown in Table III along with comparative values for other nitrate

Table III. Composition of Nitrate Reductase[a]
(g · atom or g · mole/g · mole enzyme)

	Source			
Component	Neurospora[b] crassa	Chlorella[c] vulgaris	Eschericia[d] coli K12	Micrococcus[e] denitrificans
Mo	1.0	2	~ 1.5	~ 0.4
Heme	1.0	2	0	0
Total Fe	1.3	2	~ 20	~ 8
Non-heme Fe	(0.3)	0	~ 20	~ 8
Labile S	n.d.	0	~ 20	~ 8
FAD	~.1[f]	2	0	0

[a] Ref. *43*.
[b] Based on activity: 20 units = 1 nmol enzyme.
[c] Ref. *44*.
[d] Forget (*24*), MacGregor et al. (*3*) suggest four molybdenum per molecule protein.
[e] Ref. *45*.
[f] Dialyzed enzyme; FAD estimated from A_{450} (*46*). This is an upper limit. Added FAD is required for activity.

reductases. The value for non-heme iron is the difference between the heme and total iron determinations and is probably caused by adventitiously bound iron. The low flavin content is reasonable since the activity of the dialyzed enzyme in the absence of added FAD was less than 10% of that found with excess flavin.

EPR spectra at 12°K of *N. crassa* nitrate reductase samples in the form as isolated (the oxidized form) show resonances at g = 2.98, 2.27, which we ascribe to a low spin b-type cytochrome. According to the classification of heme EPR made by Blumberg and Peisach (47), these g-values indicate that the axial ligands to the heme are both imidazole. When the same samples were observed at a very low power, an easily saturable EPR signal near $g = 1.97$, which we attribute to Mo(V), become apparent.

During anaerobic reductive titrations with NADPH, the heme resonance disappeared because of reduction of the cytochrome while the Mo(V) signal increased in magnitude and subsequently disappeared. At the same time, a resonance with two g-values near $g = 4$ and one near $g = 2$ grew in and then went away. We ascribe this to an $S = 3/2$ species, probably Mo(III) which is known to yield an $S = 3/2$ ground state in most ligand fields. These experiments all suggest that the sequence Mo(VI) → Mo(V) → Mo(IV) → Mo(III) → Mo(II) occurs during reduction of nitrate reductase (underlined species are potentially observable in EPR). No evidence of a flavin radical was seen at any time during these measurements which were conducted at equilibrium. When the enzyme was reduced with an excess of NADPH and reoxidized (anaerobically) with KNO_3, we observed the disappearance of the above species in reverse order. For these reasons we feel confident that molybdenum in at least four oxidation states, as well as the heme iron, is thermodynamically competent to play a role in the mechanism of nitrate reductase. Further studies on kinetic aspects of the mechanism await the acquisition of larger quantities of this very active, and therefore scarce, enzyme.

Acknowledgments

The authors gratefully acknowledge the support of the NIH through a training grant GM 00,236 and a research grant GM 17,170, as well as the Graduate Research Committee and the College of Agricultural and Life Sciences of the University of Wisconsin-Madison.

Literature Cited

1. Enemark, J. H., et al. ADV. CHEM. SER. (1977) **162**, 440.
2. Clark, L. J., Axley, J. H., *Anal. Chem.* (1955) **27**, 2000–2003.
3. MacGregor, C. H., Schnaitman, C. A., Normansell, D. E., Hodgins, M. G., *J. Biol. Chem.* (1974) **249**, 5321–7.
4. Brill, W. J., personal communication.
5. Bray, R. C., Swann, J. C., *Struct. Bonding* (1972) **11**, 107–144.
6. Johnson, J. L., Ward, W. R., Cohen, H. L., Rajagopalan, K. V., *J. Biol. Chem.* (1974) **249**, 5056–5061.
7. Johnson, J. L., Cohen, H. L., Rajagopalan, K. V., *J. Biol. Chem.* (1974) **249**, 5046–5055.
8. Brill, W. J., et al., ADV. CHEM. SER. (1977) **162**, 402.

9. Notton, B. A., Hewitt, E. J., *Biochim. Biophys. Acta* (1972) **275**, 355–357.
10. Pateman, J. A., Cove, D. J., Rever, B. M., Roberts, D. B., *Nature* (1964) **201**, 58–59.
11. Nason, A., Lee, K.-Y., Pan, S.-S., Ketchum, P. A., Lamberti, A., DeVries, J., *Proc. Natl. Acad. Sci. U.S.A.* (1971) **68**, 3242–3246.
12. Lee, K.-Y., Pan, S.-S., Erickson, R., Nason, A., *J. Biol. Chem.* (1974) **249**, 394.
13. Que, L., Jr., Holm, R. H., Mortenson, L. E., *J. Am. Chem. Soc.* (1975) **97**, 463–64.
14. Erbes, D. L., Burris, R. H., Orme-Johnson, W. H., *Proc. Natl. Acad. Sci. USA* (1975) **72**, 4795–4799.
15. Massy, V., Komai, H., Palmer, G., Elion, G. B., *J. Biol. Chem.* (1970) **245**, 2837–2842.
16. Bray, R. C., Meriwether, L. S., *Nature* (1966) **212**, 467–469.
17. Averill, B. A., Orme-Johnson, W. H., unpublished data.
18. Jarrett, H. S., *J. Chem. Phys.* (1965) **27**, 1298–1304.
19. Bray, R. C., *Biochem. J.* (1961) **81**, 189–197.
20. Olson, J. S., Ballou, D. P., Palmer, G., Massey, V., *J. Biol. Chem.* (1974) **249**, 4363–4382.
21. Cramer, S. P., et al., ADV. CHEM. SER. (1977) **162**, 408.
22. Tso, M.-Y., Ljones, T., Burris, R. H., *Biochim. Biophys. Acta* (1972) **267**, 600.
23. DerVartanian, D. V., Bramlett, R., *Biochim. Biophys. Acta* (1970) **220**, 443.
24. Forget, P., *Eur. J. Biochem.* (1974) **42**, 325–332.
25. Cohen, H. J., Fridovich, I., Rajogopalan, K. V., *J. Biol. Chem.* (1971) **246**, 374–382.
26. Branzzoli, U., Massey, V., *J. Biol. Chem.* (1974) **249**, 4339–4345.
27. Rajagopalan, K. V., Handler, P., *J. Biol. Chem.* (1967) **242**, 4097.
28. Hart, L. I., McGartoll, M. A., Chapman, H. R., Bray, R. C., *Biochem. J.* (1970) **116**, 851–856.
29. Edmondson, D., Ballou, D. P., Van Heuvelen, A., Palmer, G., Massey, V., *J. Biol. Chem.* (1973) **248**, 6129–6135.
30. Massy, V., Edmondson, D., *J. Biol. Chem.* (1970) **245**, 6595.
31. Bray, R. C., Knowles, P. F., *Proc. Roy. Soc. A.* (1968) **302**, 351.
32. Orme-Johnson, W. H., Davis, L. C., "Current Problems and Topics in the Enzymology of Nitrogenase," in "Iron-Sulfur Proteins," W. Lovenberg, Ed., Vol. III, Chapter 2, Academic, 1977.
33. Newton, W. E., Nyman, C. J., "Proceedings of the First International Symposium on Nitrogen Fixation," Vols. I and II, Washington State University Press, Pullman, 1976.
34. Munck, E., Rhodes, H., Orme-Johnson, W. H., Davis, L. C., Brill, W., Shah, V. K., *Biochem. Biophys. Acta* (1975) **400**, 32–53.
35. Davis, L. C., Henzl, M. T., Burris, R. H., Orme-Johnson, W. H., *Biochemistry*, in press.
36. Orme-Johnson, W. H., *Ann. Rev. Biochem.* (1973) **42**, 159–204.
37. Meriwether, L. S., Marzluff, W. F., Hodgson, W. G., *Nature* (1966) **212**, 465–467.
38. Jacob, G. S., Orme-Johnson, W. H., unpublished data.
39. Nason, A., Evans, H. J., *J. Biol. Chem.* (1953) **202**, 655–673.
40. Nicholas, D. J. D., Nason, A., McElroy, *J. Biol. Chem.* (1954) **207**, 341–352.
41. Nicholas, D. J. D., Nason, A., *J. Biol. Chem.* (1954) **207**, 253–260.
42. Garrett, R. H., Nason, A., *J. Biol. Chem.* (1969) **244**, 2870–2882.
43. Jacob, G. S., Orme-Johnson, W. H. (1977), *J. Biol. Chem.*, in press.
44. Solomonson, L. P., Lorimer, G. H., Hall, R. L., Borchers, R., Bailey, J. L., *J. Biol. Chem.* (1975) **250**, 4120–4127.

45. Forget, P., DerVartanian, D. V., *Biochem. Biophys. Acta* (1972) **256,** 600–606.
46. Beinert, H., "The Enzymes," Boyer, Lardy, Myrback, Eds., 2nd ed., Vol. 2, p. 339–416, 1960.
47. Blumberg, W. E., Peisach, J., in "Probes of Structure and Function of Macromolecules and Membranes," B. Chance, T. Yonetami, A. S. Meldva, Eds., p. 215, Academic, New York, 1971.

RECEIVED July 26, 1976.

22

Uptake and Role of Molybdenum in Nitrogen-Fixing Bacteria

PHILIP T. PIENKOS, VINOD K. SHAH, and WINSTON J. BRILL

Department of Bacteriology and Center for Studies of Nitrogen Fixation, University of Wisconsin, Madison, WI 53706

Mutant strains of nitrogen-fixing bacteria have been obtained with defects in their ability to synthesize an active molybdenum cofactor. Such strains are used to assay this cofactor. Ammonia in the medium represses the synthesis of the molybdenum cofactor. Molybdenum is transported by an energy-requiring reaction, and the metal becomes part of a molybdenum storage compound. Molybdenum in the medium affects the regulation of nitrogenase synthesis because Klebsiella pneumoniae *does not synthesize either of the two nitrogenase components when it lacks sufficient molybdenum.*

We shall outline approaches used to investigate the role of molybdenum in nitrogen fixation by bacteria. It is hoped that this type of work will make it easier for the chemist to focus on the multiple roles that molybdenum plays in these bacteria. The approach that has given us most insight into the mechanism of nitrogen fixation involves the use of mutant strains that specifically are unable to grow on nitrogen (Nif⁻ mutants). Such studies have yielded information on the active site of nitrogenase (1), how nitrogenase synthesis is regulated (2), the order of genes specific for nitrogen fixation (*nif* genes) (3, 4), and the existence of factors other than nitrogenase that are specifically required if an organism is to fix nitrogen (4). Two organisms are discussed, *Azotobacter vinelandii,* a bacterium that only fixes nitrogen aerobically, and *Klebsiella pneumoniae,* an organism that only fixes nitrogen under anaerobic conditions.

The enzyme, nitrogenase, is composed of two proteins—component I and component II. Component I has a molecular weight of about 220,000 and contains either one or two molybdenum and 24 iron atoms. Component II has a molecular weight of 60,000 and has four iron atoms. The reaction requires 12–24 ATP for each nitrogen fixed (5). We were able to prepare cell-free extracts of the Nif⁻ mutant strains and titrate each extract (1) with purified components. Thus it was possible to identify which of the components was lacking in activity. It was important to know if a mutant lacking component I activity does not produce component I at all or whether component I is synthesized, but in an inactive form. Serological tests for the components were performed using antiserum from rabbits injected with either of the two purified components isolated from the wild type. With techniques such as these, several hundred mutant strains were classified.

Molybdenum Cofactor

Ketchum et al. (6) showed that crude extracts of a mutant strain of *Neurospora crassa,* called Nit-1, that would not reduce nitrate could be activated by adding acid-treated molybdoprotein to the extract. The activating factor could be obtained from molybdoproteins such as nitrogenase component I, bovine liver sulfite oxidase, xanthine oxidase, aldehyde oxidase, and nitrate reductase from various sources (7). Kondorosi et al. (8) obtained mutant strains of *Rhizobium meliloti* that could not reduce nitrate. Some of these strains also were unable to produce effective (nitrogen fixing) alfalfa nodules, and they concluded that a common genetic determinant is required for nitrate reductase and nitrogenase activity. This determinant might be the molybdenum cofactor that seems to be common to all molybdoproteins.

The molybdenum cofactor from *Rhodospirillum rubrum* is dialyzable and is insensitive to trypsin (9). The cofactor can easily be inactivated by heat. One of the problems in purifying this cofactor is the instability and low yields from purified enzymes and crude extracts. Lee et al. (10) showed that the molybdenum cofactor is stabilized by $0.01M$ sodium molybdate and that the absence of air adds to the stability. These workers used Mo^{99} labeling to show that the molybdenum from the cofactor is found in activated nitrate reductase from the mutant strain of *N. crassa.*

Ganelin et al. (11) provided evidence that the molybdenum cofactor is a molybdenum peptide with a molecular weight of about 1000. The same properties have been claimed for the cofactor from xanthine oxidase (12).

It was important to identify mutant strains with defects in nitrogenase similar to the defect observed in nitrate reductase in the Nit-1 mutant strain of *N. crassa* (6). It was postulated that such strains would be able to synthesize active component II and an inactive component I that could be activated in vitro by the molybdenum cofactor. Cell-free extracts from strains producing inactive component I were prepared and tested with molybdenum cofactor made by acid-treating and neutralizing purified component I from the wild type organism (*13*). Strains of *A. vinelandii* (*14*) and *K. pneumoniae* (*4*) were found with inactive component I that could be activated in vitro with the molybdenum cofactor. The mutations that prevent the formation of the molybdenum cofactor have been located on the chromosome in both organisms (*3, 4*). Figure 1 shows the location (*nif* B) of the gene(s) that specify the molybdenum cofactor in *K. pneumoniae*. This gene(s) is located close to the structural genes for the two nitrogenase components.

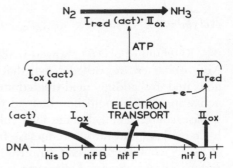

Figure 1. Map of nif *genes in* Klebsiella pneumonia. *The designation, (act), represents the molybdenum cofactor. Component II (II_{ox}) is reduced by the* nif *electron transfer system. II_{red} then binds and reduces I_{ox}(act), and the resulting complex with ATP can reduce nitrogen to ammonia.*

It is possible to make the wild type *A. vinelandii* produce active component II and an inactive component I by substituting tungsten for molybdenum in the medium (*14*). In this way, an active molybdenum cofactor is not produced. However, if acid-treated component I from cells grown on molybdenum is added to extracts from tungsten-grown cells, activation ocurs. Addition of molybdate to such extracts does not reactivate component I. We are using mutant strains defective in the molybdenum cofactor as well as tungsten-grown cells as assays to purify the molybdenum cofactor.

Regulation

Nitrogen-fixing cells have an obvious requirement for molybdenum; however, no molybdenum is required when these cells grow in a medium containing excess NH_4^+. NH_4^+ completely represses the synthesis of nitrogenase (*15*). Nitrogen fixation is quite an energy-demanding process, and so it makes sense for the organism not to produce nitrogenase when it is not needed. The synthesis of the molybdenum cofactor also is repressed by NH_4^+ (*14*). Certain mutant strains have been isolated that continue to synthesize active nitrogenase in the presence of NH_4^+ (*2, 16*). Therefore, it seems that the factors responsible for the regulation of the nitrogenase structural genes are also responsible for regulating the synthesis of active molybdenum cofactor. Some of these regulatory genes also have been mapped on the chromosome.

There are many reports (e.g., Refs. *17, 18*) in which nitrogen fixation is limited in a soil because molybdenum is deficient. Addition of molybdenum to such soils allows nitrogen fixation to proceed. This molybdenum deficiency is extremely important if a crop of nitrogen-fixing legumes is desired. In fact, farmers can buy molybdates to spread on the field or to apply to the seed before sowing. We wondered what happens to a nitrogen-fixing bacterium that is deficient in fixed nitrogen, but is unable to fix nitrogen because molybdenum is unavailable. Such a situation might cause the bacterium to synthesize active component II and an inactive component I. This condition is detrimental to the cell since it wastes energy synthesizing proteins that have no benefit to it. However, *K. pneumoniae* has a regulatory mechanism that prevents nitrogenase synthesis when there is not enough molybdenum available. When we derepress this organism in the absence of molybdenum, neither component I nor component II is synthesized, even in an inactive form (*19*). A possible mechanism for control of nitrogenase synthesis might involve a protein that activates transcription of the *nif* genes. This activator is inactivated when NH_4^+ accumulates. It is easy to introduce a mechanism for control by molybdenum by hypothesizing that this activator requires molybdenum. Mutant strains that are defective in nitrogenase regulation by molybdenum should be useful for testing this hypothesis.

Molybdenum Storage

We are now trying to understand how the molybdate that is added to the medium ultimately becomes a part of nitrogenase component I. Studies with metabolic inhibitors indicate that molybdate is taken up by an energy-demanding process in *A. vinelandii*. We were surprised to find that this organism accumulates more than 20 times more molybdenum than it actually requires for maximum nitrogenase activity. Cells that grow on NH_4^+ do not require molybdenum. *Azotobacter vinelandii* was grown to mid-log phase in a medium containing NH_4^+ but no molybdenum. The cells were washed free of NH_4^+ and then allowed to derepress for nitrogenase synthesis in the presence and absence of chloramphenicol when molybdate was added. Table I presents evidence that the protein synthesis is not required for molybdenum uptake but is required for synthesis of nitrogenase. The molybdenum uptake and storage factors then are not inducible by molybdenum.

When a crude extract was fractionated on an ion exchange column, most of the molybdenum was associated with a fraction that is not component I (*20*). This fraction contains what we have named the molybdenum-storage compound, since molybdenum can be removed from this compound and transposed to component I. In fact, cells grown in the presence of molybdenum and NH_4^+, when transferred to medium lacking both, can synthesize active nitrogenase for many generations using the molybdenum in the molybdenum-storage compound. This is a useful mechanism by which A. *vinelandii* stores molybdenum when excess is available, but then uses this excess when it finds itself in an environment lacking molybdenum. Tungsten also can be stored by the molybdenum-storage protein. We are now studying mutant strains that seem to lack the molybdenum-storage protein.

Table I.   Effects of Chloramphenicol on Molybdenum
Accumulation in *Azotobacter vinelandii*

Organism	CAM	Nitrogenase Specific Activity (nmoles acetylene reduced/min $\times$ 10^7 cells)	Mo ($\mu g/10^{11}$ cells)
A. *vinelandii*	—	0.59	5.70
A. *vinelandii*	+	0.00	6.80

There are many questions that must still be answered. How and in what form does molybdenum enter the cell? How and in what form does molybdenum get into molybdenum-storage protein? How and in what form does molybdenum get transferred to the molybdenum cofactor? How and in what form does the molybdenum cofactor get into nitrogenase component I? How and in what form does molybdenum regulate the synthesis of nitrogenase? These questions require integrated effort by chemists, enzymologists, geneticists, and bacterial physiologists. Hopefully, such work will ultimately have an impact on agriculture.

Acknowledgments

A part of this work was supported by the College of Agricultural and Life Sciences, University of Wisconsin, Madison, by NIH Grant GM22130, and by the Cellular and Molecular Biology Training Grant GM 07215.

Literature Cited

1. Shah, V. K., Davis, L. C., Gordon, J. K., Orme-Johnson, W. H., Brill, W. J., *Biochim. Biophys. Acta* (1973) **292**, 246–255.
2. Gordon, J. K., Brill, W. J., *Proc. Natl. Acad. Sci. USA* (1972) **69**, 3501–3503.
3. Bishop, P. E., Brill, W. J., Abstracts Annual Meeting American Society Microbiology, p. 163, 1976.

4. St. John, R. T., Johnston, H. M., Seidman, C., Garfinkel, D., Gordon, J. K., Shah, V. K., Brill, W. J., *J. Bacteriol.* (1975) **121**, 759–765.
5. Eady, R. R., Postgate, J. R., *Nature* (1974) **249**, 805–810.
6. Ketchum, P. A., Cambier, H. Y., Frazier, W. A., Madansky, C. H., Nason, A., *Proc. Natl. Acad. Sci. USA* (1970) **66**, 1016–1023.
7. Nason, A., Lee, K. Y., Pan, S. S., Ketchum, P. A., Lamberti, A., De Vries, J., *Proc. Natl. Acad. Sci. USA* (1972) **68**, 3242–3246.
8. Kondorosi, A., Barabas, I., Svab, Z., Orosz, L., Sik, T., Hotchkiss, R. D., *Nature* (1973) **246**, 153–154.
9. Ketchum, P. A., Sevilla, C. L., *J. Bacteriol.* (1973) **116**, 600–609.
10. Lee, K. Y., Pan, S. S., Erickson, R., Nason, A., *J. Biol. Chem.* (1974) **249**, 3941–3952.
11. Ganelin, V. L., L'vov, N. P., Sergeev, N. S., Shaposhnikov, G. L., Kretovich, V. L., *Dokl. Akad. Nauk SSSR* (1972) **206**, 1236–1238.
12. McKenna, C., L'vov, N. P., Ganelin, V. L., Sergeev, N. S., Kretovich, V. L., *Dokl. Akad. Nauk SSSR* (1974) **217**, 228–231.
13. Shah, V. K., Brill, W. J., *Biochim. Biophys. Acta* (1973) **305**, 445–454.
14. Nagatani, H. H., Shah, V. K., Brill, W. J., *J. Bacteriol.* (1974) **120**, 697–701.
15. Shah, V. K., Davis, L. C., Brill, W. J., *Biochim. Biophys. Acta* (1972) **256**, 498–511.
16. Gordon, J. K., Garfinkel, D., Brill, W. J., Abstracts Annual Meeting American Society Microbiology, p. 175, 1975.
17. Evans, H. J., Purvis, E. R., Bear, F. E., *Soil Sci.* (1951) **71**, 117–124.
18. Lobb, W. R., *N. Z. Soil News* (1953) **3**, 9–16.
19. Brill, W. J., Steiner, A. L., Shah, V. K., *J. Bacteriol.* (1974) **118**, 986–989.
20. Pienkos, P. T., Brill, W. J., Abstracts Annual Meeting American Society Microbiology, p. 164, 1976.

RECEIVED July 26, 1976.

23

Reactions of Molybdenum with Nitrate and Naturally Produced Phenolates

PAUL A. KETCHUM and DAVID JOHNSON
Department of Biological Sciences, Oakland University, Rochester, MI 48063

R. CRAIG TAYLOR, DONALD C. YOUNG, and ARTHUR W. ATKINSON
Department of Chemistry, Oakland University, Rochester, MI 48063

The reaction of $Mo(H_2O)_6{}^{3+}$ and nitrate in aqueous solutions results in the formation of $Mo_2O_4(H_2O)_6{}^{2+}$ and nitrite. Mo(III) coordinated to oxygen and nitrogen donor atoms of EDTA also reduces nitrate in aqueous solutions. The reduction of nitrate by a Mo(III)–EDTA complex results in the formation of nitrite and a Mo(V)–EDTA complex, as determined by chemical and spectrophotometric techniques. These reactions serve as models for biological nitrate reduction. In addition, molybdate coordinates to naturally produced phenolates. The molybdenum-coordinating phenolates also coordinate tungstate and ferric iron. Two of these phenolates contain threonine, glycine, alanine, and 2,3-dihydroxybenzoic acid.

A ssimilatory NADPH–nitrate reductase was first described by Evans and Nason in 1952 (*1*). Assimilatory nitrate reductases are molybdoflavoproteins with molecular weights varying from 190,000 to 356,000 daltons depending on the source of the enzyme (*2–8*). The *Neurospora* enzyme has a molecular weight of 230,000 and is a soluble electron transport enzyme with the molybdenum moiety functioning as the terminal redox site (*3*). With this information on the enzymology of NADPH–nitrate reductase, a number of groups have investigated various inorganic molybdenum models for nitrate reduction (*9, 10, 11, 12, 13*). Model systems for nitrate reduction have concentrated on Mo(V) since a Mo(V) EPR signal has been observed in purified preparations of nitrate reductase. The Mo(V) model systems reduce nitrate to nitrogen(IV) oxides in nonaqueous environments. Since nitrogen(IV) oxides disproportionate in water to nitrite and nitrate, it has been suggested (*12*) that Mo(V) exists in nitrate reductase in a hydrophobic environment and is oxidized to Mo(VI) by nitrate. The reactions of nitrate with Mo(V) in non-aqueous solutions are enlightening, however, they do not explain access of an anion to a hydrophobic active enzyme site. Furthermore recent EPR evidence indicates that oxidation states of molybdenum lower than 5+ are involved in nitrate reductases. The observation by Forget and Dervartanian (*14*) that the Mo(V) signal observed in the *Micrococcus denitrificans* nitrate reductase disappears upon reduction suggests that Mo(V) is found in the oxidized enzyme. Recent investigations on the *Neurospora* enzyme (*15*) implicate Mo(III) as the reduced form of molybdenum in this nitrate reductase.

The reaction between Mo(III) and nitrate was first reported by Haight (*16*) who demonstrated that polarographically produced Mo(III) reduces nitrate. The recent advances in the chemistry of hexaaquomolybdenum(III) by Bowen and Taube (*17, 18*), Kustin and Toppen (*19*), and Sasaki and Sykes (*20*) have led us to investigate the Mo(III)/Mo(V) couple as a model system for nitrate reduction (*21*).

Molybdenum(III) and Nitrate

The reaction between $Mo(H_2O)_6^{3+}$, prepared and purified following the procedure of Bowen and Taube (*18*), and nitrate was followed spectrophotometrically under strict anaerobic conditions with nitrate in excess. The absorption spectrum of nitrite in $1.0M$ HPTS (*p*-toluene sulfonic acid) exhibits a multicomponent band (vibrational fine structure) between 350 and 400 nm which is attributable to the $^1B_1 \leftarrow {}^1A_1$ electronic transition. Purified $Mo(H_2O)_6^{3+}$ in $1.0M$ HPTS has a low absorption at 293 nm, indicating the purity of the preparation (*18*). When $Mo(H_2O)_6^{3+}$ is mixed with nitrate (constant concentration in large excess) and the reaction is allowed to go to completion, the nitrite fine structure appears between 350 and 400, concomitant with an increase in absorbance at 293 nm. The molybdenum species resulting from the oxidation of

Table I. Observed Rate Constants for the Oxidation of
$Mo(H_2O)_6^{3+}$ by Nitrate in 1M HPTS[a]

Experiment	Total Mo ($M \times 10^4$)	$k_{obs} \times 10^5$ (s^{-1})
1	4.70	1.42
2	4.18	1.37
3	4.45	1.32
4	4.45	1.25
5	4.35	1.10

[a] $[NO_3^-] = 1.33 \times 10^{-2}M$. The data were obtained at $28.5 \pm 0.5°C$ in $1.0M$ HPTS. The average value for k_{obs} with standard deviation: $k_{obs} = 1.29 \pm 0.14$ s^{-1}.

$Mo(H_2O)_6^{3+}$ by nitrate has a spectrum between 340 and 280 nm identical to that of $Mo_2O_4(H_2O)_6^{2+}$ reported by Ardon and Pernick (22).

The kinetics of the reaction between $Mo(H_2O)_6^{3+}$ and nitrate (Table I) were determined using five different preparations of Mo(III) at 28.5 $\pm 0.5°C$ in 1.0M HPTS in the presence of $1.33 \times 10^{-2}M$ nitrate. Good first-order plots were observed in all cases, and k_{obs} was found to be $(1.29 \pm 0.14) \times 10^{-5}$ s^{-1}. Detectable nitrite was not produced from nitrate under identical conditions when $Mo_2O_4(H_2O)_6^{2+}$, prepared either by the air oxidation of $Mo(H_2O)_6^{3+}$ or by the reduction of Na_2MoO_4 by hydrazine, was substituted for Mo(III). Moreover, the spectrum of $Mo_2O_4(H_2O)_6^{2+}$ in 1.0M HPTS did not change during 1 hr of observation either in the presence or absence of $1.33 \times 10^{-2}M$ nitrate. Previous reports have demonstrated that nitrate is reduced to NO^+ by Mo(V) in the presence of Cl$^-$ in tartrate buffer (13). These results have no bearing on our conclusions since the rate of nitrate reduction by Mo(V) is approximately one tenth that of the nitrate reduction by Mo(III).

The concentrations of the products produced by the reaction between $Mo(H_2O)_6^{3+}$ and NO_3^- are consistent with one mole of dimolybdenum species and two moles of nitrite (Table II). The concentration of

Table II. Stoichiometry of the Oxidation of $Mo(H_2O)_6^{3+}$ by Nitrate[a]

Experiment	Total Mo (M × 10⁴)	[NO₂⁻] (M × 10⁴)	[Mo₂O₄²⁺] (M × 10⁴)	[NO₂⁻]/ [Mo₂O₄²⁺]
1	4.70	4.90	2.24	2.19
2	4.18	2.37	1.16	2.04
3	4.45	3.24	2.18	1.49
4	4.50	1.28	0.72	1.78
5	4.50	2.44	1.32	1.70

[a] Nitrite was measured chemically, and $Mo_2O_4(H_2O)_6^{2+}$ was measured spectrophotometrically. The average ratio and standard deviation of $[NO^-_2]/[Mo_2O_4^{2+}]$ was 1.84 ± 0.28 (21).

Nature

$Mo_2O_4(H_2O)_6^{2+}$ was measured spectrophotometrically just prior to air oxidation, and the concentration of nitrite was determined colorimetrically (4). Nitrite disappeared with time from these solutions suggesting that Mo(V) reduces nitrite, as previously observed under different conditions (9). This may account for the discrepancies in some of the nitrite values reported. When the experiment was allowed to go to completion (Experiments 1 and 3), all of the $Mo(H_2O)_6^{3+}$ present (total molybdenum) was accounted for in the $Mo_2O_4(H_2O)_6^{2+}$ formed (Table II).

Another check on the stoichiometry (Table III) was made by following the rates of appearance of NO_2^- and $Mo_2O_4^{2+}$. Nitrite was meas-

Table III. Observed Rate Constants for the Appearance of Nitrite ($k_{obs}*$) and $Mo_2O_4^{2+}$ (k_{obs}) in the Presence of Sulfanilic Acid and NEDA[a]

Experiment	Total Mo ($M \times 10^4$)	$k_{obs} \times 10^6$ (s^{-1})	$k_{obs}* \times 10^5$ (s^{-1})	$k_{obs}*/k_{obs}$
1	1.20	4.00	0.768	1.92
2	1.30	4.86	1.06	2.18
3	1.50	4.80	0.989	2.06
4	1.50	4.65	1.00	2.15

[a] Both the reference and sample cuvettes contained $5.16 \times 10^{-3}M$ sulfanilic acid and $8.3 \times 10^{-4}M$ NEDA. The reaction was initiated by adding nitrate (final concentration $= 1.33 \times 10^2 M$) to the sample cuvette. The average ratio and standard deviation of $k_{obs}*/k_{obs}$ was 2.08 ± 0.10.

ured during the reaction by adding sulfanilic acid and NEDA to the reaction mixture and recording the Δ Abs.$_{560}$ with time. The observed rate constant for the appearance of nitrite is 2.08 ± 0.10 times the observed rate constant for the appearance of $Mo_2O_4^{2+}$. The observed rate constants for $Mo_2O_4^{2+}$ were slightly higher in the presence of the nitrite-trapping reagents.

The data presented for the reaction of $Mo(H_2O)_6^{3+}$ with nitrate indicate that the reaction proceeds via Equation 1. Two basic mecha-

$$2Mo(H_2O)_6^{3+} + 2NO_3^- \xrightarrow{k} Mo_2O_4(H_2O)_6^{2+} + 2NO_2^- + 4H^+ + 4H_2O \quad (1)$$

nisms have been proposed to account for the interaction of substrate with molybdenum in molybdoenzymes—the coupled proton–electron transfer mechanism (23, 24) and the oxygen-atom transfer mechanism (21, 25). The data presented here are consistent with the oxygen-atom transfer mechanism postulated by Williams (25), even though the mechanism of the nitrate reduction by Mo(III) has not been elucidated. The oxygen-atom transfer mechanism for the enzymic reduction of nitrate by $Mo(H_2O)_6^{3+}$ is presented in Figure 1. $MoO_2(H_2O)_4^+$ produced is reduced by the enzymic electron transport chain which donates $2e^-$ and $2H^+$. This scheme is consistent with the electron transport studies on the assimilatory nitrate reductase which position molybdenum at the terminal end of the electron transport chain, and it is consistent with the physio-

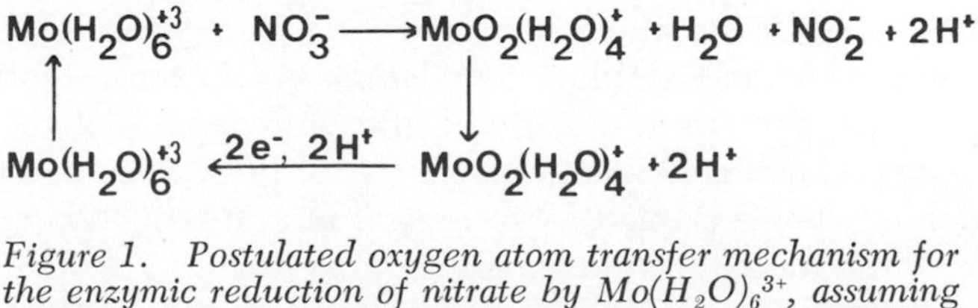

Figure 1. Postulated oxygen atom transfer mechanism for the enzymic reduction of nitrate by $Mo(H_2O)_6^{3+}$, assuming NADPH, H^+ to be the substrate for reducing $MoO_2(H_2O)_4^+$ to $Mo(H_2O)_6^{3+}$

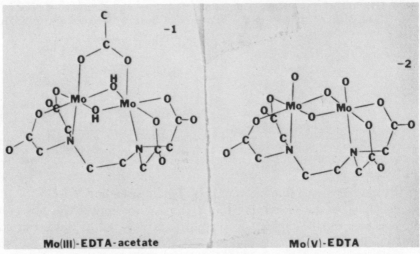

Mo(III)-EDTA-acetate   **Mo(V)-EDTA**

Figure 2.   (a) Structure of [Mo₂(OH)₂EDTA(O₂CCH₃)]⁻ as determined by Sykes et al. (29); (b) structure of [Mo₂O₄EDTA]²⁻ prepared by Haynes and Sawyer (30)

logical substrate, reduced nicotinamide adenine dinucleotide phosphate (NADPH, H⁺), which donates two electrons and two protons to the electron transport chain.

The rate observed for the inorganic reaction between Mo(III) and nitrate is approximately three orders of magnitude less than the enzymatic reduction of nitrate by the assimilatory nitrate reductase [9.4 mol/sec calculated using a specific activity of 150,000 units/mg protein (3) assuming two moles of molybdenum per mole of enzyme]. However, this is not a serious drawback to the proposed inorganic model system since the turnover rate per mole of enzyme-bound molybdenum should depend upon the ligand binding of the molybdenum atom(s) in the enzyme. Modification of the behavior of both Co(III) and Mo(III) by changes in the ligand environment of these metals has been reported. Co(III) is unstable to reduction by water to form Co(II) [$E°_{red} =$ 1.84 V] whereas the situation is reversed for the cyanide complex i.e., Co(CN)₆³⁻ + e⁻ → Co(CN)₅³⁻ + CN⁻ [$E°_{red} = -0.83$] (26). The instability of K₃MoCl₆ in water is altered by forming Mo(III) complexes which are soluble and stable to oxidation in aqueous solutions (26). Therefore, an understanding of the ligand environment of the molybdenum in molybdenum-containing enzymes is of primary importance to the elucidation of the mechanism of biological nitrate reduction and the significance of the turnover number in model systems.

One of the principal disadvantages of this system is the inability to study the rate of nitrite production by Mo(H₂O)₆³⁺ at physiological pH.

All attempts to observe this reaction at higher pH's were unsuccessful. Addition of $Mo(H_2O)_6^{3+}$ to phosphate, tris(hydroxymethyl) aminomethane, or morpholinopropane sulfonic acid (MOPS) buffer systems caused instant precipitation of an unidentified species.

EDTA Complexes of Molybdenum(III) and Molybdenum(V)

The demonstration that Mo(III) reduces nitrate to nitrite has prompted us to investigate other Mo(III) species which are stable at more alkaline pH and which contain ligands (O, N, and/or S donor atoms) which might more accurately reflect the binding of molybdenum in the enzyme.

One of the systems which fulfills these requirements is the EDTA complex of Mo(III), μ-acetato-di-μ-hydroxo-μ(N,N')ethylenediaminetetraacetatobis[molybdenum(III)]. The complex was originally prepared but incorrectly formulated by Kloubek and Podlaha (*28*). Its solid-state structure has recently been determined by Sykes et al. (Figure 2). Preliminary results in our laboratory have demonstrated that this compound

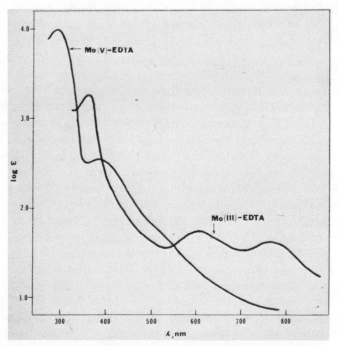

Figure 3. Optical spectra of Mo_2O_4 $EDTA^{2-}$ and $[Mo_2-(OH)_2EDTA(O_2CCH_3)]^{1-}$; the latter in acetate buffer, pH = 4.5. Oxidation of the latter compound with NO_3^- yields a spectrum identical to that obtained for $Mo_2O_4EDTA^{2-}$.

reduces nitrate to nitrite (detected colorimetrically) although at a rate considerably slower than the reduction of nitrate by $Mo(H_2O)_6^{3+}$. The spectrum of the oxidized molybdenum-containing species (Figure 3) is identical to that observed for di-μ-oxo-μ(N,N')ethylenediaminetetraacetatobis[oxomolybdate(V)], Mo_2O_4 EDTA^{2-} (30, 31). The reaction proceeds both in acetate ($3.8 <$ pH < 5.5) and phosphate buffers ($6.8 <$ pH < 7.3). Apparently, the presence of the μ-acetato or μ-phosphato bridge is necessary for the stability of the Mo(III) dimer since we have been unsuccessful in reducing Mo_2O_4 EDTA^{2-} in solutions containing high Mo(V) and low buffer concentrations. The presence of the bridging buffer moiety apparently hinders the complexation of NO_3^- which would be necessary for reduction to occur via an inner sphere mechanism. The equilibrium in Reaction 2 is plausible. In high buffer concentration

$$[Mo_2(OH)_2EDTA(O_2CCH_3)]^- + NO_3^- \rightleftarrows$$
$$[Mo_2(OH)_2EDTA(NO_3)]^- + O_2CCH_3^- \tag{2}$$

necessary for in situ reduction of Mo_2O_4 EDTA^{2-}, the ionized form of the buffer effectively competes with NO_3^- for the available coordination sites on the Mo(III) atoms. These systems are undergoing further study.

Natural Products Which Coordinate Molybdenum

Another approach to the coordination chemistry of molybdenum is to search for naturally produced molybdenum coordinating compounds. It was postulated that such compounds could have an analogous role to the siderochromes which coordinate ferric iron as discussed earlier in this symposium (see Refs. 32 and 33). There is both biochemical and genetic evidence for the existence of physiologically important low molecular weight molybdenum coordinating compounds (34–40). The in vitro restoration of NADPH–nitrate reductase using extracts of the nitrate reductase mutant of nit-1 and preparation of acid-treated molybdenum containing enzymes (34, 39) suggests the involvement of a low molecular weight molybdenum cofactor contributed by the molybdenum-containing enzymes. In Aspergillus the synthesis of the cofactor is controlled by five separate genes (40, 41). We have found that extracts of Bacillus thuringiensis are also a good source of this cofactor activity (35). Therefore B. thuringiensis was used in our initial search for naturally produced molybdenum coordinating compounds which coordinate molybdate and are excreted into the culture medium.

When B. thuringiensis is grown on an iron- and molybdenum-deficient medium (42) at pH 6.8 with added l-arginine, it produces compounds which form charge transfer complexes with molybdate. Preliminary purification and characterization of the molybdenum-reactive compounds in filtrates of *B. thuringiensis* demonstrate that they are phenolates which coordinate tungstate and ferric iron as well as molybdate.

Approximately 50% of the molybdenum-reactive compounds were extracted from acidified culture filtrates of *B. thuringiensis* with ethyl acetate. The ethyl acetate was washed with $NaHCO_3$ and then re-extracted with ethyl acetate according to the method of Tait (*43*). High voltage electrophoresis of the ethyl-acetate extractable compounds at pH 3.6 for 45 min at 5 kV and 60 ma separates out three blue-green fluorescent bands which react with molybdate to form yellow complexes. They

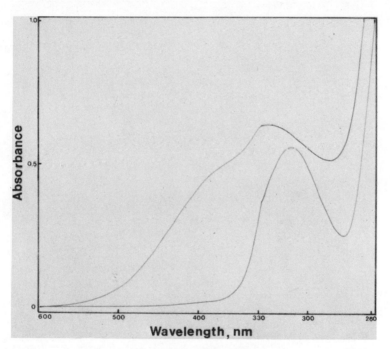

Figure 4. Absorption spectra of MoO_4^{2-} · slow band (upper trace) and the slow band (lower trace)

are designated the slow, intermediate, and fast bands. Two of the molybdenum-coordinating compounds can be differentially precipitated by adding benzene to the ethyl acetate extract, the third compound remaining in solution in 4/1 mixtures of benzene/ethyl acetate. All three bands contain 2,3-dihydroxybenzoic acid as shown by spectral data, Arnow (*44*) reactivity, and the high voltage electrophoretic mobility of the fluorescent band following acid hydrolysis. High voltage electrophoresis at pH 1.9 of acid-hydrolyzed samples of the slow and intermediate bands demonstrates the presence of threonine, alanine, and glycine. Their presence in the intermediate band has been confirmed by amino acid analysis (42). Our results suggest that *B. thuringiensis* produces two phenolates which contain threonine. These phenolates are therefore different from the previously described phenolates which contain serine (45), lysine (44), spermidine and threonine (43), or glycine (45) as the amino acid.

The uv-visible spectrum of the slow band is compared with that of its molybdate, tungstate, and Fe(III) complexes in Figures 4, 5, and 6 respectively. The dominant features of the complex spectra arise from charge transfer bands; the absorption maxima are listed in Table IV.

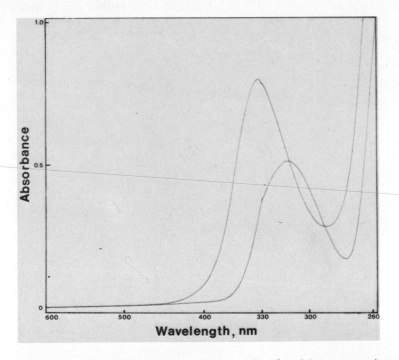

Figure 5.   Absorption spectra of $WO_4^{2-} \cdot$ slow band (upper trace) and the slow band (lower trace)

The molybdate and tungstate slow band spectra are similar to those obtained for molybdate and tungstate complexes of the fast and intermediate bands, and they are similar to the reported spectra of bisphenolate complexes of molybdate and tungstate (47, 48).

Stability constants were determined from spectrophotometric titration data treated by molar-ratio method procedures. In all cases only 1:1 complexes were assumed to form. Thus, the molar absorbtivities and the conditional stability constants in Table IV reflect this assumption.

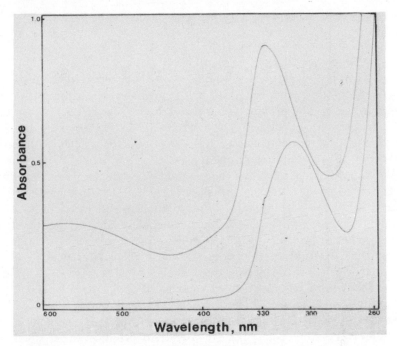

Figure 6. Absorption spectra of Fe^{3+} · slow band (upper trace) and slow band (lower trace)

For the molybdate and tungstate complexes, the conditional stability constants did not change over a pH range of 3.5–6. This suggests a pH-independent overall reaction like that formulated for other bisphenolate complexes, Reaction 3. If this reaction is valid, the stabilities in Table IV represent true formation constants rather than conditional constants. The stability of the Fe(III) complex has been determined only at pH 5. Since the molybdate and tungstate data indicate phenolate-coordinating groups, it is reasonable to assume that Reaction 4 represents the Fe(III) reaction. Hence, the Fe(III) constant is a conditional constant.

Table IV. Characteristics of Slow Band and Its Complexes

Species	λmax (nm)	ε	log K	Mol Wt
Slow band	315	8500		1280 ± 50
MoO_4^{2-} · slow band	328	9200	6.4	
	~ 370	7200		
WO_4^{2-} · slow band	333	14500	6.6	
Fe^{3+} · slow band	330	12400	6.3	
	~ 580	4000		

$$MO_4{}^{2-} + H_4L \rightarrow MO_2L^{2-} + 2H_2O \tag{3}$$

$$Fe^{3+} + H_4L \rightarrow FeL^- + 4H^+ \tag{4}$$

We are not aware of any stability constants reported for complexation of molybdate and tungstate with naturally occurring phenolates. There are several reports involving stabilities of molybdate and tungstate with simple synthetic phenolates (49, 50, 51). The range of these reported stability constants varies by about a factor of 100. Our results lie within this range for each metal species.

The physiological roles of the phenolates produced by B. thuringiensis are not known. Neither the slow band nor the intermediate band serves as the molybdenum cofactor in the in vitro restoration of NADPH–nitrate reductase. It is reasonable nevertheless to posulate that these compounds are excreted by this organism in nature and that under the appropriate environmental conditions, these phenolates will coordinate tungstate, ferric iron, or molybdate.

Selectivity of the metal ion to be coordinated may depend on the pH of the environment. Ferric iron is extremely insoluble at alkaline pH's since it is present as ferric hydroxide. This is a compelling argument for suggesting (52) that cells need to produce either phenolates or hydroxamates in order to extract iron from alkaline soils. Moreover the reaction of phenolates with ferric iron is pH dependent and favored by alkaline pH's (see Ref. 33). On the other hand, molybdenum is readily leached from acid soils, greatly reducing its availability to cells. Since coordination of molybdate to phenolates probably involves the removal of water (47, 48), this reaction should be less sensitive to pH than the coordination of iron to phenolates. The conditional binding constants of the slow band for ferric iron, molybdate, and tungstate at slightly acid pH's suggest that there would be little or no metal ion selectivity (except for concentration effects) at acid pH's. The chemistry of the interaction between phenolates and biologically important trace metals may therefore play a large role in determining the physiological function of the naturally produced phenolates.

Molybdenum is essential to the formation and activity of assimilatory nitrate reductases. Cells must assimilate molybdate from the environment, metabolize molybdenum in some manner to form active molybdenum cofactor, and then incorporate it into a large molecular weight protein so that it can perform a reversible redox reaction with nitrate. Investigations on the aqueous Mo(III) model systems for nitrate reduction and the coordination of molybdate by naturally produced phenolates will hopefully lead to an understanding of the complex process of molybdenum acquisitions by and molybdenum function in nitrate reductases.

Acknowledgments

This work was supported by grants from the National Science Foundation (GB-27490 and PCM-75-21183), the National Institute of Health Biomedical Sciences Support Services (5507-RR07131) to Oakland University, and the National Science Foundation Undergraduate Research Participation Program.

Literature Cited

1. Evans, H., Nason, A., *Arch. Biochem. Biophys.* (1952) **39**, 234.
2. Downey, R. J., *J. Bacteriol.* (1971) **105**, 759.
3. Garrett, R. H., Nason, A., *J. Biol. Chem.* (1969) **244**, 2870.
4. Nason, A., Evans, H. J., *J. Biol. Chem.* (1953) **202**, 655.
5. Nicholas, D. J. D., Nason, A., *J. Biol. Chem.* (1954) **207**, 353.
6. Payne, W. J., *Bacteriol. Rev.* (1973) **37**, 409.
7. Solomonson, L. P., Lorimer, G. H., Hall, R. L., Borchers, R., Bailey, J. L., *J. Biol. Chem.* (1975) **250**, 4120.
8. Nicholas, D. J. D., Nason, A., *J. Biol. Chem.* (1954) **211**, 183.
9. Frank, J. A., Spence, J. T., *J. Phys. Chem.* (1964) **68**, 2131.
10. Garner, C. D., Hyde, M. R., Mabbs, F. E., Routledge, V. I., *Nature* (1974) **252**, 579.
11. Garner, C. D., Hyde, M. R., Mabbs, F. E., Routledge, V. I., *J. Chem. Soc. Dalton Trans.* (1975) 1180.
12. Garner, C. D., Hyde, M. R., Mabbs, F. E., *Nature* (1975) **253**, 623.
13. Guymon, E. P., Spence, J. T., *J. Phys. Chem.* (1966) **70**, 1964.
14. Forget, P., Der Vartanian, D. V., *Biochim. Biophys. Acta* (1972) **256**, 600.
15. Orme-Johnson, W. H., Jacob, G. S., Henzl, M. T., Averill, B. A., ADV. CHEM. SER. (1977) **162**, 389.
16. Haight, G. J., Jr., *Acta Chem. Scand.* (1961) **15**, 2012.
17. Bowen, A. R., Taube, H., *J. Am. Chem. Soc.* (1971) **93**, 3287.
18. Bowen, A. R., Taube, H., *Inorg. Chem.* (1974) **13**, 2245.
19. Kustin, K., Toppen, D., *Inorg. Chem.* (1972) **11**, 2851.
20. Sasaki, Y., Sykes, A. G., *J. Chem. Soc. Dalton Trans.* (1975) 1048.
21. Ketchum, P. A., Taylor, R. C., Young, D. C., *Nature* (1976) **259**, 202.
22. Ardon, M., Pernick, A., *Inorg. Chem.* (1973) **12**, 2484.
23. Stiefel, E. I., *Proc. Natl. Acad. Sci. USA* (1973) **70**, 988.
24. Stiefel, E. N., Newton, W. E., Watt, G. D., Hadfield, K. L., Bulen, W. A., ADV. CHEM. SER. (1977) **162**, 353.
25. Williams, R. J. P., *Biochem. Soc. Trans.* (1973) **1**, 1.
26. Huheey, J. E., "Inorganic Chemistry: Principles of Structure and Reactivity," Harper & Row, 1972.
27. Mitchell, P. C. H., Scarle, R. D., *J. Chem. Soc. Dalton Trans.* (1972) 1809,
28. Kloubek, J., Podlaha, J., *J. Inorg. Nucl. Chem.* (1971) **33**, 2981.
29. Kneale, G. G., Geddes, A. J., Sasaki, Y., Shibahara, T., Sykes, A. G., *J. Chem. Soc. Chem. Commun.* (1975) 356.
30. Haynes, L. V., Sawyer, D. T., *Inorg. Chem.* (1967) **6**, 2146.
31. Pecsok, R. L., Sawyer, D. T., *J. Am. Chem. Soc.* (1956) **78**, 5496.
32. Neilands, J. B., ADV. CHEM. SER. (1977) **162**, 3.
33. Raymond, K. N., ADV. CHEM. SER. (1977) **162**, 33.
34. Ketchum, P. A., Cambier, H., Frazier, W., III, Madansky, C., Nason, A., *Proc. Natl. Acad. Sci., USA* (1970) **66**, 1016.
35. Ketchum, P. A., Swarin, R. S., *Biochem. Biophys. Res. Commun.* (1973) **52**, 1450.

36. Lee, K-Y., Pan, S-S., Erickson, R., Nason, A., *J. Biol. Chem.* (1974) **249**, 3941.
37. Lee, K-Y., Erickson, R., Pan, S-S., Jones, G., May, F., Nason, A., *J. Biol. Chem.* (1974) **249**, 3953.
38. Nason, A., Antoine, A., Ketchum, P. A., Frazier, W., III, Lee, D., *Proc. Natl. Acad. Sci., USA* (1970) **65**, 137.
39. Nason, A., Lee, K., Pan, S-S., Ketchum, P. A., Lamberti, A., DeVries, J., *Proc. Natl. Acad. Sci., USA* (1971) **68**, 2242.
40. Pateman, J., Cove, D., Rever, B., Roberts, D., *Nature* (1964) **201**, 581.
41. Arst, H., MacDonald, D., Cove, D., *Molec. Gen. Genetics* (1970) **108**, 129.
42. Ketchum, P. A., Owens, M., *J. Bacteriol.* (1975) **122**, 412.
43. Tait, G., *Biochem. J.* (1975) **146**, 191.
44. Corbin, J. L., Bulen, W. A., *Biochem.* (1969) **8**, 757.
45. Pollack, J. R., Neilands, J. B., *Biochem. Biophys. Res. Commun.* (1970) **38**, 989.
46. Ito, T., Neilands, J. B., *J. Am. Chem. Soc.* (1958) **80**, 4645.
47. Kustin, K., Liu, S. T., *J. Am. Chem. Soc.* (1973) **95**, 2487.
48. Kustin, K., Liu, S. T., *Inorg. Chem.* (1973) **12**, 2362.
49. Halmekoski, J., *Ann. Acad. Sci. Fenn.* (1959) **Ser A2**, 96.
50. Halmekoski, J., Suom. Kemistilehti, B. (1963) **36**, 46.
51. Pisko, E., *Chem. Zvesti* (1958) **12**, 95.
52. Neilands, J. B., *Struct. Bonding* (1966) **1**, 59.

RECEIVED July 26, 1976.

Preparation and X-Ray Structure of a Hydrazine-Bridged Dinuclear Molybdenum Complex [Mo(S₂CN(C₂H₅)₂)₂(CO)₂]₂N₂H₄ · CH₂Cl₂

J. A. BROOMHEAD and J. BUDGE

Department of Chemistry, Faculty of Science, Australian National University, Canberra A.C.T. 2600, Australia

J. H. ENEMARK, R. D. FELTHAM, J. I. GELDER, and P. L. JOHNSON

Department of Chemistry, University of Arizona, Tucson, AZ 85721

The reaction of hydrazine with bis(diethyldithiocarbamato)-dicarbonylmolybdenum, $Mo(S_2CN(C_2H_5)_2)_2(CO)_2$, results in formation of the dimeric complex $[Mo(S_2CN(C_2H_5)_2)_2-(CO)_2]_2N_2H_4$, (1). The ^1H NMR spectrum of 1 in $CDCl_3$ is consistent with the solid state composition. Determination of the structure of the CH_2Cl_2 solvate of 1 by x-ray diffraction has shown that the two crystallographically independent molybdenum atoms are asymmetrically bridged by the hydrazine ligand (Mo1–N5 = 2.36(1) and Mo2–N6 = 2.44(1) Å). Each molybdenum atom is seven-coordinate, but their stereochemistries are not identical. The bridging hydrazine ligand is disordered about two well resolved positions, and the N–N distance is 1.44(2) Å.

Coordinated hydrazine has frequently been proposed as an intermediate in the reduction of coordinated dinitrogen to ammonia. (*See* for example Ref. *1* and references therein.) Although hydrazine is known to form a variety of transition metal complexes (*2*) and has been shown to react with several molybdenum complexes (*3, 4*), there is little detailed structural information about discrete metal complexes of hydrazine. Two recent reports (*5, 6*) describe the preparation and structure determinations of two molybdenum complexes derived from substituted hydrazines.

This chapter describes the preparation and molecular structure of a compound, 1, in which hydrazine itself is coordinated to two molybdenum atoms.

Experimental

Preparation of [Mo(S$_2$CN(C$_2$H$_5$)$_2$)$_2$(CO)$_2$]$_2$N$_2$H$_4$. Molybdenum hexacarbonyl (1 g, 3.79 mmol) was placed in a 250-ml three-necked flask fitted with a magnetic stirrer. A nitrogen stream was then passed through the flask and was maintained throughout the preparation. All solvents used in the preparation were thoroughly deoxygenated. Dichloromethane (20 ml) was added and the suspension cooled in an acetone–solid CO$_2$ bath. Dibromine (0.61 g, 3.79 mmol) in dichloromethane (2 ml) was added dropwise to the stirred suspension. Upon warming to room temperature, carbon monoxide was evolved and dibromotetracarbonylmolybdenum(II) formed. Dichloromethane was allowed to evaporate in the nitrogen stream before adding methanol (10 ml) and then a solution of sodium diethyldithiocarbamate-3-hydrate (1.71 g, 7.58 mmol) in methanol (10 ml). Dichloromethane ($\sim$ 15 ml) was added to the precipitated mixture of di- and tricarbonyl complexes (7), and the solution was stirred for about 10 min. Hydrazine-1-hydrate 99% w/w (0.15 g, 3 mmol) in methanol (1 ml) was next added slowly. Immediately the solution became dark red, and a bright red crystalline product (1) was obtained.

The reaction vessel was transferred to a glove-box and the product removed by filtration under nitrogen. The product was washed with methanol (3 × 10 ml) and dissolved in dichloromethane (10 ml) to which a drop of hydrazine-1-hydrate had been added. The addition of methanol (50 ml) resulted in the formation of crystals of 1. After washing with methanol (3 × 10 ml) and drying at 0° for 8 hr in vacuum, the yield was 0.98 g (56% based on Mo(CO)$_6$).

Analytically calculated for C$_{24}$H$_{44}$Mo$_2$N$_6$O$_4$S$_8$:
                C, 31.03; H, 4.78; N, 9.05; S, 27.61

Found: C, 30.94; H, 4.92; N, 9.13; S, 27.83

The infrared spectrum of compound 1 (KBr discs) showed bands at 3125 (ν_{NH}); 1930(s) and 1845(s) cm^{-1}, (ν_{CO}). The C$\cdots$N, NC$_2$, and CS$_2$ bands were also present at 1500(s), 1149(m), and 1002(m) cm^{-1}. The ^{1}H NMR spectrum (100 mHz) in CDCl$_3$ (Figure 1) consists of a triplet (δ,1.26), quartet (δ,3.71), and singlet (δ,5.07) with integrated intensities of 6:4:1, in agreement with the stoichiometry [Mo(S$_2$CN-(C$_2$H$_5$)$_2$)$_2$(CO)$_2$]$_2$N$_2$H$_4$.

Crystals suitable for x-ray structure determination were grown slowly ($\sim$ 1 hr) from dichloromethane by the gradual addition of methanol until incipient crystallization. Under these conditions the crystals obtained contained one molecule of dichloromethane per molecule of 1. The dichloromethane of crystallization was lost upon standing in air or by drying in a vacuum (vide supra). The solvated crystals of 1 were

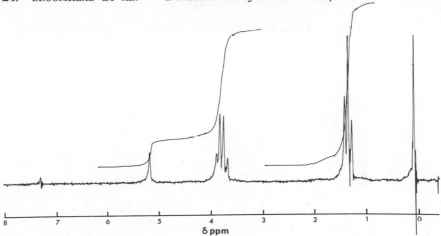

Figure 1. The proton magnetic resonance spectrum of 1 in CDCl₃

stable indefinitely when sealed in a capillary under nitrogen saturated with dichloromethane vapor.

Structure Determination. A crystal of **1** was sealed in a 0.3 mm glass capillary in nitrogen atmosphere saturated with CH_2Cl_2. Preliminary x-ray diffraction pictures showed that the crystal was triclinic with probable space group $P\bar{1}$. The data crystal also contained one molecule of CH_2Cl_2 per formula unit of **1**. The cell parameters are summarized in Table I. A total of 7578 independent reflections having $2\theta \leq 50°$ were collected using MoKα radiation. There were 6129 reflections with $F_o^2 > 3\sigma(F_o^2)$. The positions of the two molybdenum atoms were determined from a Patterson function, and the remaining atoms were located by successive least-squares refinements and difference electron density maps. These maps revealed that **1** was dimeric and contained the Mo–N–N–Mo linkage. However, the two bridging nitrogen atoms were statistically disordered among four positions. These positions were compatible with models **Ia** and **Ib**. In order to avoid a prejudicial choice of either model and to minimize computing time, an unconstrained refinement of the nitrogen atom disorder was carried out.

This refinement indicated that model **Ia** was preferred and that the relative occupancies of the N and N′ positions were 65:35. A final difference map from this refinement also revealed the presence of 31 of the 44

Ia   Ib

Table I.   Crystal Data for $[Mo(CO)_2(S_2CNEt_2)_2]_2N_2H_4 \cdot CH_2Cl_2$

a	13.872(7) Å
b	14.145(13) Å
c	13.695(9) Å
α	118.50(5)°
β	103.28(3)°
γ	65.76(4)°
Space group	$P\bar{1}$
Z	2
d_{obs}	1.56(1) g cm^{-3}
d_{calcd}	1.57 g cm^{-3}

possible hydrogen atoms of the ethyl groups. The hydrogen atoms were included as fixed contributions in the final least-squares refinement which gave $R_1 = (\Sigma||F_o|-|F_c||)/\Sigma|F_o| = 0.058$ and $R_2 = [\Sigma w(F_o-F_c)^2/\Sigma wF_o^2]^{1/2} = 0.085$, for the 6129 data with $F_o^2 > 3\sigma(F_o^2)$. The standard deviation of an observation of unit weight was 3.35. (For tables of final atomic parameters and a list of F_o and F_c, see Ref. 8).

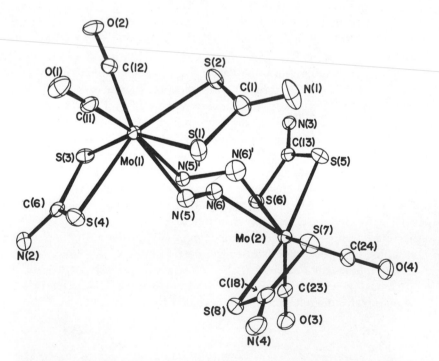

Figure 2.   Perspective view of the structure of 1. Both positions of the disordered bridging hydrazine ligand are shown. The ethyl groups of the dedtc ligands and the H atoms on the N_2H_4 groups have been omitted for clarity.

Results and Discussion

The numbering scheme and a perspective view of 1 including both positions of the disordered N_2H_4 moiety are shown in Figure 2. Each molybdenum atom of the dimer is seven-coordinate with the coordination sphere consisting of two adjacent carbonyl groups, four sulfur atoms from two diethyldithiocarbamate (dedtc) ligands, and one nitrogen atom from the bridging hydrazine ligand. The geometry about each molybdenum atom is best understood by first examining the geometry of the two crystallographically independent $Mo(dedtc)_2(CO)_2$ fragments of the dimer. Comparison of the interatomic distances and angles of the two $Mo(dedtc)_2(CO)_2$ fragments (Tables II and III) shows that their geometries are essentially identical. The four sulfur atoms about each molybdenum are coplanar. The maximum deviation of a sulfur atom from the plane S1, S2, S3, S4 is 0.028 Å; the maximum deviation of a sulfur atom from plane S5, S6, S7, S8 is 0.004 Å. However, the interligand S–Mo–S angles are grossly dissimilar as shown in II.

Each molybdenum atom is displaced ~0.4 Å from its trapezoidal plane of sulfur atoms toward the carbonyl groups, and the $Mo(CO)_2$ unit forms a plane which is nearly normal to the corresponding plane of sulfur atoms. Figure 3 presents a view of the orientation of the $Mo(CO)_2$

Table II. Selected Bond Distances in $[Mo(CO)_2(S_2CN(C_2H_5)_2)_2]_2N_2H_4 \cdot CH_2Cl_2$

A–B	Distance (Å)	A–B	Distance (Å)
Mo1–S1	2.513(3)	Mo2–S5	2.524(2)
Mo1–S2	2.515(2)	Mo2–S6	2.522(4)
Mo1–S3	2.524(3)	Mo2–S7	2.510(4)
Mo1–S4	2.508(2)	Mo2–S8	2.507(2)
Mo1–N5	2.366(9)	Mo2–N6	2.44(1)
Mo1–N5′	2.44(2)	Mo2–N6′	2.34(2)
Mo1–C11	1.910(7)	Mo2–C23	1.939(8)
Mo1–C12	1.930(7)	Mo2–C24	1.934(7)
N5–N6	1.44(2)	N5′–N6′	1.44(3)
C11–O1	1.178(8)	C23–O3	1.160(8)
C12–O2	1.175(8)	C24–O4	1.158(8)
S1–C1	1.707(7)	S5–C13	1.702(7)
S2–C1	1.692(8)	S6–C13	1.711(7)
C1–N1	1.322(9)	C13–N3	1.324(8)
N1–C2	1.55(1)	N3–C14	1.48(1)
N1–C4	1.87(2)	N3–C16	1.44(1)
C2–C3	1.51(2)	C14–C15	1.48(1)
C4–C5	1.21(2)	C16–C17	1.52(1)
S3–C6	1.709(7)	S7–C18	1.711(8)
S4–C6	1.699(7)	S8–C18	1.697(8)
C6–N2	1.324(8)	C18–N4	1.329(9)

Table III. Selected Bond Angles in
$[Mo(CO)_2(S_2CN(C_2H_5)_2)_2]_2N_2H_4 \cdot CH_2Cl_2$

$A-B-C$	Angle $(°)$	$A-B-C$	Angle $(°)$
S1–Mo1–S2	68.67(9)	S5–Mo2–S6	68.50(9)
S3–Mo1–S4	68.58(9)	S7–Mo2–S8	68.72(10)
S1–Mo1–S3	155.77(7)	S5–Mo2–S8	157.80(7)
S1–Mo1–S4	95.21(9)	S5–Mo2–S7	96.30(10)
S2–Mo1–S3	120.49(9)	S6–Mo2–S8	120.14(9)
S2–Mo1–S4	157.40(7)	S6–Mo2–S7	157.94(7)
S1–Mo1–N5	77.4(3)	S5–Mo2–N6	85.4(2)
S2–Mo1–N5	83.0(2)	S6–Mo2–N6	73.1(2)
S3–Mo1–N5	81.5(3)	S7–Mo2–N6	90.4(3)
S4–Mo1–N5	77.9(3)	S8–Mo2–N6	78.7(2)
S1–Mo1–N5′	92.9(6)	S5–Mo2–N6′	73.6(5)
S2–Mo1–N5′	77.4(5)	S6–Mo2–N6′	84.8(5)
S3–Mo1–N5′	69.5(5)	S7–Mo2–N6′	75.3(5)
S4–Mo1–N5′	88.1(5)	S8–Mo2–N6′	86.5(4)
S1–Mo1–C11	81.0(2)	S5–Mo2–C24	82.1(2)
S2–Mo1–C11	110.2(2)	S6–Mo2–C24	111.2(2)
S3–Mo1–C11	112.7(2)	S8–Mo2–C24	110.2(2)
S4–Mo1–C11	81.4(2)	S7–Mo2–C24	81.0(2)
S1–Mo1–C12	126.2(2)	S5–Mo2–C23	126.5(2)
S2–Mo1–C12	78.2(2)	S6–Mo2–C23	78.5(2)
S3–Mo1–C12	77.9(2)	S8–Mo2–C23	75.6(2)
S4–Mo1–C12	124.4(2)	S7–Mo2–C23	123.2(2)
N5′–Mo1–C11	167.2(6)	N6–Mo2–C24	163.9(3)
N5–Mo1–C11	148.4(4)	N6′–Mo2–C24	143.5(6)
N5′–Mo1–C12	120.1(7)	N6–Mo2–C23	124.0(3)
N5–Mo1–C12	139.4(4)	N6′–Mo2–C23	144.5(6)
C11–Mo1–C12	72.2(3)	C23–Mo2–C24	72.0(3)
N5–Mo1–N5′	19.4(5)	N6–Mo2–N6′	20.6(5)
Mo1–N5–N6	124.3(7)	Mo2–N6–N5	120.3(7)
Mo1–N5′–N6′	119(1)	Mo2–N6′–N5′	125(1)
Mo1–C11–O1	177.8(6)	Mo2–C24–O4	177.9(6)
Mo1–C12–O2	177.0(6)	Mo2–C23–O3	178.6(6)
Mo1–S1–C1	88.9(3)	Mo2–S5–C13	89.5(2)
Mo1–S2–C1	89.2(3)	Mo2–S6–C13	89.3(3)
Mo1–S3–C6	89.0(2)	Mo2–S7–C18	89.2(3)
Mo1–S4–C6	89.8(2)	Mo2–S8–C18	89.6(3)
S1–C1–S2	113.1(4)	S5–C13–S6	112.6(4)
S3–C6–S4	112.6(4)	S7–C18–S8	112.4(4)
S1–C1–N1	122.1(6)	S5–C13–N3	124.8(5)
S2–C1–N1	124.8(6)	S6–C13–N3	122. (6)
S3–C6–N2	124.0(5)	S7–C18–N4	124.4(6)
S4–C6–N2	123.4(5)	S8–C18–N4	123.2(6)

fragment of Mo1 with respect to its plane of sulfur atoms and shows that the fragment is cocked in such a way that C12 lies over the S2,S3 edge of the trapezoidal plane. Each of the Mo(dedtc)$_2$(CO)$_2$ fragments has approximate C$_s$ symmetry.

The unusual feature of the structure of **1** is the presence of two equivalent $Mo(dedtc)_2(CO)_2$ fragments of C_s symmetry which coordinate to the bridging hydrazine ligand to produce two different seven-coordinate stereochemistries about the molybdenum atoms with two different Mo–N distances. Figure 3 shows these two stereochemistries about Mo1. Geometries a and b occur in a 65:35 ratio for Mo1 because of the disorder of the bridging hydrazine ligand. Geometry a has Mo1–N5 = 2.37 Å. The Mo1–N5 vector approximately bisects the C11–Mo1–C12 angle and is nearly perpendicular to the plane of the four sulfur atoms. In geometry b atom N5' is more nearly trans to one of the carbonyl groups (N5'–Mo1–C11 = 167°) and Mo–N5' = 2.44(1) Å. Very similar coordination geometries are adopted by Mo2 except that for Mo2, geometries a and b occur in a 35:65 ratio rather than the 65:35 ratio found for Mo1. Thus, each molecule of **1** contains one molybdenum atom wtih coordination geometry a and one molybdenum atom with coordination geometry b, and the data crystal contains **IIIa** and **IIIb** in a 65:35 ratio.

Further evidence that **IIIa** and **IIIb** are the preferred descriptions of the disorder is provided by examining the other possible N–N vectors

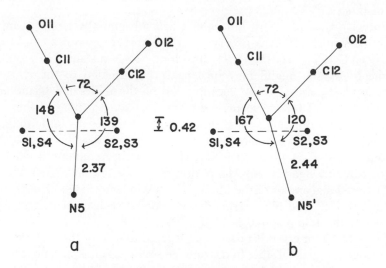

a   b

Figure 3.   Diagram of the two disordered stereochemistries of the inner coordination sphere about Mo1.

The $Mo1(CO)_2N$ plane is in the plane of the paper, and the S1–S4 plane is normal to the plane of the paper. For Mo1 geometries a and b occur in a 65:35 ratio. The coordination geometries about Mo2 are very similar. Maximum deviations from the weighted least-squares planes are: $Mo1(CO)_2N5$ (C12 = 0.007 A); $Mo1(CO)_2N5'$ (N5' = 0.143 A); $Mo2(CO)_2N6$ (N6 = 0.070 A) and $Mo2(CO)_2N6'$ (N6' = 0.021 A). The angles between planes are S1–S4: $Mo1(CO)_2N5$ = 90.2°; S1–S4: $Mo1(CO)_2N5'$ = 90.3°; S5–S8: $Mo2(CO)_2N6$ = 91.4°; S5–S8: $Mo2(CO)_2N6'$ = 91.4°. The angles about Mo2 appear in Table III; Mo2 is 0.40 Å out of the S5–S8 plane.

II

IIIa   IIIb

among the four well resolved positions for the two nitrogen atoms. The other combinations: N5–N5' = 0.81(2); N5–N6' = 1.60(2); N5–N6 = 1.02(2), and N6–N6' = 0.87(2) Å are all physically unreasonable for N–N distances.

Both kinds of Mo–N distances in 1 are substantially longer than the Mo–N distance of 2.26(2)Å observed for the Mo(II) complex $(\eta—C_5H_5)$-$Mo(CO)_2(COCH_2CH_2NH_2)$ (9). The long Mo–N distances in 1 suggest that the hydrazine ligand may be weakly coordinated to the two Mo-$(dedtc)_2(CO)_2$ units. Weak coordination of the hydrazine ligand in 1 could also account for the surprising result that the positions of the other atoms in the coordination sphere do not appear to be detectably different for the two modes of attachment of the bridging hydrazine ligand. The compound $[(\eta—C_5H_5)Mo(NO)I(NH_2NHC_6H_5)]BF_4$ contains a dihapto-phenylhydrazine ligand with Mo–N distances of 2.184 and 2.183 Å (5).

The torsional angles of the Mo1–N5–N6–Mo2 and Mo1–N5'–N6'–Mo2 moieties are 152° and 156°, respectively. The N–N distance of 1.44(2) Å is similar to the 1.46 Å found in $[N_2H_5][N_3]$ (10).

The CH_2Cl_2 of crystallization is well separated from the molybdenum and nitrogen atoms of the hydrazine complex. There are no obvious packing interactions which appear to be responsible for the two different molybdenum coordination geometries.

Finally, it is of interest to compare the stereochemistries about Mo1 and Mo2 with the idealized coordination geometries for seven-coordinate $M(bidentate)_2(unidentate)_3$ complexes (11). The two coordination geometries of the molybdenum atoms of 1 (Figure 3) are intermediate between idealized pentagonal bipyramidal stereochemistry (IVa) and idealized monocapped trigonal prismatic geometry (IVb). Generally stereochemistry IVa is favored by ligands such as dedtc with small nor-

IVa   IVb

malized bite angles (1.12). However, it was suggested (*11*) that **IVb** will become increasingly favored as the chemical differences between the unidentate and bidentate ligands increase.

The unusual structure of **1** and known reactivity of $Mo(dcdtc)_2(CO)_2$ compounds with a variety of small molecules (*7, 12*) in addition to hydrazine suggest that further chemical and structural studies of these and related Mo(II) complexes may provide additional insight concerning the binding of small molecules by low valent molybdenum compounds.

Acknowledgments

The authors thank the National Science Foundation for financial support, the University of Arizona Computer Center for a generous allocation of computer time, and the Australian–American Education Foundation for a Fulbright–Hays Grant. We thank K. F. Bizot and D. McAuley for technical assistance.

Literature Cited

1. Hardy, R. W. F., Burns, R. C., Parshall, G. W., in "Bioinorganic Chemistry," Vol. 2, G. I. Eichorn, Ed., p. 745, Elsevier, New York, 1973.
2. Bottomley, F., *Quart. Rev.* (1970) **24**, 617.
3. Mitchell, P. C. H., Scarle, R. D., *Nature* (1972) **240**, 417.
4. Sellmann, D., Brandl, A., Endell, R., *J. Organometal. Chem.* (1975) **97**, 229.
5. Kita, W. G., McCleverty, J. A., Mann, B. E., Seddon, D., Sim, G. A., Woodhouse, D. I., *Chem. Commun.* (1974) 132.
6. Bailey, N. A., Frisch, R. D., McCleverty, J. A., Walker, N. W., Williams, J., *Chem. Commun.* (1975) 350.
7. Colton, R., Scollary, G. R., Tomkins, I. B., *Aust. J. Chem.* (1968) **21**, 15.
8. Gelder, J. I., Ph.D. Thesis, University of Arizona, 1975.
9. Jones, G. A., Guggenberger, L. T., *Acta. Crystallogr.* (1975) **B31**, 900.
10. Chiglien, G., Etienne, J., Jaulmes, S., Laruelle, P., *Acta Crystallogr.* (1974) **B30**, 2229.
11. Kepert, D. L., *J. Chem. Soc., Dalton Trans.* (1975) 963.
12. McDonald, J. W., Newton, W. E., Creedy, C. T. C., Corbin, J. L., *J. Organometal. Chem.* (1975) **92**, C25.

RECEIVED July 26, 1976.

INDEX

431

G

H

The text of this book is set in 10 point Caledonia with two points of leading. The chapter numerals are set in 30 point Garamond; the chapter titles are set in 18 point Garamond Bold.

The book is printed offset on Text White Opaque 50-pound. The cover is Joanna Book Binding blue linen.

Jacket design by Sharri Harris. Editing and production by Virginia deHaven Orr.

The book was composed by Service Composition Co., Baltimore, Md., printed and bound by The Maple Press Co., York, Pa.